MSU LIBRARIES
P9-DNU-842
WITHDRAWAL

DICTIONARY OF MILITARY AND NAVAL QUOTATIONS

by the Same Author

The Defense of Wake
Marines at Midway
Soldiers of the Sea

Co-Author

The Marshalls: Increasing the Tempo
The Marine Officer's Guide

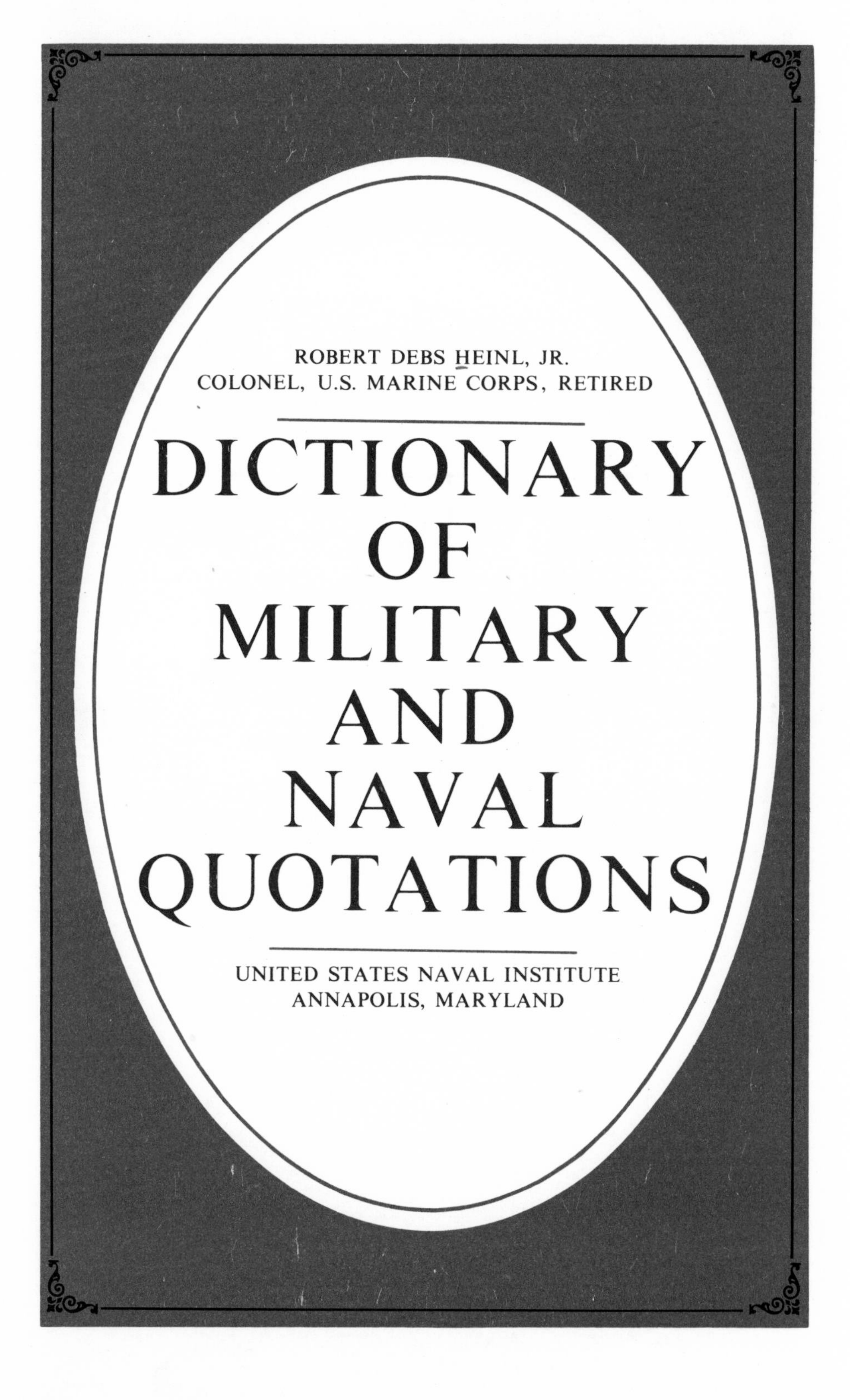

ROBERT DEBS HEINL, JR.
COLONEL, U.S. MARINE CORPS, RETIRED

DICTIONARY OF MILITARY AND NAVAL QUOTATIONS

UNITED STATES NAVAL INSTITUTE
ANNAPOLIS, MARYLAND

Library of Congress Catalogue No. 66-22342
Printed in the United States of America

For N.G.H.,
The Colonel's Lady

Nothing so comforts the military mind as the maxim of a great but dead general.

—B. W. Tuchman

PREFACE

From 1962 until my retirement in 1964, it may be remembered that my writings became the object of official solicitude on the part of the Department of Defense. As a result of experiences during this period, I was led to conclude that the most acceptable type of professional writing for an officer on active duty in today's Armed Forces might well be a dictionary of quotations. In a sense, therefore, I feel obliged to dedicate this compilation to those, in and out of uniform, who so forcibly crystallized my thoughts along these lines in early 1963.

But the idea of a dictionary of military and naval quotations goes farther back. Like most historical and topical writers, and not a few military instructors, I have tried for years to keep track of quotations which promised to be useful or interesting in connection with my work. As the stockpile grew, there arose the need for organization, and from that the idea of a book.

So far as I know, this is the first work of its kind. There are, of course, many excellent general compilations of quotations (and some bad ones, too, as I have come to find out while working on this Dictionary). Mencken, Stevenson, and Bartlett, in that order, seem to me the best, and I have rummaged freely among these and others. But the idea of collecting quotations dealing with war and related subjects and ideas is, I believe, new. I underscore this novelty not to seek credit but to avoid blame: as an innovation, this Dictionary may conceivably be allowed latitude not accorded, say, to the 13th edition of Bartlett. Yet no such work could be perfect. We have Samuel Johnson—who should know—as authority; in a preface of his own, he wrote:

Dictionaries are like watches: the worst is better than none, and the best cannot be expected to go quite right.

The quotations in this Dictionary are intended to comprehend the whole art of war—its incidents, its personalities, its participants, their weapons and equipment, their traditions and customs, the warlike virtues and failings, and the ways, techniques, and modes of war ". . . in the air, on land, and sea."

While I have striven for accuracy, particularity, and completeness in quotations and in attributory information, there remain many cases in which exactness has been impossible. The exact dates of sources cited are sometimes unknown, unavailable, or unclear (was it date of publication, of composition, or of utterance?). In a few instances we know a source by initials rather than Christian name; in a very few we have a last name only. Confronted by such entries, I include all ascertainable facts, and hope that eventually the gaps may be filled.

Specifically, for each quotation I have tried to give the individual or other source, a date, the name of the work from which taken (when applicable), and, in many cases, explanatory or clarifying notes as to context. Where an exact date cannot be established, I give birth or death dates of the source. Occasional posthumous dates may puzzle readers. These are usually found in connection with a posthumous first publication or compilation. For example, the two primary compilations of Napoleon's maxims or aphorisms are 1831 and 1848, and initial publication of Upton's *Military Policy of the United States* was in 1904, twenty years after its author's death.

On names and ranks I use whatever combination, with or without Christian name or initials, best and briefly identifies a source to the general reader. Ranks are generally omitted except where needed for identification or clarity. Readers seeking full names, ranks, and titles, can find this information in the index of sources.

Thus the well known sources—Shakespeare, Mahan, Nelson, Clausewitz—appear by last name only. Most others appear under the full name by which generally known—St. John de Crevecoeur, Henry Ford, Ferdinand Foch, Nikita Khrushchev, Mao Tse-tung—or in a few cases by initials (A. E. Housman) when this represents common usage. In one case (Stonewall Jackson) I use the soubriquet in deference to overwhelming custom. Needless to say, General Jackson appears by complete name and rank in the source index.

In works with multiple or foreign titles, I employ the title best known to English-speaking readers. In doubtful cases this means the original title in the original language (other than oriental).

A chronological, i.e., historical, order of quotations is observed within each rubric except that where a single rubric contains more than one item from a given source, all such are grouped together for the reader's convenience even though there may be some spread of dates.

Quotations which present questions of authenticity are flagged with the phrase "Attributed to . . ." Generally, where I have encountered a dubious, or even an outright bogus, item which is nevertheless well known and widely quoted, I include it with an appropriate note. This procedure serves to correct erroneous beliefs as to provenance, and protects me against being thought to have overlooked it. Further, if the quotation is intrinsically worthwhile, it can thus be kept in the public domain on its own merits (for example, the famous but spurious lines attributed to Wellington on "mere quill-driving.").

Choice of rubrics and the classification of quotations unavoidably reflect the military vocabulary of a professional soldier. With a few exceptions (such as Leadership, which is mostly a collection of maxims) I have tried to correlate quotations with the operative noun or gerund of the rubric (or vice versa). I have not hesitated to make fine shadings—for example there are headings for both Assault and Attack, for Generals and for Generalship, Command and Commander, Sailor and Seaman, Artilleryman and Gunner. Some few quotations, which defy compartmentation, are repeated under two or even three rubrics when I feel that the item answers in full force to each.

By extensive use of cross-references I have avoided an index or concordance, but, in order to permit readers a *coup d'oeuil* of what is in the Dictionary, and to suggest starting points, a table of rubrics is supplied. The index lists authors or sources, giving, to the extent known, full names, highest ranks if any, and dates of birth and death.

This Dictionary is of course an American book. Yet so much about war that is worth remembering was said before America existed that I have felt not compunction but eagerness in using every ancient or foreign source known to me. In addition to foreign sources, I have tried to avail myself fully of the wealth of British quotations which, despite events in 1775, remain part of our common heritage and domain.

While trying in every way for completeness, I have had to use and classify quotations as they came, and it may surprise readers, as I have been surprised, to discover the paucity or even lack of quotations in areas where they might be expected. Generally, the older a rubric, the more has been said or written of it. As cavalry goes back to Biblical times, there are many quotations on this subject. The tank, being barely fifty years old, has been the subject of far less quotable comment than the warhorse. Many more worthwhile things have been said about sea power than air power, and so on.

Because of the universality of war, this Dictionary possibly transcends, at least in a few places, the exact bounds of the profession of arms. I have made no attempt to draw inflexible frontiers, including, for example, even such exotics as Diplomacy, Women, and Comptrollership. This being so, the uses of the Dictionary will vary from person to person. The readers whom I have particularly kept in mind are historians, military instructors, military students (especially would-be pundits at the War Colleges), and that anonymous but swarming breed in today's Pentagon, information officers, speech-writers, and ghost-writers. In addition, I hope this work may be useful to editors, librarians, briefers, and other reference specialists. A few, perhaps, may even browse here and there for pleasure. I would like to think so.

Much of the material contained herein naturally pertains to fighting and fighting men. I trust such thoughts from fighting men of the past may lend direction to those who may have to fight in the future. The Dictionary cannot be considered as standard equipment for foxhole, seabag or cockpit, yet, hopefully, it will furnish perspective and inspiration—items never found in the allowance lists—for those who inhabit such places.

While I obviously cannot vouch for the assertions, opinions, feelings or intentions of those quoted, which comprise the greater part of a book of this kind, I avow for myself, as required by *U. S. Navy Regulations*, that the opinions or assertions contained herein are the private ones of the author and are not to be construed as official or reflecting the views of the Navy Department or the Naval Services at large.

Errors, gaps, and false attributions will almost certainly be discovered in this Dictionary. My hope is that all such will be drawn to my attention for later correction.

Finally, while my selection of quotations has been designed to illuminate each subject in the most vivid possible way rather than to convey a point of view, close reading of the work as a whole cannot but disclose what it has meant to me, over the years of a career, to have been a soldier, or, more exactly, "Soldier an' Sailor, too."

R.D. HEINL, Jr.,
Colonel, U.S. Marine Corps (Retired)

Washington, D.C.
12 August 1966

ACKNOWLEDGEMENTS

Much credit for this Dictionary must go to institutions. In the preface I have already voiced indebtedness to the Department of Defense for having given this project an appropriate and timely character I would otherwise likely have failed to apprehend. Two U.S. Naval Hospitals, those at Guantanamo Bay and Portsmouth, Virginia, provided me, in 1961 and 1963, with enforced inactivity, release from the pressures of duty, and excellent library facilities. For the Navy's doctors, nurses, hospital corpsmen—people whom Marines like and respect anyway—and particularly for the hospital librarians and for the Gray Ladies with their book carts, my warm and grateful remembrances are recorded. A very different Institution, the Royal United Service, in London, with its incomparable military library, where on chill, foggy autumn afternoons the staff, ably directed by Brigadier John Stephenson, nourished me with life-giving tea, played a role which will not soon be forgotten. Still another collective assist, and no small one, came somewhat unexpectedly when Vice Admiral John Hayes, RN, mobilized his staff, that of F.O. 2/c, Home Fleet, to help him provide valued comments on the draft ms. of this Dictionary. The extent and quality of the Home Fleet's contribution provides final proof of the capacity of a naval staff to meet any challenge, however unforeseen.

It is a convention nowadays that author-husbands testify to wifely help, sustenance, and forbearance. I so testify but must go further. Without my wife's original enthusiasm and continued, gentle pressure, this project might never have started. The very first quotation in this Dictionary—not to speak of hundreds of others—was spotted, neatly transcribed, and presented purposefully by her to me. After that, retreat was impossible. I had no direction to go but forward. With continued research and editorial help, down to the last galley proof, that is what we did.

Those who greatly forwarded this Dictionary by principal contributions, most especially but not exclusively in critical reading, and thus bore the heat and burden of the day, are listed alphabetically in the following paragraph; I owe them more than can well be expressed.

Colonel B. N. L. Ditmas; Colonel J. J. Farley; Colonel A. M. Fraser; Brigadier General S. B. Griffith, II; Major Reginald Hargreaves; General Sir James Marshall-Cornwall; Major General Orlando Ward; Brigadier General Willard Webb; and (like General Griffith, once my admired commanding officer) Brigadier B. W. Webb-Carter.

In addition to persons already mentioned, I am deeply grateful to a host of others for permission to quote, and for help particular and general, and among these I must single out: Rita Adelman, Earl F. Bartholomew; Lieutenant Commander W. J. Bingham; Pamela G. Burdick; Ambassador G. Corley-

Smith; Lieutenant Jonathan E. Davis; Rear Admiral E. M. Eller; Dr. Bernard Fall; Captain Peter LaNiece; Captain Sir B. H. Liddell Hart; First Sergeant R. F. Lee; The Reverend E. Dering Leicester; Eugenia D. Lejeune; William Lichtenwanger; Colonel J. H. Magruder, III; Ambassador Robert M. McClintock; The Honorable Paul H. Nitze; Timothy J. O'Connor; D. M. O'Quinlivan; Dr. Adam M. Parry; Dr. Forrest Pogue; Vice Admiral J. F. Shafroth; Robert Sherrod; Field Marshal Sir William Slim; General Maxwell D. Taylor; Ambassador R. L. Thurston; Gilbert G. Twiss; Mary R. Underwood; Mark Watson; Instructor-Lieutenant Commander A. S. Watt; His Grace the Seventh Duke of Wellington; and Rear Admiral J. C. Wylie.

And finally, to those in the Office of Naval History, the Office of the Chief of Military History, and in the reference services of the Library of Congress, who have answered so many troublesome inquiries, I extend my thanks.

R. D. HEINL, Jr.

FOREWORD

During the last war the United States Army distributed to official historians a mimeographed list of selected quotations on strategy, tactics, and war in general. This proved very useful to budding military historians. Colonel Heinl has profitably employed his leisure hours—which have been few and far between for a Marine officer on active duty—by compiling apt military quotations from his extensive reading, and he has wisely decided to publish the collection.

A most useful book it will be, despite the disparaging remark of Barbara Tuchman (p. 186) "Nothing so comforts the military mind as the maxim of a great but dead general." It is amazingly comprehensive, ranging from the wars of the Old Testament to Vietnam, Winston Churchill and Admiral Nimitz; and, in scope, from Air Warfare to Wounds. Persons are referred to as well as events and theories; nor has he omitted popular ballads and slogans. Sun Tzu and Xenophon are here, as well as Clausewitz, Jomini, and Captain Liddell Hart.

One of the amusing remarks preserved by Colonel Heinl (p. 236) is Admiral Ramsay's endorsement on Captain Alfred Thayer Mahan's fitness report, "It is not the business of a naval officer to write books." A distinguished naval officer thus capped it for me in 1942: "I'm against writing naval history—our enemies profited more by Mahan than we did!" Perhaps so, but whose fault was that? I hope that naval and military officers, and civilians, too, will continue to write books on history, strategy, tactics, and even politics. Every one will find Colonel Heinl's Dictionary a priceless source of quotations to illustrate the point that he is trying to make.

There are categories here not found in any other dictionary of quotations or anthology; e.g., Medicare, Initiative, Nuclear Warfare, Discipline (four pages of quotes), and Marshal Foch's apt characterization of Piecemeal Attacks. On War itself there are 13 pages of quotations starting with Homer's "Men grow tired of sleep, love, singing and dancing, sooner than of war," and including, I am glad to see, Sir John Slessor's "War is basically a conflict of wills."

Every reader will probably miss some favorite quotation, but he will find hundreds of others that are new to him; and after considerable spot checking I can assure him that Colonel Heinl has taken care to be accurate.

SAMUEL E. MORISON
Rear Admiral, U.S. Naval Reserve (Retired)

TABLE OF RUBRICS

INDEX OF SOURCES

Numbers following a given source are page-numbers on which quotations from that source may be found.

Where biographic data are missing or incomplete, I will greatly appreciate information or clues which will permit complete or more accurate identification.

The following abbreviations are used: b. = born; c. = circa, about; d. = died; fl. = floruit, i.e., is only known to have been alive on the date given; pseud. = pseudonym.

LEO VI of Byzantium, 886–911, 126

DICTIONARY OF MILITARY AND NAVAL QUOTATIONS

Action

It is even better to act quickly and err than to hesitate until the time of action is past.
Clausewitz: On War, 1832

The great end of life is not knowledge, but action.
Thomas H. Huxley, 1825–1895

Life is action and passion. I think it is required of a man that he should share the action and passion of his time at peril of being judged not to have lived.
Oliver Wendell Holmes, Jr.: Speech to Harvard Law School alumni, New York, February 1913

Action is the governing rule of war.
Ferdinand Foch: Precepts, 1919

A man who has to be convinced to act before he acts is *not* a man of action . . . You must act as you breathe.
Georges Clemenceau, 1841–1929

To take no action is to take undecided action.
Robert S. McNamara: In New York Times, 4 July 1965

(*See also* Activity.)

Action Officer

Whose sore task does not divide the Sunday from the week.
Shakespeare: Hamlet, i, 2, 1600

(*See also* Staff Officer.)

Action Report

Peccavi. (Latin = "I have sinned" = "I have Sind.")
Sir Charles James Napier: Despatch after the battle of Hyderabad, 1843, which gained the province of Sind

Many return from war who can't give an account of the battle.
Italian Proverb

(*See also* History, Report.)

Activity

This I like, active service or none!
Nelson: Letter, 1795

Activity, activity, speed! (Activité, activité, vitesse!)
Napoleon I: Order to Massena before Eckmühl, 17 April 1809

A fundamental principle is never to remain completely passive, but to attack the enemy frontally and from the flanks, even while he is attacking us.
Clausewitz: Principles of War, 1812

Activity in war is movement in a resistant medium. Just as a man immersed in water is unable to perform with ease and regularity the most natural and simplest movement, that of walking, so in war, with ordinary powers, one cannot keep even the line of mediocrity.
Clausewitz: On War, 1832

We must make this campaign an exceedingly active one. Only thus can a weaker country cope with a stronger; it must make up in activity what it lacks in strength.
Stonewall Jackson: Letter, April 1863

In war it is necessary not only to be active but to seem active.
Andrew Bonar Law, 1858–1923

(*See also* Action.)

Adjutant

An adjutant is a wit *ex officio.*
Francis Grose: Advice to the Officers of the British Army, 1782

Administration

Nothing shows a general's attention more than requiring a number of returns, particularly such as it is difficult to make with any degree of accuracy. Let your brigade-major, therefore, make out a variety of forms, the more red lines the better: as to the information they convey, that is immaterial; no one ever reads them, the chief use of them being to keep the adjutants and sergeants in employment.
Francis Grose: Advice to the Officers of the British Army, 1782

My Lord,
If I attempted to answer the mass of futile correspondence that surrounds me, I should be debarred from all serious business of campaigning. I must remind your

Lordship—for the last time—that so long as I retain an independent position, I shall see that no officer under my Command is debarred, by attending to the futile drivelling of mere quill-driving in your Lordship's office, from attending to his first duty—which is, and always has been, so to train the private men under his command that they may, without question, beat any force opposed to them in the field.

Attributed to Wellington under date of 1810 but not to be found in any authentic source, this widely known quotation can in fact only be traced to a British Middle East Command training circular, c. 1941, and is almost certainly spurious.

Do the business of the day in the day.

Wellington: in Portugal, 1811

Napoleon directed Bourrienne to leave all his letters unopened for three weeks, and then observed with satisfaction how large a part of the correspondence had thus disposed of itself, and no longer required an answer.

R. W. Emerson: Representative Men, 1850

Do not engage in any paper wars.

Florence Nightingale: Letter to Lothian Nicholson, 1853

Stick close to your desks and never go to sea,
And you all may be rulers of the Queen's Navee!

W. S. Gilbert: HMS Pinafore, 1878

There has been a constant struggle on the part of the military element to keep the end—fighting, or readiness to fight—superior to mere administrative considerations . . . The military man, having to do the fighting, considers that the chief necessity; the administrator equally naturally tends to think the smooth running of the machine the most admirable quality.

Mahan: Naval Administration and Warfare, 1903

Administration needs papers, as a mill needs grist.

Mahan: Naval Administration and Warfare, 1903

. . . detailed enquiries of that meticulous class, which is the delight of Departments and the despair of soldiers in the field.

Philip Guedalla: Palmerston, 1927

I am a strong believer in transacting official business by The Written Word.

Winston Churchill: Their Finest Hour, 1949

. . . the "Three Manys" and the "Three Fews" (much work, much responsibility, much reprimand; and little authority, few material advantages, and few awards or promotions).

Hoang Duy: Comment on military administration in the Viet Minh forces, Indo-China, c. 1952 (Duy was a staff noncommissioned officer with the Viet Minh.)

An army cannot be administered. It must be led.

Franz-Joseph Strauss: To the Bundestag, 1957

. . . there must be a clear-cut, long-term relationship established between operational intentions and administrative resources. Successful administrative planning is dependent on anticipation of requirements.

Montgomery of Alamein: Memoirs, 1958

There is far too much paper in circulation in the Army, and no one can read even half of it intelligently.

Montgomery of Alamein: Memoirs, 1958

It is difficult to know whether a man is a good administrator because he is so busy, or a bad one for the same reason.

Leo Rosten: Captain Newman, M.D., 1961

You never sign a letter you've written yourself, and never write a letter you sign yourself.

Harlan Cleveland: Speech, 9 December 1961

Paper-work will ruin any military force.

Lewis B. Puller: Marine, 1962

Administration is something to be got on with, not deified.

Michael Ramsey: Interview, October 1962

Success cannot be administered.

Arleigh Burke: Speech, 1962

Damn your writing,
Mind your fighting.

Attributed by Sir Archibald Wavell to "a British general who rose to high command."

(*See also* Bureaucracy, Civilian Employees, Official Correspondence.)

Admiral

Thou art our admiral, thou bearest the lanthorn in the poop.
Shakespeare: King Henry IV, iii, 3, 1597

An admiral has to be put to death now and then to encourage the others. (pour encourager les autres)
Voltaire: Candide, 1759

Admirals extolled for standing still
Or doing nothing with a deal of skill.
William Cowper: Table Talk, 1782

It is might which rules the world nowadays, and I am very glad to find that...I can leave the regulating of affairs to admirals. These are the men to cut the matter short!
Count Nesselrode: Reference to the battle of Navarino, 20 October 1827

No profession in England has done its duty until it has furnished a victim; even our boasted Navy never achieved a great victory until we shot an admiral.
Benjamin Disraeli: Tancred, 1847

That part of a war-ship which does the talking while the figure-head does the thinking.
Ambrose Bierce: The Devil's Dictionary, 1906

Men gointo the Navy...thinking they will enjoy it. They do enjoy it for about a year, at least the stupid ones do, riding back and forth quite dully on ships. The bright ones find that they don't like it in half a year, but there's always the thought of that pension if only they stay in...Gradually they become crazy. Crazier and crazier. Only the Navy has no way of distinguishing between the sane and the insane. Only about five percent of the Royal Navy have the sea in their veins. They are the ones who become captains. Thereafter, they are segregated on their bridges. If they are not mad before this, they go mad then. And the maddest of these become admirals.
Attributed to George Bernard Shaw, 1856–1950

It is dangerous to meddle with admirals when they say they can't do things. They have always got the weather or fuel or something to argue about.
Winston Churchill: To Secretary of the Navy Frank Knox, December 1941 (Knox had asked Churchill what should be done about Admiral Frank Jack Fletcher's abandonment of the relief of Wake on grounds of low fuel supply.)

We had Generals who were Admirals and Admirals who wanted to be Generals. Generals acting as Admirals are bad enough but it was the Admirals who wanted to be Generals who imperilled victory among the coral islands.
Holland M. Smith: Coral and Brass, 1949

Never say no to an admiral.
Naval saying

(*See also* Commander, Naval Officer.)

Advance

Success and glory are in the advance. Disaster and shame lurk in the rear.
Major General John Pope, USA: General Order to Officers and Soldiers, Army of the Potomac, 14 July 1862.

Advance Guard

The duty of an advance guard does not consist in advancing or retiring, but in maneuvering...An advance guard should consist of picked troops, and the general officers, officers, and men should be selected for their respective capabilities and knowledge. An ill-trained unit is only an embarrassment to an advance guard.
Napoleon I: Maxims of War, 1831

When you move at night, without a light, in your own house, what do you do? Do you not (though it is a ground you know well) extend your arm in front so as to avoid knocking your head against the wall? The extended arm is nothing but an advance guard.
Ferdinand Foch: Precepts, 1919

Taking the enemy by the collar is the function of the advance guard.
Ferdinand Foch: Precepts, 1919

The first duty of an advance guard is to advance.
Sir William Slim: Unofficial History, 1959

Aerial Combat

'Twixt the green sea and the azured vault
Set roaring war.
Shakespeare: The Tempest, v, 1, 1611

Hurled headlong flaming from the ethereal sky
With hideous ruin and combustion, down . . .
Milton: Paradise Lost, 1667

I dipt into the future far as human eye could see,
Saw the vision of the world, and all the wonder that would be;
Saw the heavens fill with commerce, argosies of purple sails,
Pilots of the purple twilight, dropping down with costly bales;
Heard the heavens filled with shouting, and there rain'd a ghastly dew
From the nations' airy navies grappling in the central blue.
Alfred Tennyson: Locksley Hall, 1842

With a bullet through his head, he fell from an altitude of 9,000 feet—a beautiful death.
Manfred von Richthofen: Letter telling of the death of Count von Holck on 1 May 1916, over Verdun

To conquer the command of the air means victory; to be beaten in the air means defeat.
Giulio Douhet: The Command of the Air, 1921

Cease firing, but if any enemy planes appear, shoot them down in a friendly fashion.
William F. Halsey: message to Third Fleet at sea off Tokyo, after receipt of word of Japanese surrender, 15 August 1945

(*See also* Air Force, Air Power, Aviation, Naval Aviation.)

Aggression

It is only when aggression is legitimate that one can expect prodigies of valor.
Marshall Ney, 1769–1815

The Boers broke off negotiations and made the familiar complaint of aggressors that they were about to be attacked.
Basil Collier: Brasshat, vii, 1961 (of Kruger and the Boers in 1899)

There is never a convenient place to fight a war when the other man starts it.
Arleigh A. Burke, 1901–

Aggression unchallenged is aggression unleashed.
Lyndon B. Johnson: Statement to the nation after North Vietnamese torpedo attacks on U. S. warships in international waters, Gulf of Tonkin, 4 August 1964

It is invariably the weak, not the strong, who court aggression and war.
General Thomas S. Power, USAF: Design for Survival, 1965

(*See also* Preemptive War.)

Aggressiveness

Men rise from one ambition to another; first they seek to secure themselves from attack, and then they attack others.
Niccolo Machiavelli: Discorsi, 1531

In peace, there's nothing so becomes a man
As modest stillness and humility;
But when the blast of war blows in our ears,
Then imitate the action of the tiger.
Shakespeare: King Henry V, iii, 1598

Our Country will, I believe, sooner forgive an officer for attacking his enemy than for letting it [sic] alone.
Nelson: Letter during the attack on Bastia, 3 May 1794

Strengthen your position; fight anything that comes.
W. T. Sherman: Order to Major General McPherson before Atlanta, 11 May 1864

Rough-tough, we're the stuff! We want to fight and we can't get enough!
Attributed to the Rough Riders (1st U. S. Volunteer Cavalry) while enroute to Cuba, 1898

The Third Fleet's sunken and damaged ships have been salvaged and are retiring at high speed toward the enemy.
Willian F Halsey: Message to Admiral Nimitz, on learning of

Japanese reports that the Third Fleet had sustained heavy damage and was retiring, Leyte Gulf, 12 October 1944

When the going gets tough, the tough get going.
Cadet saying, West Point

Agincourt, (25 October 1415)

This day is called the feast of Crispian;
He that outlives this day and comes safe home,
Will stand a tip-toe when this day is nam'd,
And rouse him at the name of Crispian.
He that shall live this day, and see old age,
Will yearly on the vigil feast his neighbors,
And say, "To-morrow is Saint Crispian:"
Then will be strip his sleeve and show his scars,
And say, "These wounds I had on Crispin's day."
Old men forget: yet all shall be forgot,
But he'll remember with advantages
What feats he did that day...
Shakespeare: King Henry V, iv, 3, 1598 (address to the English troops by King Henry before Agincourt)

Aide-de-Camp

An aide-de-camp is to his general what Mercury was to Jupiter, and what the jackal is to the lion.
Francis Grose: Advice to the Officers of the British Army, 1782

Let your deportment be haughty and insolent to your inferiors, humble and fawning to your superiors, solemn and distant to your equals.
Francis Grose: Advice to the Officers of the British Army, 1782

Airborne Operations

What would be the security of the good, if the bad could at pleasure invade them from the sky? Against an army sailing through the clouds, neither walls, nor mountains, nor seas, could afford any security. A flight of northern savages might hover in the wind and light at once with irresistible violence upon the capital of a fruitful region that was rolling under them.
Samuel Johnson: Rasselas, vi, 1759

Five thousand balloons, capable of raising two men each, could not cost more than five ships of the line; and where is the prince who can afford so to cover his country with troops for its defense as that 10,000 men descending from the clouds might not in many places do an infinite deal of mischief before a force could be brought together to repel them?
Benjamin Franklin: Letter to Jan Ingenhousz, 1784

It would probably astonish the reader were I able...to state the cost in manpower and material of the airborne forces of the late war, complete with all the aircraft and manpower devoted to training and carrying them...compared with their impact upon the enemy. They would certainly find no place in the early stages of another great war. They are too vulnerable.
Sir John Slessor: Strategy for the West, 1954

A parachute is merely a means of delivery but not a way of fighting.
Bernard Fall: Street without Joy, xii, 1964 ed.

Airborne—all the way!
Motto of U. S. Army airborne troops

Aircraft

The airship will revolutionize warfare.
Alexander Graham Bell: Letter, 1909

The volant or flying automata are such mechanical contrivances as have self-motion, whereby they are carried aloft in the air, like the flight of birds. Such was the wooden dove made by Archytas, a citizen of Tarentum, and one of Plato's acquaintances, and that wooden eagle framed by Regiomontanus Noremburg, which by way of triumph did fly out of the city to meet Charles V.
John Wilkins: Mechanical Magick, 1680

It [the airplane] will be a factor in war.
Wilbur Wright: Interview, St. Louis, March 1906

Aircraft Carrier

The air fleet of an enemy will never get within striking distance of our coast as long as our aircraft carriers are able to carry the preponderance of air power to sea.
Rear Admiral W. A. Moffett, USN: While Chief of Navy Bureau of Aeronautics, 1922

Scratch one flat-top.

Commander Robert Dixon, USN: Radio report as HIJMS Shoho *blew up and sank under attack by U. S. Navy carrier aircraft, Coral Sea, 7 May 1942*

Our aircraft carriers have lasted longer than a majority of our very expensive overseas air bases.

Vice Admiral W. A. Schoech, USN: Speech, 1964

(*See also* Naval Aviation, Naval Operations, Naval Warfare.)

Air Force

The Independent Air Force should embody the greatest power compatible with the resources at our disposal; therefore no aerial resources should under any circumstances be diverted to secondary purposes, such as auxiliary aviation, local air defense, and anti-aircraft defense.

Giulio Douhet: The Command of the Air, 1921

I have mathematical certainty that the future will confirm my assertion that aerial warfare will be the most important element in future wars, and that in consequence not only will the importance of the Independent Air Force rapidly increase, but the importance of the army and the navy will decrease in proportion.

Giulio Douhet: The Command of the Air, 1921

It is probable that future war will be conducted by a special class, the air force, as it was by the armored knights of the Middle Ages.

William Mitchell: Winged Defense, 1924

Never in the field of human conflict was so much owed by so many to so few.

Winston Churchill: To the House of Commons, 20 August 1940 (of the RAF in the Battle of Britain)

The Navy can lose us the war, but only the Air Force can win it. Therefore, our supreme effort must be to gain overwhelming mastery in the air.

Winston Churchill: To War Cabinet, 3 September 1940

The large ground organization of a modern air force is its Achilles' heel.

B. H. Liddell Hart: Thoughts on War, iii,1943

The function of the Army and Navy in any future war will be to support the dominant air arm.

James H. Dolittle: Speech to the Georgetown University Alumni Association, 30 April 1949

Air Power

The nation that secures control of the air will ultimately control the world.

Alexander Graham Bell: Letter 1909

In order to assure an adequate national defense, it is necessary—and sufficient—to be in a position in case of war to conquer the command of the air.

Giulio Douhet: The Command of the Air, 1921

In the development of air power, one has to look ahead and not backward and figure out what is going to happen, not too much of what has happened.

William Mitchell; Winged Defense, 1924

New weapons operating in an element hitherto unavailable to mankind will not necessarily change the ultimate character of war. The next war may well start in the air but in all probability it will wind up, as did the last war, in the mud.

Report of the President's Board to Study Development of Aircraft for the National Defense, 1925

Air power is a thunderbolt launched from an egg-shell invisibly tethered to a base.

Hoffman Nickerson: Arms and Policy, x, 1945

The power of an air force is terrific when there is nothing to oppose it.

Winston Churchill: The Gathering Storm, 1948

Air power is the most difficult of all forms of military force to measure, or even to express in precise terms.

Winston Churchill: The Gathering Storm, 1948

Today air power is the dominant factor in war. It may not win a war by itself alone, but without it no major war can be won.

Arthur Radford: Speech, 1954

U. S. Navy

JOHN PAUL JONES

1747–1792

"I have not yet begun to fight."

Modern air power has made the battlefield irrelevant.

Sir John Slessor: Strategy for the West, 1954

(*See also* Air Force, Aircraft Carrier, Aviation, Naval Aviation.)

Alamo, The

Thermopylae had its messenger of defeat; the Alamo had none.

Found written on the wall of the Alamo, San Antonio, after the Texan garrison had been wiped out by the Mexicans, March 1836

Remember the Alamo!

Texas battle cry at San Jacinto, 21 April 1836

Alexander of Macedon (365–323 B.C.)

Alexander fought many battles, and took of the strongholds of all, and slew the kings of the earth. And he went through even to the ends of the earth, and took the spoils of many nations: and the earth was quiet before him. And he gathered a power, and a very strong army; and his heart was exalted and lifted up. And he subdued countries of nations, and princes; and they became tributary to him. And after these things he fell down upon his bed, and knew that he should die.

I Maccabees, I, 2–6

Alliances

What encourages men who are invited to join in a conflict is clearly not the good will of those who summon them to their side, but a decided superiority in real power.

Address of the Athenian envoys at Melos, during the Peloponnesian Wars, 416 B.C.

Close alliances with despots are never safe for free states.

Demosthenes: Second Philippic, c. 345 B. C.

An alliance with the powerful is never to be trusted.

Phaedrus: Fables, 1st century A. D.

How can tyrants safely govern home
Unless abroad they purchase great alliance?

Shakespeare: III King Henry VI, iii, 3, 1590

When two princes undertake the conquest of a Kingdom they never agree; because each one thinks always that his companion wants to cheat him and so they distrust each other.

Blaise Montluc: Commentaires, I, 1592

Alliances, to be sure, are good, but forces of one's own are still better.

Frederick William of Brandenburg ("The Great Elector"): Political Testament, 1667

'Tis our true policy to steer clear of permanent alliances with any portion of the foreign world.

George Washington: Farewell Address, 17 September 1796

Peace, commerce, and honest friendship with all nations—entangling alliances with none.

Thomas Jefferson: Inaugural Address, 4 March 1801

Granting the same aggregate of force, it is never as great in two hands as in one, because it is not perfectly concentrated.

Mahan: Naval Strategy, 1911

Any alliance whose purpose is not the intention to wage war is senseless and useless.

Adolph Hitler: Mein Kampf, 1925

After the war is over, make alliances.

Greek Proverb

(*See also* Allies, Diplomacy, Negotiations.)

Allies

Better to have a known enemy than a forced ally.

Napoleon I: Political Aphorisms, 1848

The allies we gain by victory will turn against us upon the bare whisper of our defeat.

Napoleon I: Political Aphorisms, 1848

It is a narrow policy to suppose that this country or that is to be marked out as the eternal ally or the perpetual enemy...We have no eternal allies, and we have no eternal enemies. Our interests are eternal and perpetual, and those interests it is our duty to follow.

Lord Palmerston: To the House of Commons, 1848

In war I would deal with the Devil and his grandmother.

J. V. Stalin, 1879–1951

We are a strong nation. But we cannot live to ourselves and remain strong.

George C. Marshall: Speech to the National Cotton Council, 22 January 1948

War without allies is bad enough—with allies it is hell!

Sir John Slessor: Strategy for the West, 1954

(*See also* Alliances)

Ambassadors

Ambassadors have no warships at their disposal, or heavy infantry, or fortresses. Their weapons are words and opportunities.

Demosthenes, 385–322 B. C.

A sovereign should always regard an ambassador as a spy.

The Hitopadesa, III, c. 500

Ambassadors are the eyes and ears of states.

Francesco Guicciacardini: Storia d'Italia, 1564

An ambassador is an honest man sent to lie abroad for his country.

Henry Wotton: Inscription in the album of Christopher Fleckamore, 1604

A man-of-war is the best ambassador.

Oliver Cromwell, 1599–1658

The zeal and efficiency of a diplomatic representative is measured by the quality and not the quantity of the information he supplies.

Winston Churchill: Memorandum to Sir Alexander Cadogan, 17 February 1941

(*See also* Diplomacy, Negotiations.)

Ambition

To take a soldier without ambition is to pull off his spurs.

Francis Bacon: Essays (Of Ambition), 1597

Ambition,
The soldier's virtue.

Shakespeare: Antony and Cleopatra, iii, 1, 1606

Nothing arouses ambition so much . . . as the trumpet clang of another's fame.

Baltasar Gracian, 1601–1658

Ambush

He smote them hip and thigh.

Judges, XV, 8

An ambuscade, if discovered and promptly surrounded, will repay the intended mischief with interest.

Vegetius: De Re Militari, 378

"All quiet along the Potomac," they said,
"Except, now and then a stray picket
Is shot as he walks on his beat to and fro
By a rifleman hid in the thicket."

Ethel L. Beers: The Picket Guard, 1861

A Snider squibbed in the jungle—
Somebody laughed and fled,
And the men of the First Shikaris
Picked up their Subaltern dead,
With a big blue mark on his forehead,
And the back blown out of his head.

Rudyard Kipling: The Grave of the Hundred Head, 1892

On the eighteenth day of November,
just outside the town of Macroom,
The Tans, in their big Crosley tenders,
they hurried along to their doom,
For the boys of the column were waiting,
with hand-grenades primed on the spot,
And the Irish Republican Army,
made sh–t of the whole ----ing lot!

Irish Republican Army ballad, c. 1922 (The "Tans" were Royal Irish Constabulary, a British Force.)

American Revolution

I rejoice that America has resisted. Three millions of people, so dead to all the feelings of liberty, as voluntarily to submit to be slaves, would have been fit instruments to make slaves of the rest.

Lord Chatham (Pitt the Elder): To the House of Commons, 14 January 1766

Four or five frigates will do the business without any military force.
Lord North: To the House of Commons, 1774

I cannot but lament . . . the impending Calamities Britain and her Colonies are about to suffer . . . Passion governs and she never governs wisely.
Benjamin Franklin: Letter, 5 February, 1775

The Sun never shined on a cause of greater worth. 'Tis not the affair of a City, a County, a Province, or a Kingdom; but of a Continent—of at least one eighth part of the habitable Globe.
Thomas Paine: January 1776

These are the times that try men's souls. The summer soldier and the sunshine patriot will, in this crisis, shrink from the service of their country; but he that stands it now deserves the love and thanks of man and woman.
Thomas Paine: The Crisis, 1776

Chimney corner patriots abound; venality, corruption, prostitution of office for selfish ends, abuse of trust, perversion of funds from a national to a private use, and speculations upon the necessities of the times pervade all interests.
George Washington: Diary, 1776

A constant Naval superiority would terminate the War speedily—without it, I do not know that it will ever be terminated honorably.
George Washington: Letter to LaFayette, 15 November 1781

On this question of principle, while actual suffering was yet afar off [the American Colonies] raised their flag against a power, to which, for purposes of foreign conquest and subjugation, Rome, in the height of her glory, is not to be compared; a power which has dotted over the surface of the whole globe with her possessions and military posts, whose morning drum-beat, following the sun, and keeping company with the hours, circles the earth with one continuous and unbroken strain of the martial airs of England.
Daniel Webster: To the Senate, 7 May 1834

By the rude bridge that arched the flood,
Their flag to April's breeze unfurled,
Here once the embattled farmers stood,
And fired the shot heard round the world.
R. W. Emerson: Concord Hymn, 4 July 1837

The snow lies thick on Valley Forge,
The ice on Delaware,
But the poor dead soldiers of King George
They neither know nor care.
Rudyard Kipling: The American Rebellion, 1906

It was a blameless insurrection, founded on equity and quotations from Blackstone, a sedate rebellion, a sedition of the highest principles.
Philip Guedalla: Fathers of the Revolution, 1926

Ammunition

. . . the fatal balls of murdering basilisks.
Shakespeare: King Henry V, v, 2, 1598 (a basilisk was an early type of cannon)

Put your trust in God, my boys, and keep your powder dry.
Oliver Cromwell: To his troops at Marston Moor, 2 July 1644

If any mourn us in the workshop, say
We died because the shift kept holiday.
Rudyard Kipling: Epitaphs of the War (Batteries Out of Ammunition), 1919

Praise the Lord and pass the ammunition.
Chaplain H. M. Forgy, USN: During the Japanese attack on Pearl Harbor, 7 December 1941

. . . ammunition, life's blood of the artillery.
Colonel F. F. Parry, USMC: Marine Corps Gazette, January 1964

(*See also* Bullet, Gunpowder.)

Amphibious Operations

You are Athenians, who know by experience the difficulty of disembarking in the presence of the enemy.
Demosthenes: To the Athenian troops, Pylos, 425 B. C.

I have found that an Admiral should endeavor to run into an enemy's port immediately after he appears before it; that he should anchor the transport ships and frigates as close as he can to the land; that he should reconnoiter and observe it as quickly as possible, and lose no time in getting the troops on shore . . . On the other hand, experience shows me that, in an affair depending on vigor and despatch, the General should settle their plan of operations so that no time may be lost in idle debate and consultation when the sword should be drawn.

James Wolfe: Letter to a friend, 1758

When the army is landed, the business is half done.

James Wolfe: Letter to his father, 20 May 1758

The fleet and the army acting in concert seem to be the natural bulwark of these Kingdoms.

Thomas More Molyneux: Conjunct Expeditions, 1759

Wherever his [the Union] fleet can be brought, no opposition to his landing can be made.

Robert E. Lee: To Jefferson Davis, 1861

The question of landing in face of an enemy is the most complicated and difficult in war.

Sir Ian Hamilton: Gallipoli Diary, 1920

Difficulties of landing on beaches are serious, even when the invader has reached them; but difficulties of nourishing a lodgment when exposed to heavy attack by land, air, and sea are far greater.

Winston Churchill: Note to the Chiefs of Staff Committee, 28 June 1940

A landing against organized and highly trained opposition is probably the most difficult undertaking which military forces are called upon to face.

George C. Marshall: during planning for the Sicilian landings, 1943

In landing operations, retreat is impossible. To surrender is as ignoble as it is foolish . . . Above all else remember that we as the attackers have the initiative. We know exactly what we are going to do, while the enemy is ignorant of our intentions and can only parry our blows. We must retain this tremendous advantage by always attacking: rapidly, ruthlessly, viciously and without rest.

George S. Patton, Jr.: General Order to the Seventh Army before Sicily landings, 27 June 1943

Unloading is the world-wide difficulty of amphibious operations.

Vice Admiral T. S. Wilkinson, USN: To Samuel Eliot Morison, 1944

On land a General is in command. He can order his troops to advance or withdraw, to feint and maneuver. By his orders he can win or lose battles. But not so the man who is the titular head of an amphibious invasion force. He cannot add troops because there are no more boats to carry them. He cannot even subtract, for so interlaced are all elements of the plan that the alteration of a single part of it will throw the rest askew. In the early stages he cannot materially alter the action of his troops in time or space. Like everyone else involved, he is simply a hired man, doing the will of the Plan.

Ralph Ingersoll, 1900

(*See also* Amphibious Warfare, Amphibious Training, Beach, Landing Craft, Naval Warfare, Sea Power.)

Amphibious Training

If the Battle of Waterloo was won on the playing fields of Eton, the Japanese bases in the Pacific were captured on the beaches of the Caribbean.

Holland M. Smith: Coral and Brass, 1949

(*See also* Amphibious Operations, Amphibious Warfare.)

Amphibious Warfare

Remember one and all of you who are embarking that you are both the fleet and army of your country.

Nicias of Athens: Speech to the Athenian forces at Syrcuse, 414 B.C.

A Military, Naval, Littoral War, when wifely prepared and discreetly conducted, is a terrible Sort of War. Happy for that People who are Sovereigns enough of the Sea to put it in Execution! For it comes like Thunder and lightning to fome unprepared Part of the World.

Thomas More Molyneux: Conjunct Expeditions, 1759

It is a crime to have amphibious power and leave it unused.

Winston Churchill: Note to the Chiefs of Staff Committee, 1 December 1940

Amphibious warfare requires the closest practicable cooperation by all the combatant services, both in planning and in execution, and a command organization which definitely assigns responsibility for major decisions throughout all stages of the operation, embarkation, overseas movement, beach assault, and subsequent support of the forces ashore.

Admiral Henry K. Hewitt USN: 1887–

. . . in all probability . . . the most far-reaching tactical innovation of the War.

J.F.C. Fuller: The Second World War, 1949

The amphibious landing is the most powerful tool we have.

Douglas MacArthur: planning conference for Inchon. Tokyo, 23 August 1950

Amphibious flexibility is the greatest strategic asset that a sea power possesses.

B. H. Liddell Hart: Deterrence or Defense, 1960

(*See also* Amphibious Operations, Amphibious Training, Beach, Landing Craft, Naval Warfare, Sea Power.)

Anchor

It is always well to moor your ship with two anchors.

Publilius Syrus: Sententiae, c. 50 B.C.

Allah is Allah—but I have two anchors astern.

A Turkish admiral writing to Lady Hester Stanhope, c. 1825

Do I drag my anchors?. . . All's well.

Matthew Fontaine Maury: Dying words, 1 February 1873

Annihilation

What the country needs is the annihilation of the enemy.

Nelson, 1758–1805

Antietam (Sharpsburg), 17 September 1862

Antietam's cannon long shall boom.

Herman Melville, 1819–1891

(*See also* Sharpsburg.)

Antisubmarine Warfare

The ships destroy us above
And ensnare us beneath.
We arise, we lie down, and we move
In the belly of Death.
The ships have a thousand eyes
To mark where we come . . .
And the mirth of a seaport dies
When our blow gets home.

Rudyard Kipling: The Fringes of the Fleet, 1915

It must not be forgotten that defeat of the U-boats carries with it the sovereignty of all the oceans of the world.

Winston Churchill: Statement to the French Admiralty, November 1939

Sighted sub, sank same.

Lieutenant Donald F. Mason, USN: message report while on antisubmarine patrol, 8 January 1942

The defeat of the U-boat is the prelude to all effective aggressive operations.

Winston Churchill: To conference of the Ministers of the Crown, 11 February 1943

(*See also* Submarine Warfare.)

Anzio (22 January 1944)

I had hoped that we were hurling a wildcat onto the shore, but all we had got was a stranded whale.

Winston Churchill: Closing the Ring, 1951

Appomattox (9 April 1865)

Then there is nothing left for me but to go and see General Grant, and I would rather die a thousand deaths.

R.E. Lee: To his staff after deciding to surrender the Army of Northern Virginia at Appomattox Court House, 9 April 1865

Whatever General Lee's feelings were I do not know. As he was a man of much dignity, with an impassable face, it was impossible to say whether he felt inwardly glad that the end had finally come, or felt sad over the result and was too manly to show it.

U.S. Grant: Personal Memoirs, 1885

Approach March

When invading an enemy's territory, men should always be confident in spirit, but they should fear, too, and take measures of precaution; and thus they will be at once most valorous in attack and impregnable in defense.

Archidamus of Sparta: To the Spartan forces invading Athenian territory, 431 B.C.

There is no sight in all the pageant of war like young, trained men going up to battle. The columns look solid and business-like . . . They go on like a river that flows very deep and strong. Uniforms are drab these days, but there are points of light on the helmets and the bayonets, and light in the quick steady eyes and the brown young faces, greatly daring. There is no singing—veterans know, and they do not sing much—and there is no excitement at all; they are schooled craftsmen, going up to impose their will, with the tools of their trade, on another lot of fellows; and there is nothing to make a fuss about.

John W. Thomason, Jr.: Fix Bayonets!, 1926

Arab Revolt, 1916–1918

The textbooks gave the aim in war as "the destruction of the organized forces of the enemy" by "the one process, battle." Victory could only be purchased by blood. This was a hard saying, as the Arabs had no organized forces, and so a Turkish Foch would have no aim: and the Arabs would not endure casualties, so that an Arab Clausewitz could not buy his victory.

T. E. Lawrence: Science of Guerrilla Warfare, 1929

Armament

When the necessity for arms ceases, armaments will disappear. The basic causes of war are not armaments, but in human minds.

Mahan, 1840–1914, Armaments and Arbitration

You know, my fellow citizens, what armaments mean: great standing armies, great stores of war materials. They do not mean burdensome taxation merely, they do not mean merely compulsory military service which saps the economic strength of the nations, but they mean also the building up of a military class.

Woodrow Wilson: Speech at Topeka, Kansas, 6 September 1919

We can do without butter, but, despite all our love for peace, not without arms. One cannot shoot with butter but with guns.

Paul Joseph Goebbels: Speech in Berlin, 17 January 1936

It is war that shapes peace, and armament that shapes war.

J.F.C. Fuller: Armament and History, 1945

It is customary in democratic countries to deplore expenditures on armaments as conflicting with the requirements of the social services. There is a tendency to forget that the most important social service that a government can do for its people is to keep them alive and free.

Sir John Slessor: Strategy for the West, 1954

Today the expenditure of billions of dollars every year on weapons, acquired for the purpose of making sure we never need to use them, is essential to keeping the peace.

John F. Kennedy: Speech at American University, Washington, June 1963

A stick is a good peace-maker (Bâton porte paix).

French Proverb

(*See also* Arms, Disarmament, Weapons.)

Armed Forces

An armed, disciplined body is, in its essence, dangerous to liberty. Undisciplined, it is ruinous to society.

Edmund Burke: Speech on the Army Estimates, 1790

The Congress shall have power...To raise and support Armies, but no Appropriation of Money to that Use shall be for a longer Term than two Years; To provide and maintain a Navy; To make Rules for the Government and Regulation of the land and naval Forces; To provide for calling forth the Militia to execute the Laws of the Union, suppress Insurrections and repel Invasions...

Constitution of the United States, I, 8, 1789

The President shall be Commander in Chief of the Army and Navy of the United States, and of the militia of the several States, when called into the actual Service of the United States.

Constitution of the United States, II, i, 1789

The Services in war time are fit only for desperadoes, but in peace are fit only for fools.

Benjamin Disraeli: Vivian Grey, 1827

...a class of men set apart from the general mass of the community, trained to particular uses, formed to peculiar notions, governed by peculiar laws, marked by peculiar distinctions.

William Windham, 1750–1810

The deterrence of war is the primary objective of the armed forces.

Maxwell D. Taylor: The Uncertain Trumpet, 1960

The principal armed service of its country—in its professional attitudes, its equipment, its officer corps—is an extension, a reflection, of that country's whole society.

Correlli Barnett: The Swordbearers, vi, 1963

Nothing's too good for the Armed Forces, and that's what they usually get.

Military saying, 20th cent.

(*See also* Air Force, Army, Marines, Navy, Standing Armies.)

Armistice

This solemn moment of triumph, one of the greatest moments in the history of the world...is going to lift up humanity to a higher plane of existence for all ages of the future.

David Lloyd George: Speech in London, 11 November 1918

(*See also* World War I.)

Armor

...like unscoured armor hung by the wall.

Shakespeare: Measure for Measure, i, 2, 1604

Nought can deform the human race
Like to the armor's iron brace.

William Blake, 1757–1827: Auguries of Innocence.

Not only does economy, but naval success, dictate the wisdom and expediency of fighting with iron against wood.

Stephen R. Mallory: Letter, 1862, while the late USS Merrimack *was being reconstructed as an ironclad.*

The best armor is to keep out of gun-shot.

Italian Proverb

(*See also* Armorer, Helmet.)

Armored Cruisers

Sixteen battleships all in a line,
In Guantanamo Bay look mighty fine,
But me for a cruiser every time
In the Armored Cruiser Squadron!

The Armored Cruiser Squadron, U. S. Navy song, c. 1903

Armorer

Who was the first that forged the deadly blade?
Of rugged steel his savage soul was made.

Tibullus: 1, Elegy, XI, c. 19 B. C.

The armorers, accomplishing the knights,
With busy hammers closing rivets up,
Give dreadful note of preparation.

Shakespeare: King Henry V, iv, prologue, 1598

The first artificer of death; the shrewd,
Contriver who first sweated at the forge,
And forc'd the blunt and yet unbloodied steel
To a keen edge, and made it bright for war.

William Cowper: The Task, v, 1758

Arms

The blade itself incites to violence.

Homer: Odyssey, xvi, c. 1000 B. C.

Arms and the man I sing.
Virgil: Aeneid, i, 19 B. C.

There cannot be good laws where there are not good arms.
Niccolo Machiavelli: The Prince, xii, 1513

When princes think more of luxury than of arms, they lose their state.
Niccolo Machiavelli: The Prince, xiv, 1513

...grating shock of wrathful iron arms.
Shakespeare: King Richard II, i, 3, 1595

Perhaps more valid Armes,
Weapons more violent when next we meet,
May serve to better us, and worse our foes,
Or equal what between us made the odds,
In Nature none...
Milton: Paradise Lost, 1667

The subjects which are Protestants may have arms for their defense suitable to their conditions, and as allowed by law.
The English Bill of Rights, VII, December 1689

No glory is achieved except by arms.
(Il n'y a pas de gloire achevée sans celle d'armes.)
Marquis de Vauvenargues: Oeuvres, 1747

I do not wish to see guns in the hands of all the world, for there are other *ferae naturae* besides hares and partridges.
Horace Walpole: Letter to the Countess Ossory, 26 September 1789

A well regulated Militia, being necessary to the security of a free State, the right of the people to keep and bear Arms, shall not be infringed.
Constitution of the United States, Amendment II, 1791

The first dry rattle of new-drawn steel
Changes the world today!
Rudyard Kipling: Before Edgehill Fight, 1911

The most solid moral qualities melt away under the effect of modern arms.
Ferdinand Foch: Precepts, 1919

Only when our arms are sufficient beyond doubt can we be certain that they will never be employed.
John F. Kennedy: Inaugural Address, 20 January 1961

Arms, women, and books need to be looked at every day.
Dutch Proverb

Arms keep peace.
Latin Proverb

(*See also* Armament, Artillery, Bayonet, Cannon, Flamethrower, Gun, Mines, Musket, Sword, Torpedo, Weapons.)

Army

...terrible as an army with banners.
Song of Solomon, VI, 10.

A soldiery dull and lazy, and corrupted by the circus and theaters.
Tacitus: History, v, c. 100

To give a young gentleman right education.
The Army's the only good school in the nation.
Jonathan Swift, 1667–1745, Hamilton Brown

It is not big armies that win battles; it is the good ones.
Maurice de Saxe: Mes Rêveries, iv, 1732

An army is composed for the most part of idle and inactive men, and unless the general has a constant eye upon them... this artificial machine...will very soon fall to pieces.
Frederick the Great: Instructions to His Generals, 1747

So sensible were the Romans of the imperfections of valor without skill and practice that, in their language, the name of an Army was borrowed from the word which signified exercise.
Edward Gibbon: Decline and Fall of the Roman Empire, 1776 ("Exercitus" ="Army" in Latin.)

The qualities which commonly make an army formidable are long habits of regularity, great exactness of discipline, and great confidence in the commander.
Samuel Johnson, 1709–1784

Our object ought to be to have a good army rather than a large one.
George Washington: To the President

of Congress, 15 September 1780
(cf. Marshal Saxe, ante.)

The country must have a large and efficient army, one capable of meeting the enemy abroad, or they must expect to meet him at home.
Wellington: Letter, 28 January 1811

The first measure for a country to adopt is to form an army.
Wellington: Letter, 10 December 1811

I detest war. It spoils armies.
Grand Duke Constantine of Russia, c. 1820

The army is the people in uniform.
Benjamin Constant, 1838–1891

The army is a good book to open to study human life. There one learns to put his hand to everything, to the lowest and highest things. The most delicate and rich are forced to see poverty nearly everywhere, and to live with it, and to measure its morsel of bread and draught of water.
Alfred De Vigny: Servitudes et Grandeurs Militaires, 1835

The whole system of the army is something egregious and artificial. The civilian who lives out of it cannot understand it. It is not like other professions which require intelligence. A man one degree removed from idiocy, with brains sufficient to direct his powers of mischief and endurance, may make a distinguished solider.
W. M. Thackeray, 1811–1863, in Punch (under pseudonym, "Titmarsh.")

I have never seen so teachable and helpful a class as the Army generally.
Florence Nightingale: Letter to her sister, March 1856

...a lancet for the diplomatic surgeon.
Albrecht von Roon: Letter, 29 May 1864

The Army is the most outstanding institution in every country, for it alone makes possible the existence of all civic institutions.
Helmuth von Moltke ("The Elder"), 1800–1891

An army is an aggregation of details, a defect in any one of which may destroy or impair the whole. It is a chain of innumerable links, but the whole chain is no stronger than the weakest link.
George S. Hillard, 1808–1879, Life and Campaigns of George B. McLellan

There is a soul to an army as well as to the individual man, and no general can accomplish the full work of his army unless he commands the soul of his men as well as their bodies and legs.
W. T. Sherman: Memoirs, 1875

The army has always been the basis of power, and it is so today. Power is always in the hands of those who command it.
Lyof N. Tolstoy: The Kingdom of God Is Within You, 1893

Governments need armies to protect them against their enslaved and oppressed subjects.
Lyof N. Tolstoy: The Kingdom of God Is Within You, 1893

Back to the Army again, sergeant,
Back to the Army again.
Rudyard Kipling: Back to the Army Again, 1894

The army embodies in itself its morality, its law and its mystique; and this is not the morale nor the mystique of the nation.
Captain T. Nangis (French Army), c. 1898

The real object of having an Army is to provide for war.
Elihu Root: Annual Report of the Secretary of War, 1899

It is my royal and imperial command...that you address all your skill, and all the valor of my soldiers, to exterminate the treacherous English, and to walk over General French's contemptible little army.
Kaiser Wilhelm II: General Order, Aix, 19 August 1914 (The origin of the nickname, "Old Contemptibles" for the British Expeditionary Force in 1914)

You're in the Army now,
You're not behind a plow.
You'll never get rich,
You son of a bitch,
You're in the Army now.
American soldiers' song, c. 1917

An army is to a chief what a sword is to a soldier.
Ferdinand Foch: Precepts, 1919

The functions of an Army are: (1) to defeat the enemy's main force; (2) seize upon his vitals.

Sir Ian Hamilton: The Soul and Body of an Army, iv, 1921

An army is a crowd—a homogeneous crowd, it is true, but retaining, despite its organization, some of the general characteristics of crowds; intense emotional suggestibility, obedience to leaders, etc. These factors must be handled by commanders.

Gustave le Bon: World in Revolt, 1924

The Army should become a State within the State, but it should be merged in the State through service, in fact it should itself become the purest image of the State.

General Hans von Seeckt, 1866–1936, Thoughts of a Soldier

History shows that there are no invincible armies.

Joseph V. Stalin: Address to the Russian people, 3 July 1941

An army is an institution not merely conservative but retrogressive by nature. It has such natural resistance to progress that it is always insured against the danger of being pushed ahead too fast. Far worse and more certain...is the danger of it slipping backward. Like a man pushing a barrow uphill, if the soldier ceases to push, the military machine will run back and crush him.

B. H. Liddell Hart: Thoughts on War, v, 1944

The Army, for all its good points, is a cramping place for a thinking man.

B. H. Liddell Hart: Thoughts on War, xi, 1944

The nature of armies is determined by the nature of the civilization in which they exist.

B. H. Liddell Hart, 1895–, The Ghost of Napoleon

The Army in my day was not a nine-to-five occupation. One worked very hard for a time, then very little. Above the lowest ranks a man had certain responsibilities but how he discharged them was his business.

John Masters: in Harper's Magazine, March 1963

Whoever has an army has power, and war decides everything.

Mao Tse-tung: Selected Military Writings, 1963

(*See also* Land Warfare, Soldier.)

Army, British

The English never yield, and though driven back and thrown into confusion, they always return to the fight, thirsting for vengeance as long as they have a breath of life.

Giovanni Mocenigo: To the Doge of Venice, 8 April 1588 (Mocenigo was Venetian Ambassador in Paris)

Of all the world's great heroes,
There's none that can compare
With a tow-row-row-row-row-row—
For a British Grenadier!

The British Grenadiers. Attributed to Charles Dibdin and first performed 17 January 1780 in celebration of the attack and capture of Savannah (but a minority view dates words and music from the 17th century).

Ours [the British army in the Peninsula] is composed of the scum of the earth—the mere scum of the earth. The British soldiers are fellows who have all enlisted for drink—that is the plain fact—they have all enlisted for drink.

Wellington: Letter from Portugal, 1811

As a school of military training, the Army was nothing more or less than a gigantic Dotheboys Hall.

L. S. Amery: Of the British Army, c. 1895

The British soldier can stand up to anything except the British War Office.

George Bernard Shaw: The Devil's Disciple, 1897

The British Army should be a projectile to be fired by the British Navy.

Lord Grey, 1862–1933

The builders of this empire...were not worthy of such an army. Two centuries of persecution could not wear out its patience; two centuries of thankless toil could not abate its ardor; two centuries of conquest could not awake its insolence. Dutiful to its masters, merciful to its enemies, it clung steadfastly to its old, simple ideals—obedience, service, sacrifice.

John W. Fortescue: History of the British Army, 1899–1930

Army of Northern Virginia, CSA

Who that ever looked upon it can forget that body of tattered uniforms and bright muskets—that body of incomparable infantry which for four years carried the revolt on its bayonets...which, receiving terrible blows, did not fail to give the like; and which, vital in all its parts, died only with its annihilation?

William Swinton: Campaigns of the Army of the Potomac, 1882

Pickett's Virginians were passing through;
Supple as steel and brown as leather,
Rusty and dusty of hat and shoe,
Wonted to hunger and war and weather;
Peerless, fearless, an army's flower!
Sterner soldiers the world saw never,
Marching lightly, that summer hour,
To death and failure and fame forever.

Helen Gray Cone, 1859–1934: Greencastle Jenny.

(*See also* Confederacy, Civil War, American.)

Arnhem (17 September 1944)

"Not in vain" may be the pride of those who survived and the epitaph of those who fell.

Winston Churchill: To the House of Commons, 28 September 1944

Arsenal

This is the Arsenal. From floor to ceiling,
Like a huge organ, rise the burnished arms;
But from their pipes no anthem pealing
Startles the village with strange alarms.

Henry Wadsworth Longfellow: The Arsenal at Springfield, 1841

We must be the great arsenal of Democracy.

Franklin D. Roosevelt: Address to the American people, 29 December 1940

Artificer

Another lean unwash'd artificer.

Shakespeare: King John, iv, 2, 1596

The artificers are all thieves.

Lord St. Vincent: Letter to First Lord of the Admiralty, 1797

Artillery

And he made in Jerusalem engines invented by cunning men, to be on the towers and upon the bulwarks, to shoot arrows and great stones withal.

II Kings, XXVI, 15

Thou hast also pellets of brass which are throwne foorth with terrible noyse and fire. Thou miserable man, was it not ynough to heare the thunder of Immortal God from heaven? O crueltie joyned with pride! From the earth, also, was sent foorth unimitable lightning with thunder, as Virgil sayth, which the madness of man hath counterfeited to do the like: and that which was woont to be throwne out of the cloudes is now throwne abroad with...a devlysh device.

Petrarch: De Remediis Utrisque Fortunae, dialogue 95, 1366

And if the Turks by means of their artillery gained the victory over the Persians and the Egyptians, it resulted from no other merit than the unusual noise, which frightened the cavalry...Artillery is useful to an army when the soldiers are animated by the same valor as that of the ancient Romans, but without that it is perfectly inefficient, especially against courageous troops.

Niccolo Machiavelli: Discorsi, xvii, Bk 2, 1531

Then let us bring our light artillery,
Minions, falc'nets, and sakers, to the trench,
Filling the ditches with the walls' wide breach,
And enter in to seize the hold—

Christopher Marlowe: Tamburlaine the Great, iii, 3, 1587

O, you mortal engines, whose rude throats
The immortal Jove's dread clamors counterfeit...

Shakespeare: Othello, iii, 1604

No Artillery could be better served than ours.

George Washington: General Order to the Continental Army at Freehold, N. J., 29 June 1778

Then shook the hills with thunder riven,

Then rushed the steed, to battle driven,
And louder than the bolts of Heaven,
Far flash'd the red artillery.
Thomas Campbell: Hohenlinden, 1803

It is with artillery that war is made.
Napoleon I: After Löbau, May 1809

The best generals are those who have served in the artillery.
Napoleon I: To General Gaspard Gourgaud, St. Helena, 1815

Artillery, like the other arms, must be collected in mass if one wishes to attain a decisive result.
Napoleon I, 1769–1821

God fights on the side with the best artillery.
Attributed to Napoleon I, 1769–1821

Over hill, over dale, we have hit the dusty trail,
And those caissons go rolling along.
In and out, hear them shout, "Counter-march and right-about!"
And those caissons go rolling along.
Lieutenant E. L. Gruber USA: Field Artillery Song (to the tune of 5th Field Artillery March), 1908

I have seen war, and faced modern artillery, and I know what an outrage it is against simple men.
T. M. Kettle, The Ways of War, 1915

Our thousand-stringed artillery began to play its battle-tune.
Marshal Von Hindenburg: Out of My Life, 1920

It is of great value to an army, whether in defense or offense, to have at its disposal a mass of heavy batteries.
Winston Churchill: The Gathering Storm, 1948

Artillery without targets is in a fair way to become a target itself.
Major General Orlando Ward, USA: Letter, 1965

Artillery lends dignity to what might otherwise be a vulgar brawl.
Artillerymen's saying

Artillery conquers the ground; infantry occupies it.
Military maxim

(*See also* Artilleryman, Cannon, Fire Support, Gun, Horse Artillery.)

Artilleryman

What cannoneer begot this lusty blood?
Shakespeare: King John, ii, 2, 1596

Assumes the god,
Affects to nod,
And seems to shake the spheres.
John Dryden: Alexander's Feast, 1697

Leave the artillerymen alone. They are an obstinate lot.
Napoleon I, 1769–1821

(*See also* Gunner.)

Assault

Once more into the breach, dear friends, once more;
Or close the wall up with our English dead!
Shakespeare: King Henry V, iii, 1598

If they have the advantage of a wood, or you give them a moment's time to intrench themselves, they are a nation which will pop and pop forever at you.—There is no way but to march coolly up to them,—receive their fire, and in upon them, pell mell—Ding, dong, added Trim. —Horse and foot, said my uncle Toby. Helter skelter, said Trim. Right and left, cried my uncle Toby.—Blood and 'ounds, shouted the corporal;—the battle raged.
Laurence Sterne: Tristram Shandy, 1762

On no account should we overlook the moral effect of a rapid, running assault. It hardens the advancing soldier against danger, while the stationary soldier loses his presence of mind.
Clausewitz: Principles of War, 1812

My opinion is that there ought not to be much firing at all. My idea is that the best mode of fighting is to reserve your fire till the enemy get—or you get them—to close quarters. Then deliver one deadly, deliberate volley—and charge!
Stonewall Jackson, 1863

(*See also* Attack, Charge.)

Astronautics

...To pluck bright honor
From the pale faced moon.
Shakespeare: I King Henry IV, i, 3, 1957

Atlantic, Battle of (1941–1945)

The battle of the Atlantic was the dominating factor all through the war. Never for one moment could we forget that everything happening elsewhere...depended ultimately on its outcome.
Winston Churchill: Closing the Ring, 1951

Atrocities

Cruelty in war buyeth conquest at the dearest price.
Sir Philip Sidney, 1554–1586

The soldier smiling hears the widow's cries,
And stabs the son before the mother's eyes.
With like remorse his brother of the trade,
The butcher, fells the lamb beneath his blade.
Jonathan Swift: On Dreams, 1724

It is impossible to describe to you the irregularities and outrages committed by the troops...There is not an outrage of any description which has not been committed on a people who have received us as friends by soldiers who never yet for one moment suffered the slightest want or the smallest privation...We are an excellent army on parade, an excellent one to fight, but we are worse than an enemy in a country.
Wellington: Despatch to Lord Castlereagh, 17 June 1809

(*See also* Horrors of War.)

Attack

He smote them hip and thigh.
Judges, XV, 8

Paradise is under the shadow of our swords. Forward!
Caliph Omar Ibn Alkhattab: At the battle of Kadisiya, 637

Those expert in attack consider it fundamental to rely on the seasons and the advantages of the ground; they use inundations and fire according to the situation. They make it impossible for the enemy to know where to prepare. They release the attack like a lightning bolt from above the nine-layered heavens.
Tu Yu, 735–812

The enemy comes on in gallant show.
Shakespeare: Julius Caesar, v, 1, 1599

Decline the attack altogether unless you can make it with advantage.
Maurice de Saxe: Mes Rêveries, 1732

I approve of all methods of attacking provided they are directed at the point where the enemy's army is weakest and where the terrain favors them the least.
Frederick the Great: Instructions for His Generals, xvii, 1747

Gentlemen, the enemy stands behind his entrenchments, armed to the teeth. We must attack him and win, or else perish. Nobody must think of getting through any other way. If you don't like this, you may resign and go home.
Frederick the Great: To his officers, before the battle of Leuthen, 5 December 1757

But, in case Signals can neither be seen or perfectly understood, no Captain can do very wrong if he places his Ship alongside that of an Enemy.
Nelson: Plan of attack before Trafalgar, 9 October 1805

Up Guards, and at them!
Attributed to Wellington, as his command for counterattack by the Brigade of Guards at Waterloo, 18 June 1815, but never authenticated and probably based on his command, "Stand up, Guards!", while mounting the attack.

Strike—for your altars and your fires;
Strike—for the green graves of your sires;
God—and your native land!
Fitz-Greene Halleck: Marco Bozzaris, 1825

You appear much concerned at my attacking on Sunday. I am greatly concerned, too; but I felt it my duty to do it.
Stonewall Jackson: Letter to his wife from Kernstown, Virginia, March 1862

The enemy is there, General Longstreet, and I am going to strike him.
R. E. Lee: At Gettysburg, 3 July 1863

I was too weak to defend, so I attacked.
R. E. Lee (attributed), 1807–1870

For what is more thrilling than the sudden and swift development of an attack at dawn?
Winston Churchill: The River War, vi, 1899

Attack, whatever happens!...Victory will come to the side that outlasts the other.
Ferdinand Foch: Order during the battle of the Marne, 7 September 1914

Hard pressed on my right. My center is yielding. Impossible to maneuver. Situation excellent. I am attacking.
Ferdinand Foch: Message to Marshal Joffre, battle of the Marne, 8 September 1914

A well conducted battle is a decisive attack successfully carried out.
Ferdinand Foch: Precepts, 1919

Strength lies not in defense but in attack.
Adolf Hitler: Mein Kampf, 1925

We are so outnumbered there's only one thing to do. We must attack.
Sir Andrew Browne Cunningham: Before attacking the Italian fleet at Taranto, 11 November 1940

When the situation is obscure, attack.
Attributed to General Heinz Guderian, 1888–

Attack repeat Attack.
William F. Halsey: Signal to South Pacific Force before the battle of Santa Cruz Islands, 26 October 1942

To advance is to win. When in doubt, attack.
Military maxim.

The attack, always the attack. (Attaque, toujours l'attaque).
French Proverb

(*See also* Aggressiveness, Assault, Charge.)

Attrition

Their force is wonderful, great and strong, yet we pluck their feathers by little and little.
Sir William Howard: Of the Spanish Armada, 1588

Audacity

Fortune favors the audacious.
Erasmus: Adagia, 1508

Impetuosity and audacity often achieve what ordinary means fail to achieve.
Niccolo Machiavelli: Discorsi, xliv, Bk 3, 1531

Arm me, audacity, from head to foot.
Shakespeare: Cymbeline, i, 6, 1609

Audacity, audacity again, and audacity always! (De l'audace, encore de l'audace, et toujours de l'audace!)
Georges Danton: To the French Legislative Assembly, 2 September 1792

In audacity and obstinacy will be found safety.
Napoleon I: Maxims of War, 1831

If the theory of war does advise anything, it is the nature of war to advise the most decisive, that is the most audacious.
Clausewitz: Principles of War, 1812

Never forget that no military leader has ever become great without audacity.
Clausewitz: Principles of War, 1812

Success is the child of audacity.
Benjamin Disraeli: The Rise of Iskander, 1834

My critics...want war too methodical, too measured; I would make it brisk, bold, impetuous, perhaps sometimes even audacious.
Jomini: Précis de l'Art de la Guerre, 1838

(*See also* Boldness, Daring, Resolution.)

Authority

Authority and place demonstrate and try the tempers of men, by moving every passion and discovering every frailty.
Plutarch, 46–120, Lives

Authority without wisdom is like a heavy ax without an edge, fitter to bruise than polish.

Anne Bradstreet: Meditations Divine and Moral, c. 1670

The general story of mankind will evince that lawful and settled authority is very seldom resisted when it is well employed.
Samuel Johnson: The Rambler, 8 September 1750

In the name of the great Jehovah and the Continental Congress!
Ethan Allen: Reply to Captain Delaplace of the British Army on being asked by whose authority he demanded the surrender of Fort Ticonderoga, 10 May 1775

I never knew a sailor who found fault with the orders and ranks of the service; and if I expected to pass the rest of my life before the mast, I would not wish to have the power of the captain diminished an iota.
Richard Henry Dana, Jr.: Two Years Before the Mast, 1840

The highest duty is to respect authority.
Pope Leo XIII: Libertas Praestantissimum, 20 June 1888

It is easier to give directions than advice, and more agreeable to have the right to act, even in a limited sphere, than the privilege to talk at large.
Winston Churchill: The Gathering Storm, 1948

The best test of a man is authority.
Montenegran Proverb

Aviation

If the heavens be penetrable, and no lets, it were not amiss to make wings and fly up, and some new-fangled wits should some time or other find out.
Robert Burton: The Anatomy of Melancholy, 1621

Les Anglais, nation trop fière,
S'arrogent l'empire des mers;
Les Français, nation légère,
S'emparent de celui des airs.
(The haughty English arrogate to themselves the empire of the seas;
The French, a buoyant nation, make themselves masters of the air.)
Comte de Provence (afterward Louis XVIII): Impromptu on Montgolfier's first successful balloon ascension, 1793

It seems to me worthwhile for this government to try whether it will not work on a large enough scale to be of use in the event of war.
Theodore Roosevelt: Letter concerning experiments by Samuel Langley with steam-powered aircraft, 25 March 1898

The Navy must have that. It will be important to us.
Lieutenant George C. Sweet, USN: Report to the Navy Department after an aviation demonstration by Orville Wright, September 1908

—That's good sport, but for the Army the airplane is of no use.
Ferdinand Foch: Remark at the 1910 Circuit de l'Est

(*See also* Aircraft, Aircraft Carrier, Air Force, Air Power, Balloons, Naval Aviation.)

Awards

To those young men who, either in war or other circumstances, have deserved commendation, prizes should be given.
Plato, 428–347 B. C.

To brave men, the prizes that war offers are liberty and fame.
Lycurgus of Sparta, 396–323 B. C.

It is not titles that honor men, but men that honor titles.
Niccolo Machiavelli: Discorsi, xxxviii, Bk 3, 1531

One honor won is surety for more.
Francois de la Rochefoucauld: Reflexions, 1665

Glory is the true and honorable recompense of gallant actions.
Alain René Le Sage: Gil Blas, 1735

A soldier will fight long and hard for a bit of colored ribbon.
Napoleon I: To the Captain, HMS Bellerophon, *15 July 1815*

A clergyman, or a doctor, or a lawyer feels himself no whit disgraced if he reaches the end of his worldly labors without special note of honor. But to a soldier or a sailor such indifference to his merit is wormwood. It is the bane of the profession. Nine men out of ten who go into it must live discontented and die disappointed.

Anthony Trollope: The Three Clerks, 1858

Let him who has won the palm, bear it. (Palmam qui meruit ferat.)

Latin Proverb

(*See also* Decorations, Medals.)

Bagpipe

Some men there are...when the bagpipe sings i' the nose
Cannot contain their urine.
Shakespeare: Merchant of Venice, iv, 1, 1596

And wild and high the "Cameron's Gathering" rose!
The war-note of Lochiel, which Albyn's hills
Have heard, and heard, too, have her Saxon foes—
How in the noon of night that pibroch thrills
Savage and shrill!
Byron: Childe Harold's Pilgrimage (Eve of Waterloo), III, xxvi, 1816

General Stewart, having regard to the extraordinary loss and fatigue sustained by them, desired that the 92d (Gordon Highlanders) should not join in the charge...But this time the pipe-major was not to be denied. He struck up the charging tune of *The Haughs of Cromdale,* his comrades, seized with what in the Highlands is called *mire chath*—the frenzy of battle—without either asking or obtaining permission, not only charged but led the charge...
Lieutenant Colonel C. G. Gardyne: The Life of a Regiment, 1903 (on the Gordon Highlanders at the battle of Maya, 25 July 1813).

Balaklava (25 October 1854)

It is magnificent, but it is not war.
Pierre Bosquet: On observing the charge of the Light Brigade, 25 October 1854

Those fancy fellows in the Cavalry got themselves into a pretty pickle yesterday. The whole army is delighted.
Letter home from a subaltern of the Line, Crimea, 26 October 1854

Half a league, half a league,
Half a league onward,
All in the valley of death
Rode the six hundred.
Tennyson: The Charge of the Light Brigade, 1854

(*See also* Crimean War.)

Balance of Power

...the balance of power.
Robert Walpole: Speech to the House of Commons, 13 February 1741 (the origin of this phrase).

Balloons

How posterity will laugh at us, one way or the other! If half a dozen break their necks, and balloonism is exploded, we shall be called fools for having imagined it could be brought to use; if it should be turned to account, we shall be ridiculed for having doubted.
Horace Walpole: Letter to Horace Mann, 24 June 1785

Five thousand balloons, capable of raising two men each, could not cost more than five ships of the line; and where is the prince who can afford so to cover his country with troops for its defense as that 10,000 men descending from the clouds might not in many places do an infinite deal of mischief before a force could be brought together to repel them?
Benjamin Franklin: Letter to Jan Ingenhousz, 1784

An observer is doubtless more at his ease in a clock tower than in a frail basket floating in mid-air, but steeples are not always at hand in the vicinity of battlefields, and they cannot be transported at pleasure.
Jomini: Précis de l'Art de la Guerre, 1838

Stand by to crash.
Commander Herbert V. Wiley, USN: Last order to the crew of the dirigible, USS Akron, *as she was going down, 4 April 1933*

(*See also* Aircraft.)

Bands

The Lacadaemonians moved slowly and to the music of many flute-players, who were stationed in their ranks, and played, not as an act of religion, but in order that the army might march evenly and in true measure, and that the line might not break, as often happens in great armies when they go into battle.
Thucydides: History of the Peloponnesian Wars (of the battle of Mantinea 418 B. C.)

Shooters, not tooters, are required in this service.

D. H. Hill: Endorsement disapproving an infantryman's request for transfer to duty as a bandsman, February 1862

I am delighted at the action you have taken about bands, but when are we going to hear them playing about the streets? Even quite small parade marches are highly beneficial...In fact, wherever there are troops and leisure for it there should be an attempt at military display.

Winston Churchill: Note to Secretary of State for War, 12 July 1940

(*See also* Martial Music.)

Barbary Wars, 1805–1816

My head or yours!

Yusuf Caramanli (Bey of Tripoli): Reply to the demand that he surrender Derna to U. S. Marines, 25 April 1805.

If the Dey of Algiers should bully and fume,
Or hereafter his claim to this tribute resume,
We'll send him Decatur once more to defy him,
And his motto shall be, if you please, *Carpe Diem.*

Carpe Diem—Seize the Dey, verses composed by unknown U. S. Naval personnel on conclusion of the Barbary Wars, c. 1816

Barracks

Single men in barricks don't grow into plaster saints.

Rudyard Kipling: Tommy, 1890

Bases

Ships . . .must have secure ports to which to return, and must be followed by the protection of their country throughout the voyage.

Mahan: The Influence of Sea Power Upon History, 1890

Important naval stations should be secured against attack by land as well as by sea.

Mahan: Naval Strategy, 1911

The Navy cannot play international football without a goal-keeper.

Sir Arthur Wilson, 1842–1921

Fleets cannot operate without bases.

Major General John A. Lejeune, USMC: Testimony to House Naval Affairs Committee, 13 March 1920

Battery

Captain Bragg, it is better to lose a battery than a battle.

Zachary Taylor: In reply to Bragg's assertion that he would have to displace rearward or lose his guns, Buena Vista, 22 February 1847

It is of great value to an army, whether in defense or offense, to have at its disposal a mass of heavy batteries.

Winston Churchill: The Gathering Storm, 1948

A battery seen is a battery lost.

Artilleryman's maxim.

(*See also* Artillery, Artillerymann, Horse Artillery.)

Battle

He smote them hip and thigh, with a great slaughter.

Judges, XV, 8

Set ye Uriah in the forefront of the hottest battle.

II Samuel, XI, 15

The horseman lifteth up both the bright sword and the glittering spear: and there is a multitude of slain, and a great number of carcasses; and there is none end of their corpses; they stumble upon their corpses.

Nahum, III, 3

The race is not to the swift, nor the battle to the strong.

Ecclesiastes, IX

So ends the bloody business of the day.

Homer, c. 1000 B.C., Odyssey

Nought better can betide a martial soul
Than lawful war. Happy the warrior
To whom comes joy of battle . . .

Bhagavad-Gita, Chant of Krishna to Prince Arunja before battle, c. 5th Century, B.C.

. . . the arbitrament
Of bloody strokes and mortal staring war.
Shakespeare: King Richard III, v, 3, 1592

Fight, gentlemen of England! Fight, bold yeomen!
Draw, archers, your arrows to the head!
Spur your proud horses hard, and ride in blood;
Amaze the welkin with your broken staves!
Shakespeare: King Richard III, v, 1592

A day of battle is a day of harvest for the Devil.
William Hooke: Sermon at Taunton, Massachusetts, 1640

It is an ill battle where the Devil carries the colors.
John Ray: English Proverbs, 1670

'Tis a hard battle where none escapes.
Jeremy Collier: A Short View of the Immorality and Profaneness of the English Stage, 1698

The battle was a very pretty battle as one should desire to see, but we were all so intent upon victory that we never minded the battle.
George Farquhar: The Recruiting Officer, 1706

I do not favor battles, particularly at the beginning of a war. I am sure a good general can make war all his life and not be compelled to fight one.
Maurice de Saxe: Mes Rêveries, 1732

The battle is not to the strong alone; it is to the vigilant, the active, the brave.
Patrick Henry: To the Virginia Convention, 23 March 1775

When we enter the lists of battle, we quit the sure domain of truth and leave the decision to the caprice of chance.
William Godwin: An Enquiry Concerning Political Justice, 1793

Now's the day, and now's the hour:
See the front o' battle lour . . .
Robert Burns: Scots Wha Hae wi' Wallace Bled, 1794

How hot the day!
Only our swords to shade us!
How thick the smoke, how dark the night!
Only our guns to light us!
Song of Hamzad, the Caucasus, 18th Century

The business of an English Commander-in-Chief being first to bring an Enemy's Fleet to Battle on the most advantageous terms to himself, (I mean that of laying his Ships close on board the Enemy, as expeditiously as possible); and secondly to continue them there until the Business is decided . . .
Nelson: Excerpt from Order to the Fleet, 1805

. . . my precise object . . . a close and decisive Battle.
Nelson: Excerpt from Order to the Fleet, 1805

In the lost battle,
Borne down by the flying,
Where mingles war's rattle
With groans of the dying.
Walter Scott: Marmion, canto III, xi, 1808

Battles decide everything.
Clausewitz: Principles of War, 1812

Between a battle lost and a battle won, the distance is immense and there stand empires
Napoleon I: On the eve of the battle of Leipzig, 15 October 1813

A battle sometimes decides everything; and sometimes the most trifling thing decides the fate of a battle.
Napoleon I: Letter to Barry E. O'Meara, St. Helena, 9 November 1816

I hope to God I have fought my last battle. It is a bad thing to be always fighting. While in the thick of it I am too much occupied to feel anything; but it is wretched just after. It is quite impossible to think of glory.
Wellington: To Lady Frances Shelley after Waterloo, Brussels, 19 June 1815

Next to a battle lost, the greatest misery is a battle gained.
Wellington: To Lady Frances Shelley, Brussels, 19 June 1815

. . . Battle's magnificently stern array.
Byron: Childe Harold's Pilgrimmage, (Eve of Waterloo), III, 1816

Hand to hand, and foot to foot!
Nothing there, save death, was mute;

The Bettman Archive

WILLIAM T. SHERMAN

1820–1891

"*War is hell.*"

Stroke, and thrust, and flash, and cry
For quarter, or for victory,
Mingle there with the volleying thunder.

Byron, 1788–1824, The Siege of Corinth

The battle may therefore be regarded as War concentrated, as the center of effort of the whole war or campaign. As the sun's rays unite in the focus of a concave mirror in a perfect image, and in the fulness of their heat; so the forces and circumstances of war unite in a focus in the great battle for one concentrated utmost effort.

Clausewitz: On War, 1832

Battles, even in these ages, are transacted by mechanism; men now even die, and kill one another, in an artificial manner.

Thomas Carlyle: The French Revolution, 1837

The bursting shell, the gateway wrenched asunder,
The rattling musketry, the flashing blade,
And ever on, in tone of thunder,
The diapason of the cannonade.

Henry Wadsworth Longfellow, 1807–1882

So all day long the noise of battle roll'd
Among the mountains by the winter sea,
Until King Arthur's table, man by man,
Had fallen in Lyonnesse about their lord.

Alfred Tennyson: Morte d'Arthur, 1842

Manfully
They stood, and everywhere with gallant front
Opposed in fair array the shock of war.
Desperately they fought, like men expert in arms,
And knowing that no safety could be found,
Save from their own right hands.

Robert Southey, 1744–1843

Where a battle has been fought, you will find nothing but the bones of men and beasts; where a battle is being fought, there are hearts beating.

Henry David Thoreau, 1817–1862

I have seen battles, too—
Have waded foremost in their bloody waves,
And heard their hollow roar of dying men.

Matthew Arnold: Sohrab and Rustum, 1853

The battle is there. I am going!

Joseph E. Johnston: To Beauregard at First Bull Run, 21 July 1861

Battle is the ultimate to which the whole life's labor of an officer should be directed. He may live to the age of retirement without seeing a battle; still, he must always be getting ready for it as if he knew the hour and the day it is to break upon him. And then, whether it come late or early, he must be willing to fight—he must fight.

Brigadier General C. F. Smith, USA: To Colonel Lew Wallace, September 1861

Every battle has a turning point when the slack water of uncertainty becomes the ebb tide of defeat or the flood water of victory.

Vice Admiral Charles Turner Joy, USN, 1895–1956

Read here the moral roundly writ
For him who into battle goes—
Each soul that, hitting hard or hit,
Endureth gross or ghostly foes.
Prince, blown by many overthrows,
Half blind with shame, half choked with dirt,
Man cannot tell, but Allah knows
How much the other side was hurt!

Rudyard Kipling: Verses on Games, 1898

. . . A method of untying with the teeth a knot that would not yield to the tongue.

Ambrose Bierce: The Devil's Dictionary, 1906

It is always necessary in battle to do something which would be impossible for men in cold blood.

Colonel de Grandmaison: Lecture, Ecole de Guerre, Paris, February, 1911

As long as man dwells upon the globe, his destiny is battle.

Oliver Wendell Holmes, Jr.: Letter, 1914

A battle is a swirl of "ifs" and "ands".

Sir Ian Hamilton, 1853–1947

Tout le monde à la bataille! (All hands into battle!)

Ferdinand Foch: Attack order in July 1918

Modern battle . . . is a struggle between nations, fighting for their existence, for independence, or for some less noble interest; fighting, anyhow, with all their resources and all their passions. These masses of men and of passions have to be shaken and overthrown.

Ferdinand Foch: Precepts, 1919

Battles are won by slaughter and maneuver. The greater the general, the more he contributes in maneuver, the less he demands in slaughter.

Winston Churchill: The World Crisis, II, 1923

We love battle. If battle should at length die out of the world, then all joy would die out of life.

Busso Loewe: Creed of the German Pagan Movement, 1936

France has lost a battle. But France has not lost the war.

Charles de Gaulle: Broadcast to the French people, 18 June 1940

Battle is the most magnificent competition in which a human being can indulge. It brings out all that is best; it removes all that is base.

George S. Patton, Jr: To officers, 45th Division, before the Sicily landings, 27 June 1943

The acid test of battle brings out the pure metal.

George S. Patton, Jr.: War As I Knew It, 1947

The late M. Venizelos observed that in all her wars England—he should have said Britain, of course—always wins one battle—the last.

Winston Churchill: Speech at the Lord Mayor's Luncheon, London, 10 November 1942

Battle should no longer resemble a bludgeon fight, but should be a test of skill, a maneuver combat, in which is fulfilled the great principle of surprise by striking "from an unexpected direction against an unguarded spot."

B.H. Liddell Hart: Thoughts of War, xiv, 1944

When things are going badly in battle the best tonic is to take one's mind off one's own troubles by considering what a rotten time one's opponent must be having.

Sir A.P. Wavell: Other Men's Flowers, 1944

Battles in which no one believes should not be fought.

Sir A.P. Wavell: Unpublished "Recollections," 1946

In battle the prudent become daring, the miserly lavish, and even cowards display valor.

Jean Dutourd: Taxis of the Marne, 1957

While the battles the British fight may differ in the widest possible ways, they have invariably two common characteristics—they are always fought uphill and always at the junction of two or more map sheets.

Sir William Slim: Unofficial History, 1959

That was a hard fought battle from which no man returned to tell the tale.

Irish Proverb

(*See also* Assault, Attack, Combat, Engagement, Fighting.)

Battle, Naval

In the moment of action remember the value of silence and order, which are always important in war, especially at sea.

Phormio of Athens: Speech to the Athenian seamen and marines before action in the Crisaean Gulf, 429 B.C.

What can be more terrible than a sea fight, in which both fire and water unite for the destruction of the combatants?

Vegetius: De Re Militari, v, 14, 378

It is impossible to express unto another how a smart sea-fight elevates the spirits of a man, and makes him despise all dangers. In and after all sea-fights I have been very thirsty.

Thomas Browne, II: Letter to his father, 17 July 1666

Don't give up the ship! You will beat them off!

James Mugford: Dying words as his ship, the schooner Franklin, *was being attacked by the British in Boston Harbor, 1776*

There could be no doubt that the intention

of her commander was, if he could not conquer, to sink alongside.

Captain Richard Pearson, RN: To the court of inquiry after the loss of HMS Serapis *to USS* Bonhomme Richard, *23 September 1779*

The battle rages long and loud,
And stormy winds do blow.

Thomas Campbell: Ye Mariners of England, 1801

... No Captain can do very wrong if he places his Ship alongside that of an Enemy.

Nelson: Plan of attack before Trafalgar, 9 October 1805

The captain who is not in action is not at his post.

Vice Admiral Pierre Villeneuve: Order to the combined French-Spanish fleet before Trafalgar, October 1805

Tell the men to fire faster and not to give up the ship. Fight her till she sinks.

James Lawrence: Final order as he lay mortally wounded during the action between USS Chesapeake *and HMS* Shannon, *1 June 1813 (another version is that Capt Lawrence said, "Don't give up the ship. Blow her up.")*

The boy stood on the burning deck,
Whence all but he had fled;
The flame that lit the battle's wreck
Shone round him o'er the dead.

Felicia D. Hemans: 1793–1835: Casabianca (ode to Giacomo Casabianca, son of the captain of the French ship Orient, *battle of the Nile, 1 August 1798)*

Ship after ship, the whole night long, their high-built galleons came,
Ship after ship, the whole night long, with their battle-thunder and flame,
Ship after ship, the whole night long, withdrew with her dead and her shame.

Alfred Tennyson: Ballad of the Revenge, 1880

The unpleasant features of battle are not so apparent while the fight is on; then, one is busy, his pride is aroused, and the strain upon his nerve enables him to look upon death and bloodshed with some little indifference; but exposed to danger, seeing your comrades shot down, and idle meanwhile, is trying in the extreme. There is but one thing to do under such circumstances and that is to stand up manfully and take what comes. On board a man-of-war there is no other course.

Acting Master S.B. Coleman, USN: Paper read before Michigan Commandery, The Loyal Legion, 6 March 1889

(*See also* Battle, Naval Warfare.)

Battlefield

They caught every one his fellow by the head, and thrust his sword in his fellow's side; so they fell down together; whereof that place was called Helkath-Hazzurim.

II Samuel, II, 16

... this glorious and well-foughten field.

Shakespeare: King Henry V, iv, 6, 1598

They say it was a shocking sight
After the field was won;
For many thousand bodies here
Lay rotting in the sun;
But things like that, you know, must be
After a famous victory.

Robert Southey: After Blenheim, 1798

A battlefield is at once the playroom of all the gods and the dancehall of all the furies.

Jean Paul Richter: Titan, 1803

The village of Fuentes de Oñoro, having been the field of battle, has not been much improved by the circumstance.

Wellington: Despatch to the Secretary of State for War, after the battle, 3–5 May 1811, seeking relief funds for the homeless villagers.

In a larger sense we cannot dedicate, we cannot hallow this ground. The brave men, living and dead, who struggled here, have consecrated it far above our poor power to add or detract.

Abraham Lincoln: Gettysburg Address, 19 November 1863

It is a classical maxim that it is sweet and becoming to die for one's country; but whoever has seen the horrors of a battlefield feels that it is far sweeter to live for it.

John S. Mosby: War Reminiscences, 1887

The dominant feeling of the battlefield is loneliness.

Sir William Slim: To the officers, 10th Indian Infantry Division, June 1941

Bayonet

The onset of Bayonets in the hands of the Valiant is irresistible.

Major General John Burgoyne: Extract from orderly book, 1777

The bullet is a mad thing; only the bayonet knows what it is about.

Alexander Suvorov, 1729–1800

When you want to screw in a fresh flint, do it with your bayonet: if this notches it, it will be useful as a saw, and you will, besides, show your ingenuity in making it serve for purposes for which it never was intended: though, indeed, this weapon may be said to be the most handy of any a soldier carries. It is an excellent instrument for digging potatoes, onions, or turnips. Stuck in the ground, it makes a good candlestick; and it will on occasion serve either to kill a mudlark, or to keep an impertinent boot at a proper distance, whilst your comrades are gathering his apples.

Francis Grose: Advice to the Officers of the British Army, 1782 (boot = 18th century term for countrymen.)

Rangers of Connaught! It is not my intention to expend any powder this evening. We'll do this business with the cold iron!

Sir Thomas Picton: To the 88th Foot before the assault on Badajoz, 6 April 1812

When bayonets deliberate, power escapes from the hands of the government.

Napoleon I: Political Aphorisms, 1848

The people never chafe themselves against naked bayonets.

Napoleon I: Political Aphorisms, 1848

Have you not got your bayonets?

Sir George Cathcart: At Inkermann, 5 November 1854, when word came that his division was low on ammunition

COL B.E. BEE: General, they are beating us back!

STONEWALL JACKSON: Then, sir, we will give them the bayonet!

At the First Battle of Bull Run 21 July 1861 (as reported in the Charleston Mercury, 25 July 1861)

Under Divine blessing, we must rely on the bayonet when firearms cannot be furnished.

Stonewall Jackson: Letter accompanying requisition for 1,000 pikes, 9 April 1862

Fix bayonets and go for them.

Major General Gordon Granger, USA: At Chickamauga, on receiving word that ammunition was low and the enemy massing for attack, 19 September 1863

So long as a bare couple of yards separates men, the bullet can outreach the bayonet.

B.H. Liddell Hart: Thoughts on War, xiii, 1944

A man may build himself a throne of bayonets, but he cannot sit on it.

William R. Inge, 1860–1954

A bayonet is a weapon with a worker at each end.

Socialist slogan, early 20th century

Beach

When an advancing enemy crosses water do not meet him at the water's edge. It is advantageous to allow half his force to cross and then strike.

Sun Tzu 400–320 B.C., The Art of War

To call this thing a beach is stiff,
It's nothing but a bloody cliff.

John Churchill: Y Beach, impromptu lines written after the Gallipoli landings, 1915

If the enemy starts a landing . . . concentrate all fires on the enemy's landing point and destroy him at the water's edge.

Japanese operation order for defense of Betio Island, Tarawa Atoll, 1943, (this is the first enunciation of the well known—and wholly unsuccessful—Japanese doctrine of attempting to destroy amphibious attackers on the beach itself).

(*See also* Amphibious Operations, Amphibious Warfare, Landing Craft.)

Benbow, John (1653–1702)

In 1702 the English Admiral Benbow, a courageous man, was left almost alone by his captains during three days of fighting. With an amputated arm and leg, before dying, he had four brought to trial. One was acquitted, three were hanged (sic); and from that instant dates the inflexible English severity towards commanders of fleets and vessels, a severity necessary in order to force them to fight effectively.

Ardant du Picq 1821–1870, Battle Studies, (this was the action of 19 August 1702, off Santa Marta, Colombia; actually, only two of the recreant captains were shot).

. . . that traditional seaman of the olden time, Benbow.

Mahan: The Influence of Sea Power Upon History, 1890

Bennington (16 August 1677)

There, my boys, are your enemies, red-coats and Tories. We beat them today—or Molly Stark's a widow tonight.

John Stark: To the Continental troops before engaging the British at Bennington, Vermont, 16 August 1777

Billeting

No Soldier shall, in time of peace be quartered in any house, without the consent of the Owner, nor in time of war, but in a manner to be prescribed by law.

Constitution of the United States, Amendment III, 1791

Bivouac

The hum of either army stilly sounds,
That the fixed sentinels almost receive
The secret whispers of each other's watch;
Fire answers fire, and through their paly flames
Each battle sees the other's umbered face;
Steed threatens steed, in high and boastful neighs
Piercing the night's dull ear; and from the tents
The armorers, accomplishing the knights,
With busy hammers closing rivets up,
Give dreadful note of preparation.

Shakespeare: King Henry V, iv, prologue, 1598

The first rule is always to occupy the heights; the second, that if you have a river or a stream in front of the camp, not to move more than half a musket-shot's distance from it.

Frederick The Great: Instructions for His Generals, 1747

The art of encamping in position is the same as taking up the line in order of battle in this position. To this end, the artillery should be advantageously placed, ground be selected which is not commanded or liable to be turned, and, as far as possible, the guns should command and cover the surrounding country.

Napoleon I: Maxims of War, 1831

Encampments of the same army should always be formed so as to protect each other.

Napoleon I: Maxims of War, 1831

On Fame's eternal camping-ground
Their silent tents are spread,
And Glory buards, with solemn round,
The bivouac of the dead.

Theodore O'Hara: The Bivouac of the Dead, 1847 (commemorating the American dead at Buena Vista, 22 February 1847)

Blitzkrieg

We had seen a perfect specimen of the modern Blitzkrieg; the close interaction on the battlefield of army and air force; the violent bombardment of all communications and of any town that seemed an attractive target; the arming of an active Fifth Column; the free use of spies and parachutists; and above all, the irresistible forward thrust of great masses of armor.

Winston Churchill: The Gathering Storm, 1948

Mobility, Velocity, Indirect Approach . . .

Heinz Guderian, 1888–, Definition of the blitzkrieg method as quoted by Liddell Hart, 1950

Blockade

If they are kept off the seas by our superior strength, their want of practice will make them unskillful and their want of skill, timid.

Pericles, 490–429 B.C.

In time of war . . . it is lawful for the one party to intercept the assistance and succors sent to the other.
Elizabeth I, 1533–1603

Methought the hindering of their trade the best provocation to make the enemy's fleet come out.
Lord Sandwich: After outbreak of the Second Dutch War, 1655

A battle is really nothing to the fatigue and anxiety of such a life as we lead. It is now thirteen months since I let go an anchor, and, from what I see, it may be as much longer.
Cuthbert Collingwood: Letter while on blockade station off Cadiz, 1806

Blockades in order to be binding must be effective.
Declaration of Paris, Art. 4, 1856

. . . That most hopeless form of hostilities, an inadequate commercial blockade and a war on seaborne trade.
Sir Julian Corbett: The Successors of Drake, 1900

It [blockade] is a belligerent measure which touches every member of the hostile community, and, by thus distributing the evils of war, as insurance distributes the burden of other losses, it brings them home to every man.
Mahan: Some Neglected Aspects of War, 1907

Blood

What coast knows not our blood? (Quae caret ora cruore nostro?)
Horace: Odes, Bk. ii, ode 1, 23 B.C.

Blood is the god of war's rich livery.
Christopher Marlowe: Tamburlaine the Great, iii, 2, 1587

. . . The purple testament of bleeding war.
Shakespeare: King Richard II, iii, 3, 1596

I believe that sanguine God is rather thirsty for human gore.
General Anthony Wayne: Letter, June 1779

Blood is the price of victory.
Clausewitz: On War, 1832

Blood is thicker than water.
Josiah Tatnall: In the Peiho River, China, when bringing his flagship into action to support a hard pressed British force, 25 June 1859

If blood be the price of admiralty,
Lord God, we ha' paid in full!
Rudyard Kipling: The Song of the English, 1893

(*See also* Casualties, Dead, Death, Wound, Wounded.)

Blood and Guts

War will be won by Blood and Guts alone.
George S. Patton, Jr.: Remarks to the officers, 2d Armored Division, Fort Benning, Georgia, 1940 (from this talk dated Patton's nickname, "Old Blood and Guts")

Blood and Iron

Not by speechifying and counting majorities are the great questions of the time to be solved—that was the error of 1848 and 1849—but by iron and blood.
Otto von Bismarck: To the Prussian Diet, 30 September 1862

Historical experience is written in blood and iron.
Mao Tse-tung: On Guerrilla Warfare, 1937

Blücher, Gebhardt L. von (1742–1819)

Gneisenau, if I had only learned something, what might not have been made of me! . . . But I put off everything I should have learned. Instead of studying, I have given myself to gambling, drink, and women; I have hunted, and perpetrated all sorts of foolish pranks. That's why I know nothing now. Yes, the other way I would have become a different kind of fellow, believe me.
Blücher: To Gneisenau, his chief of staff, during the campaign of 1813

. . . that drunken hussar (Cet ivrogne de de hussard.)
Napoleon I, 1769–1821

Would God that night or Blücher would come.
Attributed to the Duke of Wellington on the afternoon of Waterloo, 18 June 1815

Blunder

An enemy that commits a false step . . . is ruined, and it comes on him with an impetuosity that allows him no time to recover.
Cuthbert Collingwood, 1748–1810 (writing after Copenhagen of Nelson's quickness to profit by enemy blunders)

It is more than a crime; it is a blunder. (C'est plus qu'un crime, c'est une faute.)
Joseph Fouché: Comment on the murder of the Duc d'Enghien, 1804 (also frequently attributed to Talleyrand).

Some one had blundered.
Alfred Tennyson: The Charge of the Light Brigade, 1854

The deployments at Aboukir Bay and Trafalgar will never be seen in the Polaris Age, but any classic naval blunder may be substantially reenacted if the fleet to be defeated is so conditioned by the standard of command thinking.
Donald Neil Duncan, February 1962

To blunder twice is not allowed in war. (Bis peccare in bello non licet.)
Latin Proverb

(*See also* Mistake.)

Board

Living movements do not come out of committees.
John Henry Cardinal Newman, 1801–1890

I have always noticed that a board is long, wooden, and narrow.
George W. Goethals, 1858–1928

(*See also* Administration, Court Martial.)

Boer War (1899–1903

Fair chaos reigns here.
Sir Henry Wilson: Diary entry shortly after landing in South Africa, 25 November 1899

You entered into these two republics for philanthropic purposes and remained to commit burglary.
Lloyd George: To the House of Commons, 25 July 1900

In the Boer War . . . we owned nothing beyond the fires of our camps and bivouacs, whereas the Boers rode where they pleased all over the country.
Winston Churchill: Their Finest Hour, ii, 4, 1949.

(*See also* Colenso.)

Boldness

The gods favor the bold.
Ovid: Metamorphoses, x, c. 5 A.D.

Great empires are not maintained by timidity.
Tacitus: Histories, c. 115

Be bold, be bold, and everywhere be bold.
Edmund Spenser: The Faerie Queene, iii, 1609

Boldness be my friend!
Arm me, audacity, from head to foot!
Shakespeare: Cymbeline, i, 6, 1609

A wight man never wanted weapon. (Wight = bold)
David Fergusson: Scottish Proverbs, 1641

The measure may be thought bold, but I am of opinion the boldest are the safest.
Nelson: To Sir Hyde Parker, urging immediate, vigorous action against the Danes and Russians, 24 March 1801

Desperate affairs require desperate remedies.
Nelson, 1758–1805

Perhaps I should not insist on this bold maneuver, but it is my style, my way of doing things.
Napoleon I: Letter to Prince Eugene, 1813

"Be bold! be bold!" and everywhere—
"Be bold!"—be not too bold . . .
Henry Wadsworth Longfellow,

1807–1882, Morituri Salutamus (cf. Spenser, ante.)

Bold decisions give the best promise of success.
Erwin Rommel, 1891–1944, Rules of Desert Warfare

"Safety first" is the road to ruin in war.
Winston Churchill: Telegram to Anthony Eden, 3 November 1940

The bold are always lucky.
Danish Proverb

(*See also* Audacity, Daring, Resolution.)

Bombing, Aerial

Bombardment from the air is legitimate only when directed at a military objective, the destruction of which could constitute a distinct military disadvantage to the belligerent.
Hague Convention of Jurists, 1923

A people who are bombed today as they were bombed yesterday, and who know that they will be bombed again tomorrow and see no end to their martyrdom, are bound to call for peace at length.
Giulio Douhet: Command of the Air, 1922

The enthusiasm for baby-killing under the nicer name of "strategic bombing" is all there is to the argument for [a separate air] force.
Hoffman Nickerson: Arms and Policy, x, 1945

The bomber is the primary agent of air mastery.
Sir John Slessor: Strategy for the West, 1954

Nobody has yet found a way of bombing that can prevent foot soldiers from walking.
Walter Lippmann: in Washington Post, 18 February 1965

All experience goes to show that wars cannot be won by bombing alone.
Walter Lippmann: In Washington Post, 22 June 1965

(*See also* Air Force, Air Power.)

Boulanger, Georges (1837–1891)

Boulanger has the soul of a subaltern.
Said of Boulanger, French general and Minister of War, by Saussier, Military Governor of Paris, during the crisis of January 1889

Boyne (July 1690)

Within four yards of our fore-front,
before a shot was fired,
A sudden sniff they got that day,
which little they desired;
For horse and man fell to the ground,
and some hung in their saddle:
Others turned up their forked ends,
which we call *coup de ladle.*
Prince Eugene's regiment was the next,
on our right hand advanced,
Into a field of standing wheat,
where Irish horses pranced—
But the Brandy ran so into their heads,
their senses all did scatter,
They little thought to leave their bones
that day at the Boyne Water.
The Boyne Water (author unknown), 1690

Brass Hat

A brass hat is anybody at least one rank senior to you.
Soldier saying, c. 1920

Curse the Brass Hats: poor reptiles.
T.E. Lawrence: Letter to Ernest Thurtle, 2 May 1930

(*See also* Rank, Seniority.)

Bravery

Men of Athens, there is not much time for exhortation, but to the brave a few words are as good as many.
Hippocrates of Athens: Address to the Athenian troops before the battle of 424 B.C.

Fortune favors the brave. (Fortes fortuna adiuvat.)
Terence: Phormio, c. 160 B.C.

Only the brave enjoy noble and glorious deaths.
Dionysius of Halicarnassus: Antiquities of Rome, c. 20 B.C.

God himself helps the brave (Audentes Deus ipse iuvat.)
Ovid: Metamorphoses, c. 5 A.D.

Few men are born brave; many become so through training and force of discipline.
Vegetius: De Re Militari, iii, 378

Few men are brave by nature, but good order and experience make many so. Good order and discipline in any army are more to be depended upon than courage alone.
Niccolo Machiavelli: Art of War, 1520

A braver soldier never couchèd lance.
Shakespeare: I King Henry VI, iii, 2, 1591

What's brave, what's noble,
Let's do it after the high Roman fashion,
And make death proud to take us.
Shakespeare: Antony and Cleopatra, iv, 1606

Brave actions never want a Trumpet.
Thomas Fuller 1608–1691, Gnomologia.

Brave men are brave from the first blow.
Pierre Corneille: The Cid, 1636

A brave man never dies.
Owen Feltham, d. 1688, Resolves ("Of Fame")

None but the brave deserves the fair.
John Dryden: Alexander's Feast, 1697

Women are partial to the brave, and they think every man handsome who is going to the camp or the gallows.
John Gay: The Beggar's Opera, 1728

He (Chevalier Folard) supposes all men to be brave at all times and does not realize that the courage of the troops must be reborn daily, that nothing is so variable, and that the true skill of the general consists in knowing how to guarantee it.
Maurice de Saxe: Mes Rêveries, 1732

Who combats bravely is not therefore brave,
He dreads a death-bed like the meanest slave.
Alexander Pope: Moral Essays, 1733

How sleep the brave who sink to rest
By all their country's wishes blest.
William Collins: Ode Written in the Year 1746

The best hearts, Trim, are ever the bravest, replied my Uncle Toby.
Laurence Sterne: Tristram Shandy, 1762

The brave man is not he who feels no fear,
For that were stupid and irrational;
But he, whose noble soul its fear subdues,
And bravely shares the danger nature shrinks from.
Joanna Baillie, 1762–1851, Basil

Bravery is a quality not to be dispensed with in the officers—Like Charity, it covers a great many defects.
Benjamin Stoddert: Letter to James Simons, 13 December 1798 (Stoddert was the first Secretary of the Navy).

That man is not truly brave who is afraid either to seem or to be, when it suits him, a coward.
Edgar Allen Poe: Marginalia, 1844–1849

Bravery never goes out of fashion.
Thackeray: The Four Georges, 1860

At the bottom of a good deal of the bravery that appears in the world there lurks a miserable cowardice. Men will face powder and steel because they cannot face public opinion.
George Chapin, 1826–1880

The bravest are the tenderest,
The loving are the daring.
Bayard Taylor: The Song of the Camp, 1864

Oh, what brave people! (O, les braves gens!)
King Wilhelm of Prussia: As the French cavalry charged the Prussian lines at Gravelotte, 18 August 1870

I'll try, Sir.
Trumpeter Calvin P. Titus, USA: In reply to a call for volunteers to scale the Tartar Wall, Peking, under heavy fire, 14 August 1900 (Titus planted the Colors on the wall, was first man up, and won a Medal of Honor).

Bravery is not an individual, a racial, a national quality, in which some excel

others *per se*. It is an accident of circumstances.
Michael J. Dee: Conclusions, 1917

Oh who would not sleep with the brave?
A.E. Housman: Lancer, 1922

A brave man may fall, but he cannot yield. (Fortis cadere, cedere non potest.)
Latin Proverb

It is easy to be brave behind a castle wall.
Welsh Proverb

The world belongs to the brave.
German Proverb

(*See also* Audacity, Courage, Daring.)

Britain, Battle of (1940)

Let us therefore brace ourselves to our duties, and so bear ourselves that, if the British Empire and its Commonwealth last for a thousand year, men will say, "This was their finest hour."
Winston Churchill: To the House of Commons, 18 June 1940

Far out on the grey waters of the North Sea and the Channel coursed and patrolled the faithful, eager flotillas peering through the night. High in the air soared the fighter pilots, or waited serene at a moment's notice around their excellent machines. This was a time when it was equally good to live or die.
Winston Churchill: Their Finest Hour, 1949

The bombs have shattered my churches,
have torn my streets apart,
But they have not bent my spirit
and they shall not break my heart.
For my people's faith and courage
are lights of London town
Which still would shine in legends though
my last broad bridge were down.
Greta Briggs: London Under Bombardment, 1941

Budget

The purse and the sword ought never to get into the same hands.
George Mason, 1725–1792 (also attributed to James Madison and Alexander Hamilton, both of whom used a similar metaphor)

Popular governments are not generally favorable to military expenditure, however necessary.
Mahan: The Influence of Sea Power Upon History, 1890

The determination of United States strategy has become a more or less incidental byproduct of the administrative process of the defense budget.
Maxwell D. Taylor: The Uncertain Trumpet, 1960

The budget of the Defense Department, like the national budget, is in a sense a composite of pressures.
Hanson W. Baldwin: in New York Times, 23 November 1958

(*See also* Civil-Military Relations, Comptrollership, Cost-Consciousness, Finance.)

Bugeaud de la Piconnerie, Thomas Robert (1784–1849)

L'as-tu vue,
La casquette, la casquette,
L'as-tu vue
La casquette de Père Bugeaud?
(Have you seen it? Have you seen the helmet, the helmet of Old Man Bugeaud?)
French soldiers' song during the conquest of Algeria, after Marshal Bugeaud turned out during an enemy night attack wearing an enormous nightcap, c. 1831

Bugle

One blast upon his bugle horn
Were worth a thousand men.
Walter Scott: Lady of the Lake, 1810

And high above the fight, the lonely bugle grieves!
Grenville Mellen: Ode on the Celebration of the Battle of Bunker Hill, 17 June 1825

Blow, bugle, blow; set the wild echoes
flying.
Blow, bugle, blow; answer, echoes, dying,
dying, dying.
Alfred Tennyson: The Princess, 1850

Bring the good old bugle, boys! We'll sing
another song—
Sing it with a spirit that will start the
world along—

Sing it as we used to sing it, fifty thousand strong,
While we were marching through Georgia.
Henry Clay Work: Marching Through Georgia, 1865

What are the bugles blowin' for?" said Files-on-Parade.
"To turn you out, to turn you out," the Color-Sergeant said.
Rudyard Kipling: Danny Deever, 1890

Blow out, you bugles, over the rich dead . . .
Rupert Brooke: The Dead, 1914

(*See also* Field Music, Trumpet.)

Bull Run, First Battle of (21 July 1861)

The army which went forth to Bull Run, freighted with the hope of a loyal people, was simply a chain of weak links.
Emory Upton: The Military Policy of the United States, 1904

It was one of the best-planned battles of the war, but one of the worst-fought.
W.T. Sherman: Memoirs, 1875

Bullets

As swift as pellet out of gonne
Whan fyr is in the poudre ronne . . .
Geoffrey Chaucer, House of Fame, c. 1380

O you leaden messengers,
That ride upon the violent speed of fire,
Fly with false aim.
Shakespeare: All's Well That Ends Well, iii, 2, 1602

That shall be my music in the future!
Charles XII of Sweden: On first hearing the whistle of bullets in battle, Copenhagen, August 1700

I heard the bullets whistle; and believe me, there is something charming in the sound.
George Washington: Letter to his mother after the battle of Great Meadows, 3 May 1754

Every bullet hath its billet.
John Wesley: Journal, 6 June 1765 (Wesley called this "the odd saying of King William.")

The flying bullet down the Pass,
That whistles clear, "All flesh is grass."
Rudyard Kipling: Arithmetic on the Frontier, 1886

There is nothing more democratic than a bullet or a splinter of steel.
Wendell L. Willkie: An American Program, 1944

(*See also* Ammunition, Gunpowder.)

Bureaucracy

It [bureaucracy] consequently tends to overvalue the orderly routine and observance of the system by which it receives information, transmits orders, checks expenditures, files returns, and in general, keeps with the Service the touch of paper; in short, the organization which has been created for facilitating its own labors.
Mahan: Naval Administration and Warfare, 1903

The British bureaucrat has managed to transform inertia from a negative into a positive force. Bureaucracy is one huge "sit tight" club.
Sir Ian Hamilton: The Soul and Body of an Army, ix, 1921

The reason why there is this crabbing . . . is of course the warfare which proceeds between Air Ministry and Ministry of Aircraft Production. They regard you as a merciless critic, and even enemy. They resent having had the M.A.P. functions carved out of their show, and I have no doubt they pour out their detraction by every channel open. I am definitely of opinion that it is more in the public interest that there should be sharp criticism and counter-criticism between the two departments—than that they should be handing each other out ceremonious bouquets.
Winston Churchill: Note to Lord Beaverbrook, 15 December 1940

(*See also* Administration, Civil-Military Relations, Civilian Employees.)

Burial

In peace, sons bury their fathers; in war, fathers bury their sons.
Herodotus, 484–424 B.C., History

With all respect and rites of burial,
Within my tent his bones tonight shall lie,
Most like a soldier, order'd honorably.
Shakespeare: Julius Caesar, v, 5, 1599

I'll hide my master from the flies, as deep
As these poor pick-axes can dig.
Shakespeare: Cymbeline, iv, 2, 1609

. . . her own clay shall cover, heaped and pent,
Rider and horse,—friend, foe—in one red burial blent!
Byron: Childe Harold's Pilgrimage (The Eve of Waterloo), 1816

Not a drum was heard, not a funeral note,
As his corse to the rampart we hurried;
Not a soldier discharged his funeral shot
O'er the grave where our hero we buried.
Charles Wolfe: The Burial of Sir John Moore at Corunna, 1817

In Flanders fields the poppies blow
Between the crosses, row on row.
John McRae: In Flanders Fields, 1917

(*See also* Dead, Death, Grave.)

Caesar, Caius Julius (102–44 B.C.)

I came. I saw. I conquered.
Despatch to the Roman Senate after the battle of Zela, 47 B.C.

When Caesar says "Do this," it is perform'd.
Shakespeare: Julius Caesar, i, 1, 1599

Calm

Imperturbable calm in the Commander is essential above all things.
Sir Ian Hamilton: Excerpt from message to Lord Kitchener from Gallipoli, 12 May 1915

Camouflage

Let every soldier hew him down a bough
And bear't before him: thereby shall we shadow
The numbers of our host, and make discovery
Err in report of us.
Shakespeare: Macbeth, v, 1605

Camp-Follower

'Tis hard for a poor woman to lose nine husbands in a war, and no notice taken; nay, three of 'em, alas, in the same campaign.
Richard Steele: The Funeral, 1701

You, Sir, have an army—we have a travelling brothel.
Remark by a captured French officer to the Duke of Wellington in Spain, 1810

(*See also* Dependents, Love and War, Women.)

Cannon

... terrible as a dragon and huge as a mountain; destroyer even of the strongholds of heaven, and a weapon like a fire-raining monster.
Inscription etched on Zam-Zammah, said to be the first cannon cast in India

Men call me a breaker of tower and wall.
Through hill and dale I hurl my ball.
Sponge me dry and keep me clean,
And I'll fire a shot to Calais Green.
Inscription etched on "Queen Elizabeth's Pocket Pistol," 16th century English field piece

As when that devilish iron engine, wrought
In deepest hell, and fram'd by fury's skill,
And ramm'd with bullet round, ordain'd to kill,
With windy nitre and quick sulphur fraught,
Conceiveth fire, the heavens it doth fill.
Edmund Spenser: The Faerie Queene, i, 7, 1595

By East and West let France and England mount,
Their battering cannon chargèd to the mouths.
Shakespeare: King John, ii, 2, 1596

The cannons have their bowels full of wrath,
And ready mounted are they to spit forth
Their iron indignation.
Shakespeare: King John, ii, 1596

The thunder of my cannon shall be heard.
Shakespeare: King John, i, 1, 1596

I must report
They were as cannons overcharged with double cracks.
Shakespeare: Macbeth, i, 2, 1605

Ultima ratio regum. (The final argument of kings.)
Motto inscribed on French cannon by order of Louis XIV, 1638–1715

We are the boys
That fears no noise
Where the thundering cannons roar.
Oliver Goldsmith: She Stoops to Conquer, 1773

And nearer, clearer, deadlier than before!
Arm! Arm! it is—it is—the cannon's opening roar!
Byron: Childe Harold's Pilgrimage (Eve of Waterloo), 1816

The cannon's breath
Wings far the hissing globe of death.
Byron, 1788–1824: The Siege of Corinth

And rounder, rounder, rounder, roared the iron six-pounder,
Hurling death!

G.H. Mc Master: Carmen Bellicosum, 1849

Cannon to the right of them,
Cannon to the left of them,
Cannon in front of them
Volley'd and thunder'd.
Alfred Tennyson: The Charge of the Light Brigade, 1854

I heard the hoarse-voiced cannon roar,
The red-mouthed orators of war.
Joaquin Miller, 1841–1913

Cannon his name,
Cannon his voice, he came.
George Meredith: Napoleon, 1891

(*See also* Artillery, Cannonade, Gun.)

Cannonade

The briskest cannonade occurred on both sides—the finest music I ever heard.
Observer at Monmouth Court House, 28 June 1778

And ever on, in tone of thunder,
The diapason of the cannonade.
Henry Wadsworth Longfellow, 1807–1882

(*See also* Artillery, Cannon, Gun.)

Capabilities

We must consult our means rather than our wishes; and not endeavor to better our affairs by attempting things, which, for want of success may make them worse.
George Washington: To LaFayette, 1780

Capitulation

All generals, officers, and soldiers who capitulate in battle to save their own lives should be decimated. He who gives the order and those who obey are alike traitors and deserve capital punishment.
Napoleon I: Maxims of War, 1831

(*See also* Surrender.)

Captain

. . . the thunder of the captains and the shouting.
Job, XXXIX, 25

A brave captain is a root, out of which, as branches, the courage of his soldiers doth spring.
Sir Philip Sidney, 1554–1586

Who can look for modestie and sobrietie in the souldiers, where the captaine is given to wine, or women, and spendeth his time in riot and excesse?
Mathew Sutcliffe: The Practice, Proceedings, and Lawes of Armes, 1593

That in the captain's but a choleric word
Which in the soldier is flat blasphemy.
Shakespeare: Measure for Measure, ii, 2, 1604

As many are soldiers that are not captains
So many are captains that are not soldiers.
Nathaniel Field: A Woman Is a Weather-Cock, 1609

I take a bold Step, a rakish toss, a smart Cock, and an impudent Air to be the principal ingredients in the Composition of a Captain.
George Farquhar: The Recruiting Officer, 1706

The union of wise theory with great character will constitute the great captain.
Jomini: Précis de l'Art de la Guerre, 1838

The captain, in the first place, is lord paramount. He stands no watch, comes and goes when he pleases, and is accountable to no one, and must be obeyed in everything.
Richard Henry Dana, Jr.: Two Years Before the Mast, iii, 1840

The Captain had brushed his project aside in a way that captains have.
Winston Churchill: The Grand Alliance, 1950

(*See also* Command, Leadership, Officer.)

Career Management

Be a constant attendant at the General Officer's levees. If you get nothing else by it, you may at least learn how to scrape and bow, to simper and to display a handsome set of teeth, by watching closely the conduct of the aide-de-camps.
Francis Grose: Advice to the Officers of the British Army, 1782

(asked for)
I never (refused) anything.
(resigned)
Sir John Fisher: Memories, 1919 (Admiral Fisher attributed this to Cardinal Rampolla.)

Carnot, Lazar Nicolas Marguerite (1753–1823)

Carnot has organized victory.
Statement by one of Carnot's supporters during debate in the French Convention, 28 May 1795

Casualties

There and then our best men were killed.
Homer: Odyssey, iii, c. 1000 B.C.

I think the slain
Care little if they sleep or rise again;
And we, the living, wherefore should we ache
With counting all our lost ones?
Aeschylus: Agamemnon, c. 460 B.C.

How are the mighty fallen in the midst of battle.
II Samuel, I, 25

When life was gone, almost every man covered with his body the position he had taken at the beginning of the battle.
Sallust, 86–34 B.C., Bellum Catalinae

There were slain in this battle about 6,000 men, which, to people that are unwilling to lie, may seem very much; but in my time I have been in several actions, where, for one man that was really slain, they reported a hundred, thinking by such an account to please their masters.
Philip de Commines: Memoirs, 1498

A piteous corse, a bloody piteous corse . . .
Shakespeare: Romeo and Juliet, iii, 2, 1594

There are few that die well in a battle.
Shakespeare: King Henry V, iv, 1598

The third part of an army must be destroyed, before a good one can be made out of it.
Marquess of Halifax, 1633–1695

If you wish to be loved by your soldiers, do not lead them to slaughter . . . When you seem to be most prodigal of the soldier's blood, you spare it, however, by supporting your attacks well and by pushing them with the greatest vigor to prevent time from augmenting your losses.
Frederick The Great: Instructions for His Generals, 1747

There are six or seven thousand of the human species less than there were a month ago, and that seems to me to be all.
Lord Chesterfield: Letter to his son, 20 November 1757

What millions died—that Caesar might be be great!
Thomas Campbell: The Pleasures of Hope, 1799

Every individual nobly did his duty; and it is observed that our dead . . . were lying, as they fought, in ranks, and every wound was in the front.
Sir William Beresford: Report after Albuera, 16 May 1811

You could not be successful in such an action without a large loss. We must make up our minds to affairs of this kind sometimes, or give up the game.
Wellington: Letter to Sir William Beresford, 19 May 1811, after the Albuera

. . . one Paris night will replace them.
Attributed to Napoleon I in reference to the 75,000 French and Russian casualties at Borodino, 7 September 1812

I had never heard of a battle in which everybody was killed; but this seemed likely to be an exception, as all were going by turns.
Captain John Kincaid, fl. 1815, Adventures with the Rifle Brigade (reminiscence of Waterloo)

Let us not hear of generals who conquer without bloodshed. If bloody slaughter is a horrible sight, then that is a ground for paying more respect to war.
Clausewitz: On War, 1832

The blood of the soldier makes the glory of the general.
H.G. Bohn: Handbook of Proverbs, 1855

Well, well, General, bury these poor men and let us say no more about it.

U. S. Navy

ALFRED THAYER MAHAN

1840–1914

"Communications dominate war."

R.E. Lee: To Major General A.P. Hill after the battle of Bristoe Station, 14 October 1863

"You know," he said gravely, "it's one of the most serious things that can possibly happen to one in a battle—to get one's head cut off!"

C.L. Dodgson (Lewis Carroll): Through the Looking Glass, 1872

Casualties? What do I care about casualties?

Attributed to Major General Hunter-Weston: Gallipoli, 1915

They claim that to fire human grapeshot at the enemy without preparation, gives us moral ascendancy. But the thousands of dead Frenchmen, lying in front of the German trenches, are instead those who are giving moral ascendancy to the enemy.

Abel Ferry: Memorandum to the Cabinet of French Premier Viviani, 1916

The most fatal heresy in war, and, with us, the most rank, is the heresy that battles can be won without heavy loss.

Sir Ian Hamilton: Gallipoli Diary, 1920

Pile the bodies high at Austerlitz and
Waterloo,
Shovel them under and let me work—
I am the grass; I cover all.

Carl Sandburg, 1878–1964, Grass

A big butcher's bill is not necessarily evidence of good tactics.

Sir A.P. Wavell: Reply to Churchill, August 1940, when reproached for having evacuated British Somaliland with only 260 casualties

In battle, casualties vary directly with the time you are exposed to effective fire. Your own fire reduces the effectiveness and volume of the enemy's fire, while rapidity of attack shortens the time of exposure. A pint of sweat will save a gallon of blood!

George S. Patton, Jr.: War As I Knew It, 1947

It is always rather a pitiful business seeing men you have shot, even enemies in war.

Sir William Slim: Unofficial History, 1959

(*See also* Dead, Wounded.)

Causes of War

Wars spring from unseen and generally insignificant causes, the first outbreak being often but an explosion of anger.

Thucydides: History of the Peloponnesian Wars, ii, c. 404 B.C.

In war, important events result from trivial causes.

Julius Caesar: De Bello Gallico, 51 B.C.

The same reasons that make us quarrel with a neighbor cause war between two princes.

Michel De Montaigne: Essays, 1580

In the nature of man we find three principal causes of quarrel. First, competition; secondly, diffidence; thirdly, glory. The first maketh men invade for gain; the second, for safety; the third, for reputation.

Thomas Hobbes: Leviathan, 1651

The human heart is the starting-point in all matters pertaining to war.

Maurice de Saxe: Mes Rêveries, 1732

Troops always ready to act, my well-filled treasury, and the vivacity of my disposition—these were my reasons for making war upon Maria Theresa.

Frederick The Great: To Voltaire, discussing his invasion of Silesia, 1741

Vice foments war; virtue fights. Were there no virtue, we should have peace forever.

Marquis de Vauvenargues: Oeuvres, 1747

Each government accuses the other of perfidy, intrigue and ambition, as a means of heating the imagination of their respective nations, and incensing them to hostilities. Man is not the enemy of man, but through the medium of a false system of government.

Thomas Paine: The Rights of Man, 1791

War nourishes war.

Schiller: The Piccolomini, 1799

The well-spring of war is in the human heart.

Stephen B. Luce, 1827–1917

He drew down war by suffering himself to have an undue horror of it.

A.W. Kinglake: The Invasion of the Crimea, 1877 (of Lord Aberdeen)

States which are weak from a military standpoint, and which are surrounded by stronger neighbors invite war, and if they neglect their military organizations from false motives, they court this danger by their own supineness.

Colmar von der Goltz: Nation in Arms, 1883

War is the usual condition of Europe. A thirty years' supply of causes of war is always on hand.

Prince Kropotkin: Paroles d'un Revolté, 1884

When a war is waged by two opposing groups of robbers for the sake of deciding who shall have a freer hand to oppress more people, then the question of the origin of the war is of no real economic or political significance.

V.I. Lenin: Article in Pravda, 26 April 1917

Is there any man here or any woman—let me say, is there any child—who does not know that the seed of war in the modern world is industrial and commercial rivalry?

Woodrow Wilson: Speech in St. Louis, 5 September 1919

It takes at least two to make a peace, but one can make a war.

Neville Chamberlain: Speech in Birmingham, 28 January 1939

The urge to gain release from tension by action is a precipitating cause of war.

B.H. Liddell Hart: Thoughts on War, i, 1944

The greatest wars have the most trivial causes. (Maxima bella ex levissimis causis.)

Latin Proverb

He who has land has war.

Italian Proverb

(*See also* Declaration of War, Diplomacy, Disarmament, Militarism, National Policy, Pacifism, War.)

Cavalry

Their horses are swifter than the leopards, and are more fierce than the evening wolves; and their horsemen shall come from afar; they shall fly as the eagle that hasteth to eat.

Habakkuk, I, 8

The horse is the strength of the army. The horse is a moving bulwark.

The Hitopadesa, III, c. 500

Ten thousand cavalry only amount to ten thousand men. No one has ever died in battle through being bitten or kicked by a horse. It is men who do whatever gets done in battle.

Xenophon: Speech to the Greek army after the defeat of Cyrus at Cunaxa, 401 B.C.

The wanton troopers riding by . . .

Andrew Marvell, 1620–1678: Nymph Complaining for the Death of Her Faun.

If you choose godly, honest men to be captains of Horse, honest men will follow them.

Oliver Cromwell: Letter to Sir William Springer, September 1643

Altogether, cavalry operations are exceedingly difficult, knowledge of the country is absolutely necessary, and ability to comprehend the situation at a glance, and an audacious spirit, are everything.

Maurice de Saxe: Mes Rêveries, 1732

Old troopers and old horses are good, and recruits of either are absolutely useless.

Maurice de Saxe: Mes Rêveries, 1732

Where ever gentlemen can hunt, there can cavalry act.

Sir Banastre Tarleton: To the House of Commons, 1810

And there was mounting in hot haste: the steed,
The mustering squadron, and the clattering car,
Went pouring forward with impetuous speed,
And swiftly forming in the ranks of war.

Byron: Childe Harold's Pilgrimage (Eve of Waterloo), 1816

Boot, saddle, to horse, and away!

Robert Browning: Boot and Saddle, 1846

It makes men imperious to sit a horse.
Oliver Wendell Holmes, Sr.: Elsie Venner, 1861

Well, General, we have not had many dead cavalrymen lying about lately.
Major General Joseph Hooker USA: To Major General Averell, Chief of Cavalry, Army of the Potomac, November 1862

If you want to smell hell—
If you want to have fun—
If you want to catch the devil—
Jine the Cavalry!
Song in the Cavalry Division (J.E.B. Stuart's), Army of Northern Virginia, c. 1862

Many opportunities will be afforded to the cavalry to harass the enemy, cut off his supplies, drive in his pickets, etc. Those who have never been in battle will thus be enabled to enjoy the novel sensation of listening to the sound of hostile shot and shell, and those who have listened a great way off will be allowed to come some miles nearer, and compare the sensation caused by distant cannonade with that produced by the rattle of musketry.
D.H. Hill: Address to troops on assumption of command in North Carolina, February 1863

I have not seen in this war a cavalry command of 1,000 that was not afraid of the sight of a dozen infantry bayonets.
W.T. Sherman: To Major General J.B. Steedman, January 1864

You can't guess how spendid it is. A cavalry charge! think of that!
G.B. Shaw: Arms and the Man, i, 1894

Cavalry will never be scrapped to make room for the tanks; in the course of time cavalry may be reduced as the supply of horses in this country diminishes. This depends greatly on the life of fox-hunting.
Article by a British officer, Journal of the Royal United Service Institution, February 1921

Some entusiasts today talk about the probability of the horse becoming extinct and prophesy that the aeroplane, the tank, and the motor-car will supersede the horse in future wars . . . I am sure that as time goes on you will find just as much use for the horse—the well bred horse—as you have done in the past.
Sir Douglas Haig: Interview, 1925

Bullets can't stop cavalry.
Sir Douglas Haig, 1861–1928

A horseman unarmed is like a bird without wings.
Moroccan Proverb

(*See also* Tank, Warhorse.)

Censorship

The censorship laid upon [military] experts was, and remains, more severe than in any other Army in the world; stricter than even in the Japanese Army. There could hardly be a better proof of lack of interest. The freest people and press in the world are the most hoodwinked by military regulations for secrecy made in their own War Office! If the public had been keen about their Army they would never for a moment have stood orders framed to shield the War Office from independent criticism.
Sir Ian Hamilton: The Soul and Body of an Army, i, 1921

(*See also* Propaganda, Public Information, Public Relations.)

Centurion

I am a man under authority, having soldiers under me: and I say to this man, Go, and he goeth: and to another, Come, and he cometh.
Matthew, VIII, 9

He is to be vigilant, temperate, active, and readier to execute the orders he receives than to discuss them; strict in exercising and keeping up proper discipline among his soldiers, in obliging them to appear clean and well dressed and to have their arms constantly polished and bright.
Vegetius: De Re Militari, 378

(*See also* Gunnery Sergeant, Noncommissioned Officer.)

Ceremonies

The pride, pomp and circumstance of glorious war.
Shakespeare: Othello, iii, 1604

Before the battle joins afar,
The field yet glitters with the pomp of war.
John Dryden, 1631–1700

Bad luck to this marching,
Pipe-claying and starching;
How neat one must be to be killed by the French!
Soldiers' song, Peninsular War, 1811

What makes a regiment of soldiers a more noble object of view than the same mass of mob? Their arms, their dress, their banners, and their art and artificial symmetry of their position and movements.
Byron: Letter to John Murray, 7 February 1821

Wherever there are troops and leisure for it, there should be an attempt at military display.
Winston Churchill: Note to Secretary of State for War, 12 July 1940

No one who has participated in it or seen it well done should doubt the inspiration of ceremonial drill.
Sir A.P. Wavell: Soldiers and Soldiering, 1953

Chairborne Warfare

Forsooth, a great arithmetician . . .
That never set a squadron in the field,
Nor the division of a battle knows
More than a spinster . . .
. . . mere prattle, without practice,
Is all his soldiership.
Shakespeare: Othello, i, 1, 1604

The habit of the arm-chair easily prevails over that of the quarter-deck; it is more comfortable.
Mahan: Naval Administration and Warfare, 1903

Chance

To a good general luck is important.
Livy: History of Rome, c. 110

Now good or bad, 'tis but the chance of war.
Shakespeare: Troilus and Cressida, prologue, 1601

In the reproof of chance
Lies the true proof of men.
Shakespeare: Troilus and Cressida, i, 3, 1601

We should make war without leaving anything to chance, and in this especially consists the talent of a general.
Maurice de Saxe: Mes Rêveries, 1732

In war something must be allowed to chance and fortune seeing it is, in its nature, hazardous and an option of difficulties.
James Wolfe: Letter to a friend, 5 November 1757

Something must be left to chance; nothing is sure in a Sea Fight beyond all others.
Nelson: Plan of Attack, before Trafalgar, 9 October 1805

There is no human affair which stands so constantly and so generally in close connection with chance as War.
Clausewitz: On War, 1832

The affairs of war, like the destiny of battles, as well as of empires, hang upon a spider's thread.
Napoleon I: Political Aphorisms, 1848

Countless and inestimable are the chances of war.
Winston Churchill: The River War, vii, 1899

(*See also* Fortunes of War, Luck.)

Chancellorsville (1 May 1863)

I trust God will grant us a great victory. Keep closed on Chancellorsville.
Stonewall Jackson: Final orders to J.E.B. Stuart, 1 May 1863

Change

The Service isn't what it used to be—and never was.
Service saying.

Change of Command

I deliver to you a fleet that is mistress of the seas.
Lysander of Sparta: On turning over command of the Spartan fleet to Callicratidas, 406 B.C.

To remove a General in the midst of a campaign—that is the mortal stroke.
Duke of Marlborough, 1650–1722

When promoted to the command of a regiment from some other corps, show them that they were all in the dark before, and, overturning their whole routine of discipline, introduce another as different as possible . . . If you can only contrive to vamp up some old exploded system, it will have all the appearance of novelty to those, who have never practiced it before: the few who have, will give you credit for having seen a great deal of service.

Francis Grose: Advice to the Officers of the British Army, 1782

Whether I am to command the army or not, or am to quit it, I shall do my best to insure its success.

Wellington: Letter to Lord Castlereagh after receiving word that he might be superseded in command, 1 August 1808

There will always be found at the beginning of a war or upon change of commanders, a restless impatience to do something, to make a showing of results, which misleads the judgment of those in authority, and commonly ends, if not in failure, at least in barren waste of powder and shot.

Mahan, 1840–1914

Change of Station

Pay, pack, and follow.

Sir Richard Burton: Instructions to his wife, on being abruptly ordered home from duty in Damascus, 1870

Channels of Command

For when the king is on the field nothing is done without him; he in person gives general orders to the polemarchs, which they convey to the commanders of divisions; these again to the commanders of fifties, the commanders of fifties to the commanders of *enomoties*, and these to the *enomoty*. In like manner any more precise instructions are passed down through the army, and quickly reach their destination. For almost the whole Lacadaemonian army are officers who have officers under them, and the responsibility of executing an order devolves upon many.

Thucydides: Peloponnesian Wars, 422 B.C.

I was informed that all the causes of delay had been reported through the "usual channels," but as far as those on the spot were aware nothing very much seems to have happened. It would seem best therefore to start from the other end of the "usual channels" and sound backwards to find where the delay in dealing with the matter has occurred.

Winston Churchill: Note for General Ismay, 26 January 1941

Chaplain

He has the canonical smirk and filthy clammy palm of a chaplain.

William Wycherley: The Country Wife, c. 1673

He taught them how to live and how to die.

William Somerville, 1675–1742, In the Memory of the Rev. Mr. Moore

The chaplain is a character of no small importance in a regiment, though many gentlemen in the army think otherwise . . . Never preach any practical morality to the regiment. That would be only throwing away your time. To a man, they all know, as well as you do, that they ought not to get drunk or commit adultery: but preach to them on the Trinity, the attributes of the Deity, and other mystical and abstruse subjects, which they may never before have thought or heard of. This will give them a high idea of your learning: besides, your life might otherwise give the lie to your preaching.

Francis Grose: Advice to the Officers of the British Army, 1782

You may indulge yourself in swearing, and talking bawdy as much as you please; this will show you are not a stiff high priest.

Francis Grose: Advice to the Officers of the British Army, 1782

Theology gives sure rules for spiritual government, but not for the government of armies.

Napoleon I: Political Aphorisms, 1848

(*See also* Christian Soldier.)

Charge

Charge, and give no foot of ground.

Shakespeare: III King Henry VI, i, 4, 1590

Advance your standards, draw your willing swords.

Shakespeare: King Richard II, v, 3, 1592

I am your king. You are a Frenchman. There is the enemy. Charge!

Henry IV of France: At the battle of Ivry, 14 March 1590

The enemy say that Americans are good at a long shot but cannot stand the cold iron. I call upon you to give a lie to the slander. Charge!

Winfield Scott: To the 11th Infantry at Chippewa, Canada, 5 June 1814

Come on, you sons of bitches—do you want to live forever?

Gunnery Sergeant Daniel Daly, USMC: to his platoon at Belleau Wood, 6 June 1918

Goddamn it, you'll never get the Purple Heart hiding in a foxhole! Follow me!

Lieutenant Colonel Henry P. Crowe, USMC: Guadalcanal, 13 January 1943

Chariot

For thou shalt drive out the Canaanites, though they have iron chariots, and though they be strong.

Joshua, XVII, 18

Thy walls shall shake at the noise of the horsemen, and of the wheels, and of the chariots.

Ezekiel, XXVI, 10

Charles XII of Sweden (1682–1718)

Ten years of unbroken success—and two hours of mismanagement!

Duke of Marlborough: on hearing of Charles XII's defeat at Poltava, 27 June 1709

He left the name at which the world grew pale,
To point a moral or adorn a tale.

Samuel Johnson, 1709–1784 (of Charles XII)

Chemical Warfare

The effects of the successful gas attack were horrible. I am not pleased with the idea of poisoning men. Of course the entire world will rage about it first and then imitate us. All the dead lie on their backs with clenched fists; the whole field is yellow.

Rudolph Binding: A Fatalist at War, 1915 (of the first German use of lethal gas, Vijfwege, Belgium, April 1915)

Even in theory the gas mask is a dreadful thing. It stands for one's first flash of insight into man's measureless malignity against man.

Reginald Farrer: The Void of War, 1918

Chief of Staff

The leading qualifications which should distinguish an officer for the head of the staff are, to know the country thoroughly; to be able to conduct a reconnaissance with skill; to superintend the transmission of orders promptly; to lay down the most complicated movements intelligibly, but in a few words and with simplicity.

Napoleon I: Maxims of War, 1831

Gneisenau makes the pills which I administer.

Gebhardt von Blücher, 1742–1819: Of Gneisenau, his chief of staff

If the relations between the General and his Chief of Staff are what they ought to be, the boundaries are easily adjusted by soldierly and personal tact and the qualities of mind on both sides.

Paul Von Hindenburg: Out of My Life, 1920

Chippewa River (5 July 1814)

By God, these are Regulars.

Major General Phineas Riall: On observing the assault of Winfield Scott's brigade under heavy British fire, 5 July 1814

(*See also* Charge.)

Chivalry

. . . for Christian service and true chivalry.

Shakespeare: King Richard II, i, 1, 1595

I shall maintain and defend the honest adoes and quarrels of all ladies of honor, widows, orphans, and maids of good fame.

The Oath of Knighthood (written out by Sir William Drummond of Hawthornden, 1619)

None but the brave deserves the fair.

John Dryden: Alexander's Feast, 1697

The age of chivalry is gone; and that of sophisters, economists, and calculators, has succeeded.

Edmund Burke: Reflections on the Revolution in France, 1790

So faithful in love, and so dauntless in war,
There never was knight like the young
Lochinvar.

Walter Scott: Marmion, v (Lochinvar), 1808

As a soldier, preferring loyal and chivalrous warfare to organized assassination if it be necessary to make a choice, I acknowledge that my prejudices are in favor of the good old times when the French and English Guards courteously invited each other to fire first—as at Fontenoy—preferring them to the frightful epoch when priests, women, and children throughout Spain plotted the murder of isolated soldiers.

Jomini: Précis de l'Art de la Guerre, 1838

In one sense the charge that I did not fight fair is true. I fought for success and not for display. There was no man in the Confederate army who had less of the spirit of knight-errantry in him, or who took a more practical view of war than I did.

John S. Mosby: War Reminiscences, 1887

It is the essence of chivalry to interfere.

Philip Guedalla: Fathers of the Revolution, 1926

Generals think war should be waged like the tourneys of the Middle Ages. I have no use for knights. I need revolutionaries.

Adolf Hitler: To Hermann Rauschning, 1940

Christian Soldier

A good Christian will never make a bad soldier.

Gustavus Adolphus: After the landing at Usedom, 24 June 1630

If you choose godly, honest men to be captains of Horse, honest men will follow them.

Oliver Cromwell: Letter to Sir William Springer, September 1643

The soldier at the same time may shoot out his prayer to God, and aim his pistol at his enemy, the one better hitting the mark for the other.

Thomas Fuller, 1608–1661: Good Thoughts for Bad Times.

You appear much concerned at my attacking on Sunday. I am greatly concerned, too; but I felt it my duty to do it.

Stonewall Jackson: Letter to his wife, Kernstown, Virginia, March 1862

Our gallant little army is increasing in numbers, and my prayer is that it may be an army of the living God as well as of its country.

Stonewall Jackson: Letter to his wife, 7 April 1862

Onward, Christian soldiers,
Marching as to war,
With the cross of Jesus
Going on before!

Sabine Baring-Gould: Hymn, Onward Christian Soldiers, 1864

We are on our last legs, owing to the delay in the expedition. However, God rules all, and as He will rule to His glory and our welfare, His will be done . . . I am quite happy, thank God, and like Lawrence, I have "*tried* to do my duty."

Charles George Gordon: Last letter from Khartoum, 14 December 1884

A soldier is Christ's warrior. As such he should regard himself, and so he should behave.

General M. E. Dragomirov: Notes for Soldiers, c. 1890

If the historian of the future should deem my service worthy of some slight reference, it would be my hope that he mention me not as a commander engaged in campaigns and battles, even though victorious to American arms, but rather as one whose sacred duty it became, once the guns were silenced, to carry to the land

of our vanquished foe the solace and hope and faith of Christian morals.
Douglas MacArthur, 1880–1964

(*See also* Prayer Before Action.)

Churchill, Sir Winston (1874–1965)

Winston is back.
General message from the Admiralty to all ships and shore stations, 3 September 1939, announcing Churchill's reappointment as First Lord of the Admiralty, the post he had held in World War I.

Civil-Military Relations

To make appointments is the province of the sovereign; to decide on battle, that of the general.
Wang Ling ("Master Wang"): fl. c. 200

A uniform coat and a cocked hat are sufficient reasons for the inhabitants why they will not assist or relieve the soldier's distress.
Colonel Michael Jackson: Letter to General Henry Knox, 1777

Though soldiers are the true supports,
The natural allies of Courts,
Woe to the Monarch who depends
Too much on his red-coated friends;
For even soldiers sometimes think—
Nay, Colonels have been known to
reason—
And reasoners, whether clad in pink,
Or Red, or blue, are on the brink
(Nine cases out of ten) of treason.
Thomas Moore, 1779–1852

Military affairs should be left to military men.
James Monroe: After the battle of Bladensburg, 24 August 1814 (Much but not all of the extensive blame for this disaster could be laid at Monroe's door for meddling in the measures for the defense of Washington).

The particulars of your plan I neither know nor seek to know...I wish not to obtrude any constraints or restraints on you.
Abraham Lincoln: Letter to U. S. Grant, March 1864

The business routine of even the most military department...is in itself more akin to civil than to military life: but it by no means follows that those departments would be better administered under men of civil habits of thought than by those of military training. The method exists for the result, and an efficient fighting body is not to be attained by weakening the appreciation of military necessities at the very fountain head.
Mahan: Naval Administration and Warfare, 1903

In its proper manifestation the jealousy between civil and military spirits is a healthy symptom.
Mahan: Naval Administration and Warfare, 1903

An army cannot be directed in war nor commanded in peace under the immediate authority of a civilian. There must be a military commander, the obedient servant of the Government supported by the Government in the exercise of his powers...and sheltered by the Government against all such criticism as would weaken his authority.
Spencer Wilkinson: The Brain of an Army, preface to 1913 edition

Civilians have had no instruction, training or experience in the principles of war, and to that extent are complete amateurs in the methods of waging war. It is idle, however, to pretend that intelligent men whose minds are concentrated for years on one task learn nothing about it by daily contact with its difficulties and the way to overcome them.
Lloyd George: War Memoirs, VI, 1937

Military philosophies, bred and crystallized in the crucible of war against the elements and other adversaries, may not convincingly register on mentalities trained and experienced in totally different circumstances.
Admiral R. B. Carney, USN: Address to the Naval War College, 31 May 1963

(*See also* Civilian Interference, Civilian Supremacy, Home Front, Noncombatants, Politico-Military Affairs.)

Civil War

Deep are the wounds that civil strife inflicts. (Alta sedent civilis vulnera dextrae.)
Lucan: Bellum Civile, c. 48.

Make us enemies of every people on earth, but save us from civil war.
Lucan: Bellum Civile, c. 48

There is nothing unhappier than a civil war, for the conquered are destroyed by, and the conquerors destroy, their friends.
Dionysius of Halicarnassus: Antiquities of Rome, c. 20 BC

Civil dissension is a viperous worm.
Shakespeare: I King Henry VI, iii, 1, 1591

Civil blood makes civil hands unclean.
Shakespeare: Romeo and Juliet, prologue, 1594

The war of all against all.
Thomas Hobbes: De Cive, 1642

In such battles [civil wars] those who win are established as loyalists, the vanquished as traitors.
Calderon, 1600–1681

From hence, let fierce contending nations know
What dire effects from civil discord flow,
Joseph Addison: Cato, 1713

One of the greatest difficulties in civil war is, that more art is required to know what should be concealed from our friends than what ought to be done against our enemies.
Lord Chesterfield: Miscellaneous Pieces, 1777

Civil wars strike deepest of all into the manners of the people. They vitiate their politics; they corrupt their morals; they pervert even the natural taste and relish of equity and justice. By teaching us to consider our fellow-citizens in a hostile light, the whole body of our nation becomes gradually less dear to us.
Edmund Burke: Letter to the Sheriffs of Bristol, 3 April 1777

I am one of those who have probably passed a longer period of my life engaged in war than most men, and principally in civil war; and I must say this, that if I could avoid, by any sacrifice whatever, even one month of civil war in the country to which I was attached, I would sacrifice my life in order to do it.
Wellington: To the House of Lords, March 1829

The proudest capitals of Western Europe have streamed with civil blood.
T. B. Macaulay: History of England, 1848

A foreign war is a scratch on the arm; a civil war is an ulcer which devours the vitals of a nation.
Victor Hugo: Ninety-Three, 1879

Civil wars end in dictators.
Alan Moorehead: The Blue Nile, ix, 1962

(*See also* Insurgency, Partisan Warfare, Rebellion, Revolt, Revolution.)

Civil War, American (1861–1865)

No sir, you dare not make war on cotton. No power on earth dares make war upon it. Cotton is king.
James Henry Hammond: To the Senate, March 1858

...an irrepressible conflict between opposing and enduring forces.
William H. Seward: Speech at Rochester, N. Y., 25 October 1858, forecasting war

You had better, all you people of the South, prepare yourselves for a settlement of this question. It must come up for settlement sooner than you are prepared for it, and the sooner you commence that preparation, the better for you. You may dispose of me very easily: I am nearly disposed of now; but this question is still to be settled—this negro question, I mean.
John Brown: After his capture at Harper's Ferry, 19 October 1859

The time may come when your state may need your services; and if that time does come, then draw your swords and throw away your scabbards!
Stonewall Jackson: To the cadets, Virginia Military Institute, March 1861

In your hands, my dissatisfied fellow-countrymen, and not in mine, is the momentous issue of civil war. The government will not assail you. You can have no conflict without being yourselves the aggressors. *You* have no oath registered in heaven to destroy the government, while *I* shall have the most solemn one to "preserve, protect, and defend it."

Abraham Lincoln: Inaugural Address, 4 March 1861

Since my interview with you on the 18th inst., I have felt that I ought not longer to retain my commission in the Army...Save in the defense of my native State, I never desire again to draw my sword.

R. E. Lee: Letter to Winfield Scott, 20 April 1861

A reckless and unprincipled tyrant has invaded your soil. Abraham Lincoln, regardless of all moral, legal and constitutional restraints, has thrown his Abolition hosts among you, who are murdering and imprisoning your citizens, confiscating and destroying your property, and committing other acts of violence and outrage too shocking and revolting to humanity to be enumerated.

P. G. T. Beauregard: Proclamation to the people of Virginia, 1 June 1861

This war was never really contemplated in earnest. I believe if either the North or the South had expected that their differences would result in this obstinate struggle, the cold-blooded Puritan and the cock-hatted Huguenot and Cavalier would have made a compromise.

George E. Pickett: Letter to his fiancee, 27 June 1862

I worked night and day for twelve years to prevent the war, but I could not. The North was mad and blind, and would not let us govern ourselves, and so the war came. Now it must go on until the last man of this generation falls in his tracks and his children seize his musket and fight our battles.

Jefferson Davis: To J. F. Jaquess and J. R. Gilmore, 17 July 1864

...rebellion, begun in error and perpetuated in pride.

W. T. Sherman: Letter to Abraham Lincoln from Atlanta, 17 September 1864

If I had the power I would arm every wolf, panther, catamount and bear in the mountains of America, every crocodile in the swamps of Florida, every Negro in the South, every devil in Hell, clothe him in the uniform of the Federal Army, and then turn them loose on the rebels of the South and exterminate every man, woman and child south of Mason and Dixon's line.

Parson W. P. Brownlow: Speech in New York, 1866

In the South the war is what A. D. is elsewhere; they date from it.

S. L. Clemens (Mark Twain): Life on The Mississippi, 1883

Through our good fortune, in our youth our hearts were touched with fire...We have seen with our own eyes beyond and above the gold fields, the snowy heights of honor, and it is for us to bear the report to those who come after us.

Oliver Wendell Holmes, Jr.: Memorial Day address, 1884

[The Civil War] was a fearful lesson, and should teach us the necessity of avoiding wars in the future.

U. S. Grant: Personal Memoirs, 1885

The troops on both sides were American, and united they need not fear any foreign foe.

U. S. Grant: Personal Memoirs, I, 1885

The generations to come will celebrate the glory. This generation knows the cost.

Thomas B. Reed: Speech at the Portland, Me., Centennial, 1886

Under the sod and the dew,
Waiting the judgment-day;
Under the laurel, the Blue,
Under the willow, the Grey.

Francis Miles Finch, 1827–1907, The Blue and the Grey

Had the South had a people as numerous as it was warlike, and a navy commensurate with its other resources as a sea power, the great extent of its sea-coast and its numerous inlets would have been elements of great strength. [The Union blockade] was a great feat, a very great feat; but it would have been an impossible feat had the Southerners been more numerous, and a nation of seamen.

Mahan: The Influence of Sea Power Upon History, 1890

Never did sea power play a greater or more decisive part than in the contest which determined that the course of world history would be modified by the existence of one great nation, instead of several rival states, on the North American continent.

Mahan: The Influence of Sea Power Upon History, 1890

The transcending facts of the American Civil War are the military genius of Robert

E. Lee and the naval superiority of the North. Behind all the blood and sacrifice, behind the movements of armies and the pronouncements of political leaders, the war was essentially a contest between these two strategic forces. Lee's tactical opponent was the Army of the Potomac, but his strategic rival was the Union Navy.

Rear Admiral John D. Hayes: Sea Power in the Civil War, 1961

(*See also* Confederacy, Southern.)

Civilian Employees

Away with him, I say! Hang him with his pen and ink-horn about his neck.

Shakespeare: II Henry VI, ii, 1590

We have more machinery of government than is necessary, too many parasites living on the labor of the industrious.

Thomas Jefferson: Letter to William Ludlow, 1824

The clerk combines the vices of a courtier with all the bad habits of the poverty in which he vegetates.

Stendhal (Henri Beyle): A Life of Napoleon, 1818

Now landsmen all, whoever you may be,
If you want to rise to the top of the tree,
If your soul isn't fettered to an office stool,
Be careful to be guided by this golden rule—
Stick close to your desks and *never go to sea*,
And you all may be Rulers of the Queen's Navee!

W. S. Gilbert: HMS Pinafore, 1878

Every man who takes office in Washington either grows or swells.

Woodrow Wilson: Address, Washington, 15 May 1916

Clerks in the War Office . . . always on the nibble at any specialty in custom or dress upon which corps took a particular pride.

Sir Ian Hamilton: The Soul and Body of an Army, vi, 1921

...a set of blear-eyed clerks.

Winston Churchill: My Early Life, 1930

War hath no fury like a non-combatant.

C. E. Montague: Disenchantment, 1922

(*See also* Bureaucracy.)

Civilian Interference

There are three ways in which a ruler can bring misfortune upon his army:
When ignorant that the army should not advance, to order an advance, or ignorant that it should not retire, to order a retirement...
When ignorant of military affairs, to participate in their administration...
When ignorant of command problems to share in the exercise of responsibilities.

Sun Tzu, 400–320 B. C., The Art of War

In every circle and at truly every table there are people who lead armies in Macedonia, who know where the camp ought to be placed; what ports ought to be occupied by the troops; and when and through what pass that territory should be entered; where magazines should be established; how provisions should be conveyed by land and by sea; when it is proper to engage the enemy; and when to lie quiet. And they not only determine what is best to be done, but if anything is done in any other manner than they have proposed, they arraign the consul as if he were on trial before them... If therefore, anyone thinks himself qualified to give advice respecting the war I am to conduct, let him come with me into Macedonia...but if he thinks this too much trouble, and prefers the repose of city life to the toils of war, let him not on the land, assume the office of pilot.

Lucius Aemilius Paulus: Speech in Rome prior to departing to take command in Macedonia, 168 B.C.

If one ignorant of military affairs is sent to participate in the administration of the army, then in every movement there will be disagreement and mutual frustration and the entire army will be hamstrung.

Wang Hsi, c. 1100

The Romans left the commanders of their armies entirely uncontrolled in their operations...The republics of the present day, such as the Venetians and the Florentines, act very differently, so that, if their generals...wish merely to place a battery of artillery, they want to know and direct it...and which has brought them to the condition in which they now find themselves.

Niccolo Machiavelli: Discorsi, 1531

To a service of a very special character, involving special exigencies, calling for special aptitudes, and consequently demanding special knowledge of its requirements in order to deal wisely with it, were applied the theories of men wholly ignorant of those requirements—men who did not even believe that they existed.

Mahan: Life of Nelson, 1897 (of the French Navy under Napoleon)

...a popular outcry will drown the voice of military experience.

Mahan: Naval Strategy, 1911

Your order...has been received and promptly complied with. With such interference with my command I cannot expect to be of much service in the field.

Stonewall Jackson: To Judah P. Benjamin, Confederate Secretary of War, 31 January 1862

Politics must always exercise an extreme influence on strategy; but it cannot be gain-said that interference with the commanders in the field is fraught with the gravest danger.

Colonel G. F. R. Henderson: Stonewall Jackson, 1898

The statesman has nothing in his gift but disaster so soon as he leaves his own business of creating or obviating wars, and endeavors to conduct them.

Sir Ian Hamilton: A Staff Officer's Scrapbook, 1905

What soldier relishes the sight of a civilian flourishing a sword?

Philip Guedalla: Wellington, 1931

(*See also* Civil-Military Relations.)

Civilian Supremacy

Let the soldier give way to the civilian. (Cedant arma togae.)

Cicero: Orationes Philippicae, c. 60 B. C.

A man without one scar to show on his skin, that is smooth and sleek with ease and home-keeping habits, will undertake to define the office and duties of a general.

Plutarch, 46–120, Lives (Aemilius Paulus)

In all cases the military should be under strict subordination to and governed by the civil power.

Virginia Declaration of Rights, 12 June 1776

We shall wait in vain for the awakening in our country of that public spirit which the English and the French and other peoples possess, if we do not imitate them in setting for our military leaders certain bounds and limitations which they must not disregard.

Carl Zuckmayer, 1896, Deutschland, ein unendlicher Prolog

Nothing is more absolutely indispensable to a good soldier than perfect subordination and zealous service to him whom the national will may have made the official superior for the time being.

Major General J. M. Schofield, USA, 1831–1906

When a dispute arises between the civil head of a department and its great professional expert, in that dispute the civil head is judge. If a man goes to the Admiralty as First Lord, and has been there only a week or a fortnight, he can dismiss at his own will his expert advisers, and let the people know that when it comes to a question whether an indispensable expert, or an ignorant civilian, who has only been there a fortnight, is to go, the decision rests with the civilian.

Sir Edward Carson: To the House of Commons, 1916

It is impossible for a country that has had forty-two war ministers in forty-three years to fight effectively.

Hans Delbrück: Krieg und Politik, 1919

Our arms must be subject to ultimate civilian control and command at all times, in war as well as peace. The basic decisions on our participation in any conflict and our response to any threat—including all decisions relating to the use of nuclear weapons, or the escalation of a small war into a large one—will be made by the regularly constituted civilian authorities.

John F. Kennedy, 1917–1963

(*See also* Bureaucracy, Civilian Interference, Civil-Military Relations, Politico-Military Affairs.)

Clausewitz, Karl von (1780–1831)

My way always takes me across a great

battlefield; without my entering upon it, no permanent happiness will come to me.
Letter by Clausewitz to his fiancee, 1806

This fellow has a common sense that borders on wittiness.
Karl Marx: Of Clausewitz in a letter to Friedrich Engels, 1858

...the Mahdi of mass and mutual massacre.
Sir B. H. Liddell Hart, 1895–, The Ghost of Napoleon

Coast Defense

The raison d'etre of the defense of our military ports is to give a free hand to our navy, and to enable them to take the sea with the confident assurance that their bases of operations can be safeguarded in their absence.
Major General Henry Schaw, 1829–1902

One gun on land is worth ten at sea.
French maxim, 19th century

(*See also* Shore Bombardment.)

Cold War

The whole art of war is being transformed into mere prudence, with the primary aim of preventing the uncertain balance from shifting suddenly to our disadvantage and half-war from developing into total war.
Clausewitz: On War, viii, 1832

Although the war is over, we are in the midst of a cold war which is getting warmer.
Bernard M. Baruch: To the Senate War Investigating Committee, 24 August 1948 (Baruch had coined and used the phrase, "cold war," earlier but this seems to be the first documentation of it.)

...All mischief short of war.
Winston Churchill, 1874–1964

Cold wars cannot be conducted by hotheads. Nor can ideological conflicts be won as crusades or concluded by unconditional surrender.
Walter Lippmann: The Russian-American War, 1949

We can lose the world, one parcel of real estate after another, while we wait for a shot that may never be fired.
Arthur W. Radford: Speech, 1956

We think of war in terms of killing; [the communists] think of war in terms of subverting.
J. F. C. Fuller: Letter to Journal, Royal United Service Institution, 16 February 1960

Colenso

It was a very trying day.
Sir Redvers Buller: Of the battle of Colenso (15 December 1899), when General Buller, attempting to relieve Ladysmith, was badly beaten, losing ten guns, many prisoners, and heavy casualties.

(*See also* Boer War.)

Collision

A collision at sea can ruin your entire day.
Attributed to Thucydides, 5th century B. C.

Colonel

When giving orders to your regiment on the parade, or marching at the head of it, you will doubtless feel as bold as a cock, and look as fierce as a lion; yet, when the commander-in-chief, or any other general officer approaches, it must all subside into the meekness of the lamb and the obsequiousness of the spaniel.
Francis Grose: Advice to the Officers of the British Army, 1782

There are no bad regiments; there are only bad colonels.
Napoleon I, 1769–1821

For soldiers sometimes think—
Nay, Colonels have been known to
reason...
Thomas Moore, 1779–1852

Colors

Our colors do return
In those same hands that did display them
when
We first marched forth.
Shakespeare: King John, ii, 2, 1596

The soldiers should make it an article of faith never to abandon their standard. It should be sacred to them; it should be respected; and every type of ceremony should be used to make it respected and precious.

Maurice de Saxe: Mes Rêveries, 1732

They shall not be lost.

Anthony Wayne: Letter to Secretary of War Knox, requesting the issues of of colors and standards for Wayne's Legion, 1793

Stood for his country's glory fast,
And nail'd her colors to the mast!

Walter Scott: Marmion, i, 1808

The Colors must never be struck.

Lieutenant W. W. Burrows, Jr., USN: While dying on board his ship, USS Enterprise, *during the action in which she took HMS* Boxer, *5 September 1813*

Every means should be taken to attach the soldier to his colors.

Napoleon I: Maxims of War, 1831

We will sink with our colors flying.

Lieutenant George U. Morris, USN: When called on to strike the colors of the sinking frigate, USS Cumberland, *by the commander of the Confederate ironclad,* Virginia, *Hampton Roads, 8 March 1862*

(*See also* Flag.)

Combat

And let thy blows, doubly redoubled,
Fall like amazing thunder on the casque
Of thy adverse, pernicious enemy.

Shakespeare: King Richard II, i, 3, 1595

Scarce could they hear, or see their foes,
Until at weapon-point they close.
They close, in clouds of smoke and dust,
With sword-sway, and with lance's thrust.

Walter Scott: Marmion, vi, 1808

Combats may be quite independent of scientific combinations; they may become essentially dramatic; personal qualities and inspirations and a thousand other things frequently are the controlling elements.

Jomini: Precis de l'Art de la Guerre, 1838

Under the sky there is no uglier spectacle than two men with clenched teeth and hellfire eyes, hacking one another's flesh; converting precious living bodies, and priceless living souls, into nameless masses of putrescence useful only for turnip-manure.

Thomas Carlyle: Past and Present, 1843

...the dearest school on earth.

W. T. Sherman, 1820–1891

The men of the tattered battalion which fights till it dies,
Dazed with the dust of the battle, the din and the cries,
The men with the broken heads and the blood running into their eyes.

John Masefield, 1878–
A Consecration

We call Japanese soldiers fanatics when they die rather than surrender, whereas American soldiers who do the same thing are heroes.

Robert M. Hutchins: Address at University of Chicago, June 1945

Human beings, like plans, prove fallible in the presence of those ingredients that are missing in maneuvers—danger, death, and live ammunition.

Barbara W. Tuchman: The Guns of August, 1962

(*See also* Battle.)

Combat Efficiency

Chariots strong, horses fast, troops valiant, weapons sharp—so that when they hear the drums beat the attack they are happy, and when they hear the gongs sound the retirement they are enraged. He who is like this is strong.

Chang Yu, fl 1100

One regiment of 3,000 rifles, if well cared for, represents after a few days of campaigning, 2,800 rifles; less well managed, it will no longer include more than 2,000. The variations in morale of a force are at least as ample. How then compare two regiments with each other?

Ferdinand Foch: Precepts, 1919

Casualties many; percentage of dead not known; combat efficiency: we are winning.
David M. Shoup: Situation report from the beach, Betio Island, 21 November 1943

Combat Experience

It is right to learn, even from the enemy. (Fas est et ab hoste doceri.)
Ovid: Metamorphoses, iv, c. 8 A. D.

Men who are familiarized to danger, meet it without shrinking, whereas those who have never seen Service often apprehend danger where no danger is.
George Washington: Letter to President of Congress, 9 February 1776

OCTAVIUS: But he's a tried and valiant soldier.
ANTONY: So is my horse, Octavius, and for that I do appoint him store of provender.
Shakespeare: Julius Caesar, iv, i, 1599

War is a singular art. I assure you that I have fought sixty battles, and I learned nothing but what I knew when I fought the first one.
Napoleon I: To General Gaspar Gourgaud, St. Helena, 1815

I love a brave soldier who has undergone the baptism of fire, whatever nation he may belong to.
Napoleon I: To Barry O'Meara, St. Helena, 1816

Use makes a better soldier than the most urgent considerations of duty,—familiarity with danger enabling him to estimate the danger. He sees how much is the risk, and is not afflicted with imagination; knows practically Marshal Saxe's rule, that every soldier killed costs the enemy his weight in lead.
R. W. Emerson, 1803–1882

Nothing is more exhilarating than to be shot at without result.
Winston Churchill: The Malakand Field Force, 1898

Combat Fatigue

The greatest weapon against the so-called "battle fatigue" is ridicule.
George S. Patton, Jr.: War As I Knew It, 1947

Combined Arms

It is not so much the mode of formation as the proper combined use of the different arms which will insure victory.
Jomini: Précis de l'Art de la Guerre, 1838

(*See also* Fire Support.)

Command

Learn to obey before you command.
Solon of Athens, 638–559 B. C.

A general good at commanding troops is like one sitting in a leaking boat or lying under a burning roof. For there is no time for the wise to offer counsel nor the brave to be angry. All must come to grips with the enemy.
Wu Ch'i, 430–381 B. C.

By command I mean the general's qualities of wisdom, sincerity, humanity, courage, and strictness.
Sun Tzu, 400–320 B. C., The Art of War

I object to saying things twice.
Plautus, 254–184 B. C., Pseudolus

I am a man under authority, having soldiers under me: and I say to this man, Go, and he goeth; and to another, Come, and he cometh.
Matthew, VIII, 9

A prince should therefore have no other aim or thought, nor take up any other thing for his study, but war and its organization and discipline, for that is the only art that is necessary to one who commands.
Niccolo Machiavelli: The Prince, 1513

He who wishes to be obeyed must know how to command.
Niccolo Machiavelli: Discorsi, 1531

Who hath not served cannot command.
John Florio: First Fruites, 1578

You have that in your countenance which I would fain call master.
Shakespeare: King Lear, i, 4, 1605

There is great force hidden in a sweet command.
George Herbert: Outlandish Proverbs, 1640

If a man faint under the burden of such tediousness as usually attendeth upon warlike designments, he is in no way fit for enterprise; because the two chief parts of a soldier are Valor and Sufferance; and there is as much honour gained by suffering wants patiently in war, as by fighting valiantly, and as great achievements effected by the one, as by the other...and yet it is an easier matter to find men that will offer themselves willingly to death, than such as will endure Labour with patience.
George Monk, Duke of Albemarle, 1608–1670

No man is fit to command another that cannot command himself.
William Penn: No Cross, No Crown, 1669

The higher we rise, the more isolated we become; and all elevations are cold.
Duc de Boufflers, 1644–1711

The success of my whole project is founded on the firmness of the conduct of the officer who will command it.
Frederick the Great: Instructions for His Generals, 1747

If you are deficient in knowledge of your duty, the word of command given in a boatswain's tone of voice, with a tolerable assurance...will carry you through till you get a smattering of your business.
Francis Grose: Advice to the Officers of the British Army, 1782

I can no longer obey; I have tasted command, and I cannot give it up.
Napoleon I: In conversation with Miot de Melito, 1798

You cannot conceive how few men are qualified to command ships-of-the-line as they ought to be.
Lord St. Vincent: Letter to Earl Spenser, 22 June 1800

I command—or I hold my tongue.
Napoleon I: Political Aphorisms, 1848

Nothing is so important in war as an undivided command.
Napoleon I: Maxims of War, 1831

I consider it a great advantage to obtain command young, having observed as a general thing that persons who come into authority late in life shrink from responsibility, and often break down under its weight.
David G. Farragut: Journal entry, 1819

There are Field Marshals who would not have shone at the head of a cavalry regiment, and *vice versa.*
Clausewitz: On War, 1832

The best means of organizing the command of an army...is to: (1) Give the command to a man of tried bravery, bold in the fight and of unshaken firmness in danger. (2) Assign as his chief of staff a man of high ability, of open and faithful character, between whom and the commander there may be perfect harmony.
Jomini: Précis de l'Art de la Guerre, 1838

If officers desire to have control over their commands, they must remain habitually with them, industriously attend to their instruction and comfort, and in battle lead them well.
Stonewall Jackson: Letter of Instruction to Commanding Officers, Winchester, Virginia, November 1861

I cannot trust a man to control others who cannot control himself.
Attributed to R. E. Lee, 1807–1870

To be at the head of a strong column of troops, in the execution of some task that requires brain, is the highest pleasure of war.
W. T. Sherman: Personal Memoirs, II, xxv, 1875

Thousands of moralists have solemly repeated the old saw that only he can command who has learnt to obey. It would be nearer the truth to say that only he can command who has the courage and the initiative to disobey.
William McDougall, 1871–1938, Character and Conduct of Life

The reward of the general is not a bigger tent, but command.
Oliver Wendell Holmes, Jr.: Law and the Court, 1913

I don't know who won the Battle of the Marne, but if it had been lost, I know who would have lost it.
Joseph J. C. Joffre: To the Briey Parliamentary Commission, 1919

...that gift of command which can still animate the troops at the last stage of exhaustion.
Ferdinand Foch: Precepts, 1919

The power to command has never meant the power to remain mysterious.
Ferdinand Foch: Precepts, 1919

In action it is better to order than to ask.
Sir Ian Hamilton: Gallipoli Diary, 1920

Superiority of material strength is given to a commander gratis. Superior knowledge and superior tactical skill he must himself acquire. Superior morale, superior cooperation, he must himself create.
Admiral Joseph Mason Reeves, 1872–1948

A commander must accustom his staff to a high tempo from the outset, and continuously keep them up to it. If he once allows himself to be satisfied with norms, or anything less than an all-out effort, he gives up the race from the starting post, and will sooner or later be taught a bitter lesson.
Erwin Rommel, 1891–1944

I do not play at war. I shall not allow myself to be ordered about by commanders-in-chief. I shall make war. I shall determine the correct moment for attack. I shall shrink from nothing.
Adolf Hitler: Regarding the Russian campaign, 1943

It is sad to remember that, when anyone has fairly mastered the art of command, the necessity for that art usually expires—either through the termination of the war or through the advanced age of the commander.
George S. Patton, Jr.: War as I Knew It, 1947

At the top there are great simplifications. An accepted leader has only to be sure of what it is best to do, or at least have his mind made up about it. The loyalties which center upon number one are enormous. If he trips, he must be sustained. If he makes mistakes, they must be covered. If he sleeps, he must not be wantonly disturbed. If he is no good, he must be pole-axed.
Winston Churchill: Their Finest Hour, 1949

Command doth make actors of us all.
John Masters: The Road Past Mandalay, 1961

It is pleasant to command, though it be only a flock of sheep.
Spanish Proverb

(*See also* Commander, Generalship, Leadership, Unity of Command.)

Command Relations

Land Officers have no pretense to command at sea. Sea Officers have no pretense to command on land.
Royal Navy Signal Book, 1744

Whereas the Success of this Expedition will very much depend upon an entire Good Understanding between Our Land and Sea Officers, We do hereby strictly enjoin and require you...to maintain and cultivate such a good Understanding and Agreement ...As the Commander-in-Chief of Our Squadron is instructed, on his part to entertain and cultivate the same good Understanding and Agreement.
William Pitt: Secret instructions, drafted by Pitt for King George II, to General Wolfe for the expedition against Quebec, 1759

The Earl of Chatham with sword drawn
Was waiting for Sir Richard Strachan;
Sir Richard, longing to be at 'em,
Was waiting for the Earl of Chatham.
George Canning: On lack of cooperation between Navy and Army at Walcheren, July 1809, where Chatham and Strachan were joint commanders

We had Generals who were Admirals and Admirals who wanted to be Generals. Generals acting as Admirals were bad enough but it was the Admirals who wanted to be Generals who imperilled victory among the coral islands.
Holland M. Smith: Coral and Brass, 1949

Commander

The commander of an army is one in whom civil and martial acumen are combined. To unite resolution with resilience is the business of war.

Wu Ch'i, 430–381 B. C., Art of War

Seeing so many vertues are required in a captaine, and so small faultes lay him open to the enemy, it is no marvell if perfect Generalls be so rare and hard to find.

Mathew Sutcliffe: The Practice, Proceedings, and Laws of Armes, 1593

The troops are scattered, and the commanders very poor rogues.

Shakespeare: All's Well That Ends Well, iv, 3, 1602

The Commander of an Army neither requires to be a learned explorer of history nor a publicist, but he must be well versed in the higher affairs of state; he must know and be able to judge correctly of traditional tendencies, interests at stake, the immediate questions at issue, and the characters of leading persons; he need not be a close observer of men, a sharp dissecter of human character, but he must know the character, the feelings, the habits, the peculiar faults and inclinations, of those whom he is to command. He need not understand anything about...the harness of a battery horse, but he must know how to calculate exactly the march of a column...These are matters only to be gained by the exercise of an accurate judgment in the observation of things and men.

Clausewitz: On War, 1832

Oh ye kind Heavens, there is in every Army a fittest, wisest, bravest, best; whom, could we find and make Commander, all were in truth well—by what art discover him—for our need of him is great!

Thomas Carlyle: Sartor Resartus, 1834

Our army would be invincible if it could be properly organized and officered. There were never such men in an army before. They will go anywhere and do anything if properly led. But there is the difficulty—proper commanders.

R. E. Lee: To Stonewall Jackson, 1862

The military commander is the fate of the nation.

Helmuth Von Moltke ("The Elder"), 1800–1891

Great results in war are due to the commander. History is therefore right in making generals responsible for victories—in which case they are glorified; and for defeats—in which case they are disgraced.

Ferdinand Foch: Precepts, 1919

The be-medalled Commander, beloved of the throne,
Riding cock-horse to parade when the bugles are blown.

John Masefield, 1878– , A Consecration

Sorting out muddles is really the chief job of a commander.

Sir A. P. Wavell: The Training of the Army for War, Journal, Royal United Service Institution, February 1933

It is in the minds of the commanders that the issue of battle is really decided.

B. H. Liddell Hart: Thoughts on War, iii, 1944

A commander should have a profound understanding of human nature, the knack of smoothing out troubles, the power of winning affection while communicating energy, and the capacity for ruthless determination where required by circumstances. He needs to generate an electrifying current, and to keep a cool head in applying it.

B. H. Liddell Hart: Thoughts on War, xi, 1944

There are more tired corps and division commanders than there are tired corps and divisions.

George S. Patton, Jr.: War As I Knew It, 1947

All very successful commanders are prima donnas, and must be so treated.

George S. Patton, Jr.: War As I Knew It, 1947

The commander must be the prime mover of the battle and the troops must always have to reckon with his appearance in personal control.

Erwin Rommel: The Rommel Papers, ix, 1953

One of the most valuable qualities of a commander is a flair for putting himself in the right place at the vital time.
Sir William Slim: Unofficial History, 1959

. . . that melting-point of warfare—the temperament of the individual commander.
Barbara W. Tuchman: The Guns of August, 1962

The commander's will must rest on iron faith; faith in God, in his cause, or in himself.
Correlli Barnett: The Swordbearers, iii, 1963

(*See also* Commander-in-Chief, General, High Command.)

Commander-in-Chief

A commander-in-chief, whose power and dignity are so great and to whose fidelity and bravery the fortunes of his countrymen, the defense of their cities, the lives of soldiers, and the glory of the state, are entrusted, should not only consult the good of the army in general, but extend his care to every private soldier in it.
Vegetius: De Re Militari, 378

A commander-in-chief is greater than a sultan; for if he is not the Lord's viceregent, he is the King's, which in the idea of a military man is much better.
Francis Grose; Advice to the Officers of the British Army, 1782

A commander-in-chief is as infallible as the Pope, and, being the King's representative, he can do no wrong, any more than his royal master.
Francis Grose: Advice to the Officers of the British Army, 1782

The President shall be Commander in Chief of the Army and Navy of the United States, and of the militia of the several States, when called into the actual Service of the United States.
Constitution of the United States, II, 1, 1789

The character of the man is above all other requisites in a commander-in-chief.
Jomini: Precis de l'Art de la Guerre, 1838

(*See also* Commander, High Command.)

Communications

If you are able to hold critical points on his strategic roads the enemy cannot come. Therefore Master Wang said: "When a cat is at the rat hole, ten thousand rats dare not come out; when a tiger guards the ford, ten thousand deer cannot cross."
Tu Yu, 735–812

When masses of troops are employed, certainly they are widely separated, and ears are not able to hear acutely nor eyes to see clearly. Therefore officers and men are ordered to advance or retreat by observing the flags and banners and to move or stop by signals of bells and drums. Thus the valiant shall not advance alone, nor shall the coward flee.
Chang Yu, fl 1000

Communications dominate war; broadly considered, they are the most important single element in strategy, political or military.
Mahan: The Problem of Asia, 1900

(*See also* Lines of Communication.)

Comptrollership

. . . an arrogant controller.
Shakespeare: II King Henry VI, iii, 2, 1590

Nothing is more hurtefull to the proceedings of warres than miserable niggardise.
Mathew Sutcliffe: The Practice, Proceedings, and Lawes of Armes, 1593

War and economy are things not easily reconciled, and the attempt of leaning towards parsimony in such a state may be the worst economy in the world.
Edmund Burke, 1729–1797

War is business, to which actual fighting is incidental.
Attributed to Mahan, 1840–1914

We have at this moment to destinguish carefully between running an industry or a profession, and winning the war.
Winston Churchill: Memorandum for First Sea Lord, 8 October 1939

You can't run a government solely on a business basis.
Herbert H. Lehman, 1878–1963

The Armed Forces will never show a dollar-and-cents profit.
Observation by unidentified officer, c. 1950, quoted in The Professional Soldier, Janowitz

(*See also* Cost Consciousness, Finance.)

Comradeship

We few, we happy few, we band of brothers...
Shakespeare: King Henry V, iv, 3, 1598

It is not for myself, or on my own account chiefly, that I feel the sting of disappointment. No! It is for my brave officers, for my noble-minded friends and comrades, Such a gallant set of fellows! Such a band of brothers! My heart swells at the thought of them!
Nelson, 1758–1805

There's never a bond, old friend, like this—
We have drunk from the same canteen!
Charles Graham Halpine (Miles O'Reilly), 1829–1868, The Canteen

...Haunted by that spirit of camaraderie which, in the army, has done us so much harm, certain high officials hesitated to apply the red-hot iron where it was needed.
General Jean Mordacq: Le Ministère Clemenceau, I, 1920

Concentration

It is better to be on hand with ten men than to be absent with ten thousand.
Tamerlane, 1336–1405

One should never risk one's whole fortune unless supported by one's entire forces.
Niccolo Machiavelli: Discorsi, 1531

If you meet two enemies, do not each attack one. Combine both on one of the enemy; you will make sure of that one, and you may also get the other afterwards; but, whether the other escape or not, your country will have won a victory, and gained a ship.
Nelson, 1758–1805: Instructions to two frigate captains sent off on detached service

The essence of strategy is, with a weaker army, always to have more force at the crucial point than the enemy.
Napoleon I, 1769–1821

When you have resolved to fight a battle, collect your whole force. Dispense with nothing. A single battalion sometimes decides the day.
Napoleon I: Maxims of War, 1831

An army should always keep its columns so united as to prevent the enemy from passing between them with impunity.
Napoleon I: Maxims of War, 1831

Concentration is the secret of strength in politics, in war, in trade, in short in all management of human affairs.
R. W. Emerson: The Conduct of Life, 1860

I always make it a rule to get there first with the most men.
Nathan Bedford Forrest, 1821–1877 (Widely misquoted as, "I git thar fustest with the mostest men.")

...that power of concentration which is the strength of armies and of minorities.
Mahan: The Interest of America in Sea Power, 1893

The proverbial weakness of alliances is due to inferior power of concentration.
Mahan: Naval Strategy, 1911

In any military scheme that comes before you, let your first question to yourself be, Is this consistent with the requirement of concentration?
Mahan: Naval Strategy, 1911

The principles of war could, for brevity, be condensed into a single word—"Concentration."
B. H. Liddell Hart: Thoughts on War, ix, 1944

True concentration is the product of dispersion.
B. H. Liddell Hart: Strategy, 1954

March divided and fight concentrated.
Military maxim

Confederacy, Southern

The young bloods of the South: sons of planters, lawyers about towns, good billiard-players and sportsmen, men who never did work and never will...War suits them, and the rascals are brave, fine riders, bold to rashness, and dangerous subjects in every sense. They care not a sou for niggers, land, or any thing. They hate Yankees, *per se*, and don't bother their brains about the past, present, or future. As long as they have good horses, plenty of forage, and an open country, they are happy.

W. T. Sherman: Letter to General Halleck, 17 September 1863

You who, in the midst of peace and prosperity, have plunged a nation into war—dark and cruel war—who dared and badgered us to battle, insulted our flag, seized our arsenals and forts that were left in the honorable custody of peaceful ordnance sergeants, seized and made "prisoners of war" the very garrisons sent to protect your people against negroes and Indians, long before any overt act was committed by the (to you) hated Lincoln government; tried to force Kentucky and Indiana into rebellion, spite of themselves; falsified the vote of Louisiana; turned loose your privateers to plunder unarmed ships; expelled Union families by the thousands, burned their houses, and declared, by an act of your Congress, the confiscation of all debts due Northern men for goods had and received! Talk thus to the Marines, but not to me, who have seen these things.

W. T. Sherman: Letter to General John B. Hood, CSA, from Atlanta, 10 September 1864

An angel's heart, an angel's mouth,
Not Homer's, could alone for me
Hymn well the great Confederate South,
Virginia first, and *Lee*!

Philip Stanhope Worsley: Verses inscribed in a presentation copy of his translation of the Iliad, to R. E. Lee, January 1864

(*See also* Civil War, American.)

Confidence

In confidence and quietness shall be your strength.

Isaiah, XXX, 15

Fields are won by those who believe in winning.

Thomas Wentworth Higginson, 1823–1911

The most vital quality a soldier can possess is *self-confidence*, utter, complete, and bumptious.

George S. Patton, Jr.: Letter to Cadet George S. Patton, III, USMA, 6 June 1944

Confusion

Remember that war is always a far worse muddle than anything you can produce in peace, and that sorting out muddles is really the chief job of a commander and his staff;...but find out afterward how that particular muddle occurred and, if possible, don't let it happen again.

Sir A. P. Wavell: Training of the Army for War, Journal, Royal United Service Institution, 15 February 1933

Congress

The Congress shall have power...To raise and support Armies, but no Appropriation of Money to that Use shall be for a longer Term than two Years; To provide and maintain a Navy; To make Rules for the Government and Regulation of the land and naval Forces; To provide for calling forth the Militia to execute the Laws of the Union, suppress Insurrections and repel Invasions...

Constitution of the United States, I, 8, 1789

Congress is the great commanding theater of this nation.

Thomas Jefferson: Letter to William Wirt, 1808

When a nation is at war, the presence of a deliberative body is injurious and often fatal.

Napoleon I: Political Aphorisms, 1848

Though the President is commander-in-chief, Congress is his commander; and, God willing, he shall obey.

Thaddeus Stevens: To the House of Representatives, 3 January 1867

Congress provides; the President commands.

Charles Evans Hughes, C. J., 1862–1948 (on the division of power over the military establishment between the executive and legislative branches)

McNamara doesn't think Congress is a necessary evil; he thinks it is an unnecessary evil.

F. Edward Hébert, 1901–__.

(*See also* Commander-in-Chief, Politicians.)

Conquer

Conquered, we conquer, (Victi vincimus.)
Plautus: Casina, c. 200 B.C.

To conquer without danger is to triumph without glory.
Pierre Corneille: The Cid, i, 1636

Arise, go forth and conquer as of old.
Alfred Tennyson: The Passing of Arthur, 1872

Divide and conquer.
Ancient military maxim.

(*See also* Conqueror, Conquest, Victory.)

Conqueror

. . . A conqueror of conquerors (Victor victorum)
Plautus: Trinummus, c. 190 B.C.

I came, I saw, I conquered. (Veni, vidi, vici.)
Julius Caesar: Despatch to the Roman Senate after the battle of Zela, 47 B.C.

The conqueror would rather burst a city gate than find it open to admit him; he would rather ravage the land with fire and sword than overrun it without protest from the husbandmen. He scorns to advance by an unguarded road or to act like a peaceful citizen.
Lucan, Bellum Civile, c. 48

The vanquished never yet spake well of the conqueror.
Samuel Daniel: A Defence of Rhyme, 1602

I came, I saw, God conquered.
John III Sobieski of Poland: Message to the Pope after defeating the Turks at Vienna, 12 September 1683

The fame of a conqueror is a cruel fame.
Lord Chesterfield: Letter to his son, 30 September 1757

The English conquered us, but they are far from being our equals.
Napoleon I: Letter to General Gaspard Gourgaud, St. Helena, 1815

Rats and conquerors must expect no mercy in misfortune.
C.C. Colton: Lacon, 1820

A conqueror, like a cannon-ball, must go on. If he rebounds, his career is over.
Attributed to the Duke of Wellington, 1769–1852

The greatest conqueror is he who overcomes the enemy without a blow.
Chinese Proverb

(*See also* Conquer, Conquest.)

Conquest

To rejoice in conquest is to rejoice in murder.
Lao Tze: The Tao-Teh King, c. 500 B.C.

In the practical art of war, the best thing of all is to take the enemy's country whole and intact; to shatter and destroy it is not so good . . . Hence, to fight and conquer in all your battles is not supreme excellence; supreme excellence consists in breaking the enemy's resistance without fighting.
Attributed to Sun Tzu, 400-320 B.C.

Have they not divided the prey; to every man a damsel or two?
Judges, V, 30

Moab is my washpot: over Edom will I cast out my shoe; over Philistia will I triumph.
Psalm CVIII

When we conquer enemies by kindness and justice we are more apt to win their submission than by victory in the field. In the one case, they yield only to necessity; in the other by their own free choice.
Polybius: Histories, V, c. 125 B.C.

Whoever conquers a free town and does not demolish it commits a great error and may expect to be ruined himself.
Niccolo Machiavelli: The Prince, v, 1513

So many goodly cities ransacked and razed; so many nations destroyed and made desolate; so infinite millions of harmless people of all sexes, states and ages massacred, ravaged, and put to the sword; and the richest, the fairest, and best part of the world overturned, and defaced for the traffic of pearls and pepper: Oh, base conquest!
Michel de Montaigne: Essays, iii, 1588

We go to gain a little patch of ground,
That hath in it no profit but the name.
Shakespeare: Hamlet, iv, 1600.

Those possessions short-liv'd are,
Into which we come by war.
Robert Herrick: Hesperides, 1648

There is no such conquering weapon as the necessity of conquering.
George Herbert: Iacula Prudentum, 1651

A conquest made by a democracy is always odious to the subject states. It becomes thereby monarchical by a fiction, but it is always more oppressive than a monarchy, as the experience of all times and ages shows.
Montesquieu: The Spirit of the Laws, x, 1748

If there be one principle more deeply rooted than any other in the mind of every American, it is that we should have nothing to do with conquest.
Thomas Jefferson: Letter to William Short, 1791

There is no state, the chief of which does not desire to secure to himself a constant state of peace by the conquest of the whole universe, if it were possible.
Immanual Kant: Perpetual Peace, Supplement I, 1795

You are badly fed and all but naked . . . I am about to lead you into the most fertile plains in the world. Before you are great cities and rich provinces; there we shall find honor, glory, and riches.
Napoleon I: Proclamation to the Army of Italy, April 1796

Conquest is the most debilitating fever of a nation and the rudest of glories . . . The sword does not plow deep.
Francis Lieber, 1800–1872

I beg to present you as a Christmas-gift the city of Savannah.
W.T. Sherman: Telegram to President Lincoln, 22 December 1864

Conquest brings self-conceit and intolerance, the reckless inflation and dissipation of energies. Defeat brings prudence and concentration; it ennobles and fortifies.
Henry Havelock Ellis, 1859–1939, The Task of Social Hygiene

What is denied to amity, the fist must take.
Adolf Hitler: Mein Kampf, 1925

We sure liberated the hell out of this place.
Remark by unidentified U.S. soldier in a French village, 1944

The conquered mourns, the conqueror is undone. (Flet victus, victor interiit.)
Latin Proverb

(*See also* Conquer, Conqueror, Victory.)

Conscientious Objector

And but for these vile guns,
He would himself have been a soldier.
Shakespeare: I King Henry IV, i, 3, 1597

Conscript

Food for powder, food for powder; they'll fill a pit as well as better.
Shakespeare: I King Henry IV, iv, 2, 1597

Done give myself to Uncle Sam
Now I ain't worth a good goddam,
I don't want no mo' camp;
Lawd I want to go home.
American negro song, c. 1918

(*See also* Conscription, Manpower. Universal Military Training.)

Conscription

One volunteer is worth ten pressed men.
Military saying, 18th century

A military force cannot be raised in this manner but by means of a military force.
Daniel Webster: To the House of Representatives, 9 December 1814

Conscription is the vitality of a nation, the purification of its morality, and the real foundation of all its habits.
Napoleon I: Political Aphorisms, 1848

A rich man's war and a poor man's fight.
Slogan of the New York City draft rioters, July 1863 (Anyone who could pay $300 for a substitute was exempt from the draft.)

This republic has no place for a vast military service and conscription.

Democratic National Platform, 1900

Every man in the draft age must work or fight.

Major General Enoch H. Crowder, USA: Instructions to draft boards, 1917 (Crowder, as Provost Marshal General of the Army, was in charge of selective service.)

In the event of war in which the man-power of the nation is drafted, all other resources should likewise be drafted. This will tend to discourage war by depriving it of its profits.

Democratic National Platform, 1924

In order fully to protect the achievements of the workers' and peasants' revolution, the Russian Socialist Federated Republic declares it to be the duty of all the citizens of the Republic to protect their Socialist country, and introduces conscription. The honorable right to defend the revolution by force of arms is bestowed only on the workers. Non-working elements have to carry out other military tasks.

Constitution of the USSR, 1924

The system of conscription has always tended to foster quantity at the expense of quality.

B.H. Liddell Hart: The Untimeliness of a Conscript Army, 1950

All the countries which collapsed under the shock of the German *blitzkrieg* in 1939, 1940, and 1941 relied on long-established conscript armies for their defense.

B.H. Liddell Hart: Defense of the West, 1950

Conscription has been the cancer of civilization.

B.H. Liddell Hart: Conscription—the basic Questions, 1950

(*See also* Conscript, Manpower, Universal Military Training.)

Constitution, USS

Long may she ride, our Navy's pride,
And spur to resolution;
And seamen boast, and landsmen toast,
The Frigate Constitution.

The Frigate Constitution, *Navy song, c. 1813*

Nail to the mast her holy flag,
Set every threadbare sail,
And give her to the god of storms,
The lightning and the gale!

Oliver Wendell Holmes, Sr.: Old Ironsides, 1830 (written in protest against the proposed scrapping of USS Constitution

Contact

Wherever the enemy goes, let our troops go also.

U.S. Grant: Despatch to General Halleck, concerning Sheridan's operations in the Shenandoah Valley, 1 August 1864

At Flores in the Azores Sir Richard
Grenville lay,
And a pinnace, like a flutter'd bird, came
flying from far away;
"Spanish ships at sea! we have sighted
fifty-three!"

Alfred Tennyson: The Ballad of the Revenge, 1878

. . . *contact* (a word which perhaps better then any other indicates the dividing line between tactics and strategy) . . .

Mahan: The Influence of Sea Power Upon History, 1890

Continental Army

In their ragged regimentals
Stood the old Continentals,
Yielding not.

G.H. McMaster: Carmen Bellicosum, 1849

An army all of captains, used to pray
And stiff in fight, but serious drill's despair,
Skilled to debate their orders, not obey.

James Russell Lowell: Under the Old Elm (ode read by Lowell at Cambridge on the centennary of Washington's assumption of command of the Continental Army, 3 July 1875)

Controversy

An opinion can be argued with; a conviction is best shot.

T.E. Lawrence, 1888–1935

There is one thing about the warfare between the Air Ministry and Ministry of Aircraft Procurement which is helpful to the public interest, namely, that I get a fine view of what is going on and hear both sides of the case argued with spirit . . . I await with keen interest further developments of your controversy.

Winston Churchill: Note to Secretary of State for War, 14 December 1940

Cooks

God sends meat but the Devil sends cooks.

Thomas Deloney: Works, 1600

(*See also* Mess Call, Rations, Subsistence.)

Copenhagen, Battle of (2 April 1801)

It is warm work; and this day may be the last to any of us at a moment. But mark you! I would not be elsewhere for thousands.

Nelson: At the height of action against the Danish fleet

Of Nelson and the North
Sing the glorious day's renown.

Thomas Campbell, 1777–1844, Battle of the Baltic

Corregidor, Defense of (December 1941–May 1942)

With broken heart and with head bowed in sadness but not in shame I report . . . that today I must arrange terms for the surrender of the fortified islands of Manila Bay—Corregidor, Fort Hughes, Fort Drum, and . . . (end of message).

Lieutenant General Jonathan M. Wainwright, USA: Last message from Corregidor, 6 May 1942

Cossacks

The king who rules the Cossacks, rules the world.

Stendhal (Henri Beyle): A Life of Napoleon, 1818

Take away his horse and he is no longer a Cossack.

Russian Proverb

(*See also* Cavalry.)

Cost Consciousness

The army is a school in which the miser becomes generous, and the generous prodigal; miserly soldiers are like monsters, very rarely seen.

Cervantes: Don Quixote, i, 39, 1604

If you serve under a ministry, with whom economy is the word, make a great bustle and parade about retrenchment; it will be prudent for you, likewise, to put it in some measure, into practice; but not so far as to extend to you own perquisites, or those of your dependents. Those savings are best made out of the pay of the subaltern officers and private soldiers; who being little able to bear it, will of course make much complaint of it, which will render your regard to economy the more conspicuous.

Francis Grose: Advice to the Officers of the British Army, 1782

Cost-accounting has its limitations in an arena [warfare] where there is no salable product and no profit criterion.

Morris Janowitz: The Professional Soldier, xii, 1960

(*See also* Comptrollership.)

Council of War

It is the duty and interest of the general frequently to assemble the most prudent and experienced officers of the different corps of the army and consult with them on the state both of his own and the enemy's forces.

Vegetius: De Re Militari, iii, 378

If a man consults whether he is to fight, when he has the power in his own hands, it is certain that his opinion is against fighting.

Nelson: Letter to Viscount Sidmouth, August 1801

Whenever I hear of Councils of War being called, I always consider them as "cloaks for cowardice"—so said the brave Boscawen, and from him I imbibed this sentiment.

Lord St. Vincent: Speech at the Opening of Parliament, 1809

The same consequences which have uniformly attended long discussions and councils of war will follow at all times.

They will end in adoption of the worst course, which in war is always the most timid, or, if you will, the most prudent. The only true wisdom in a general is determined courage.

Napoleon I: Maxims of War, 1831

A council always results in the most pusillanimous decision—in war the most disastrous.

Napoleon I: Maxims of War, 1831

Call no council of war. It is proverbial that councils of war never fight.

Major General H.W. Halleck: Telegram to General Meade after Gettysburg, 13 July 1863

If the commander . . . feels the need of asking others what he ought to do, the command is in weak hands.

Helmuth von Moltke ("The Elder"): Letter to Spenser Wilkinson, 20 January 1890

A council of war never fights, and in a crisis the duty of the leader is to lead and not to take refuge behind the generally timid wisdom of a multitude of councilors.

Theodore Roosevelt: Autobiography, 1913

A conference of subordinates *to collect ideas* is the resort of a weak commander.

Montogmery of Alamein: Memoirs, 1958

Counterattack

Have your dispositions well supported. Do not forget your reserves. Take advantage of the terrain and then attack brusquely and should be able to hope for the most brilliant success.

Frederick The Great: Instructions for His Generals, xii, 1747

We have retired far enough for today; you know I always sleep upon the field of battle!

Napoleon I: To retreating French troops before the counter stroke at Marengo, 14 June 1800

A general of ordinary talent occupying a bad position and surprised by a superior force, seeks his safety in retreat; but a great captain makes good all his deficiencies by his courage, and marches boldly to meet the attack . . . By this determined conduct he maintains the honor of his army, the first essential to military superiority.

Napoleon I: Maxims of War 1831

When you are occupying a position which the enemy threatens to surround, collect all your force immediately, and menace *him* with an offensive movement.

Napoleon I: Maxims of War, 1831

Counterattack is the soul of defense. Defense is in a passive attitude, for that is the negation of war. Rightly conceived it is an attitude of alert expectation. We wait for the moment when the enemy shall expose himself to a counterstroke, the success of which will so far cripple him as to render us relatively strong enough to pass to the offensive ourselves.

Julian Corbett: Some Principles of Maritime Strategy, 1911

To every blow struck in war there is a counter.

Winston Churchill: Memorandum for the War Cabinet, 16 December 1939

(*See also* Offensive.)

Counterinsurgency

It is by combined use of politics and force that pacification of a country and its future organization will be achieved. Political action is by far the more important.

Joseph Simon Gallieni: Marshal Gallieni's instructions to the French forces occupying Madagascar, 22 May 1898

If in taking a native den one thinks chiefly of the market that he will establish there on the morrow, one does not take it in the ordinary way.

Lyautey: The Colonial Role of the Army, Revue des Deux Mondes, 15 February 1900

The result can be achieved more rapidly than most people think. It will advance not by columns, nor by mighty blows, but as a patch of oil spreads, through a step-by-step progression.

Lyautey: Letter to Galliéni, 14 November 1903 (This is the origin of the famous "oil stain" (tâche d'hûile) phrase and strategy in counterinsurgency.)

I believe it must be the policy of the United States to support free peoples who are resisting attempted subjugation by armed minorities or by outside pressures.

Harry S. Truman: Message to Congress, enunciating the "Truman Doctrine," 12 March 1947

(*See also* Guerrilla, Guerrilla Warfare, Insurgency, Partisan Warfare.

Country

Our country! In her intercourse with foreign nations, may she be always in the right; but our country, right or wrong.

Stephen Decatur: Toast at Norfolk, 1816

Damn me if I ever love another country!

Ascribed to a demobilized Confederate soldier after Appomattox, 1865

(*See also* Patriotism.)

Coup-de-Main

The success of a coup-de-main depends absolutely upon luck rather than judgment.

Napoleon I: Political Aphorisms, 1848

(*See also* Insurrection, Revolt, Revolution.)

Coup d'Oeuil

The *coup d'oeuil* is a gift of God and cannot be acquired; but if professional knowledge does not perfect it, one only sees things imperfectly and in a fog, which is not enough in these matters where it is so important to have a clear eye . . . To look over a battlefield, to take in at the first instance the advantages and disadvantages is the great quality of a general.

Chevalier Folard: Nouvelles Decouvertes sur la Guerre, 1724

I engage and after that I see what to do. (Je m'engage et après ça je vois.)

Napoleon I: Remark during the Italian campaign, 1796

There is a gift of being able to see at a glance the possibilities offered by the terrain . . . One can call it the *coup d'oeuil militaire* and it is inborn in great generals.

Napoleon I, 1769–1821, Mémoires

A vital faculty of generalship is the power of grasping *instantly* the picture of the ground and the situation, of relating the one to the other, and the local to the general.

B.H. Liddell Hart: Thoughts on War, xi, 1944

Courage

Oh friends, be men, and let your hearts be strong,
And let no warrior in the heat of fight
Do what may bring him shame in others' eyes;
For more of those who shrink from shame are safe
Than fall in battle, while with those who flee
Is neither glory nor reprieve from death.

Homer: The Iliad, v (Bryant's translation), c. 1000 B.C.

To see what is right and not to do it is want of courage.

Confucius, 551–479 B.C.

Far better it is to have a stout heart always and suffer one's share of evils, than to be ever fearing what may happen.

Herodotus: History, c. 444 B.C.

Courage may be taught as a child is taught to speak.

Euripides: The Suppliant Women, 421 B.C.

Courage is equivalent to a rampart.

Sallust: Conspiracy of Catiline, lviii, 44 B.C.

O strong of heart, go where the road
Of ancient honor climbs.
Bow not your craven shoulders.
Earth conquered gives the stars.

Boethius, 480–524 A.D.

The strongest, most generous, and proudest of all virtues is true courage.

Michel de Montaigne: Essays, 1580

'Tis true, that we are in great danger;
The greater therefore should our courage be.

Shakespeare: King Henry V, iv, 1, 1598

Courage, in soldiers, is a dangerous profession they follow to earn their living.
La Rochefoucauld: Maxims, 1665

One can't answer for his courage when he has never been in danger.
La Rochefoucauld: Maxims, 1665

Courage is a quality so necessary for maintaining virtue that it is always respected, even when it is associated with vice.
Samuel Johnson, 1709–1784

Often the test of courage is not to die but to, live.
Vittorio Alfieri, 1749–1803

When soldiers brave death, they drive him into the enemy's ranks.
Napoleon I: To a regiment of chasseurs before Jena, 14 October 1806

. . . two o'clock in the morning courage.
Napoleon I: Quoted by Las Cases, 1823

The patient courage which waits for the opportunity it cannot create.
Epitaph of Admiral Cuthbert Collingwood, RN (ob. 1810)

I would define true courage to be a perfect sensibility of the measure of danger, and a mental willingness to incur it.
W.T. Sherman: Personal Memoirs, II, xxv, 1875

. . . not the courage which throws away the scabbard, much less that which burns its ships.
Mahan: The War in South Africa, 1900

Courage disdains fame and wins it.
Royal Cortissoz: Inscription for Memorial Hall, Yale University, 1915

It is ideas that inspire courage. (Ce qui donne du courage, ce sont les idées.)
Georges Clemenceau, 1841–1929

Take her down.
Commander Howard W. Gilmore, USN: When mortally wounded on the deck of his ship, USS Growler, *submarine, during surface action with a Japanese warship.* Growler *submerged as ordered and Commander Gilmore was lost.*

There are only two classes who, as categories, show courage in war—the front-line soldier and the conscientious objector.
B.H. Liddell Hart: Thoughts on War, v, 1944

No sane man is unafraid in battle, but discipline produces in him a form of vicarious courage.
George S. Patton, Jr.: War as I Knew It, 1947

Everyone admires courage and the greenest garlands are for those who possess it.
John F. Kennedy: in notes for Profiles in Courage, c. 1955

There is nothing like seeing the other fellow run to bring back your courage.
Sir William Slim: Unofficial History, 1959

It is better to live one day as a lion than a hundred years as a sheep.
Italian Proverb

In a fight, anger is as good as courage.
Welsh Proverb

(*See also* Audacity, Bravery, Daring, Gallantry, Herosm, Valor.)

Court Martial

The charge is prepar'd; the lawyers are met;
The judges all ranged,—a terrible show!
John Gay: The Beggar's Opera, iii, 11, 1728

Talking of a court-martial that was sitting upon a very momentous public occasion [Dr. Johnson] expressed much doubt of an enlightened decision; and said that perhaps there was not a member of it who in the whole course of his life had ever spent an hour by himself in balancing probabilities.
James Boswell: Life of Johnson, 1791

The popular conception of a court martial is half a dozen bloodthirsty old Colonel Blimps, who take it for granted that anyone brought before them is guilty . . . and who at intervals chant in unison, "Maximum penalty—death!" In reality courts martial are almost invariably

composed of nervous officers, feverishly consulting their manuals; so anxious to avoid a miscarriage of justice that they are, at times, ready to allow the accused any loophole of escape. Even if they do steel themselves to passing a sentence, they are quite prepared to find it quashed because they have forgotten to mark something "A" and attach it to the proceedings.
Sir William Slim: Unofficial History, 1959

(*See also* Law, Law Specialists, Uniform Code of Military Justice.)

Courtesy

Although all *words of command* should be given in an *authoritative and firm tone,* it does not follow that *drill manners* should accompany the officer into private society.
Colonel J.G.D Tucker: Hints to Young Officers, 1826

My men can hardly hold their ground. Would you object to bringing yours into line with ours? I had the pleasure of being introduced to you at Lady Palmerston's last summer.
One British officer to another at Inkerman, 5 November 1854

Dip my broad pennant to my old messmate.
Commodore Josiah Tattnall, CSN: To his quartermaster at the opening of the Battle of Port Royal Sound, as USS Wabash *led in the Union squadron, flying the flag of RAdm Samuel F. DuPont, USN, an "Old Navy" friend and messmate of pre-war days.*

A compliance with the minutiae of military courtesy is a mark of well disciplined troops
John A. Lejeune: Letter to All Officers of the Marine Corps, 1919

But after all when you have to kill a man, it costs nothing to be polite.
Winston Churchill: The Grand Alliance, 1950

Covenants

Covenants without swords are but words.
Thomas Hobbes: Leviathan, 1651

Open covenants of peace, openly arrived at, after which there shall be no private international understandings of any kind, but diplomacy shall proceed always frankly and in the public view.
Woodrow Wilson: Address to Congress, 8 January 1918 (Point I of the Fourteen Points)

(*See also* Diplomacy, Negotiations.)

Coward

Cowards do not count in battle; they are there, but not in it.
Euripides, 480–406 B.C., Meleager

Cowards falter, but danger is often overcome by those who nobly dare.
Elizabeth I, 1553–1603

A plague of all cowards, I say, and a vengeance, too.
Shakespeare: I King Henry IV, ii, 4, 1597

Cowards die many times before their deaths;
The valiant never taste of death but once.
Shakespeare: Julius Caesar, ii, 2, 1599

Plenty and peace breed cowards;
Hardness ever of hardiness is mother.
Shakespeare: Cymbeline, iii, 6, 1609

A coward's fear can make a coward valiant.
Owen Feltham 1602–1688, Resolves

It is vain for the coward to fly; death follows close behind; it is by defying it that the brave escape.
Voltaire, 1694–1778

A Coward, when taught to believe, that if he breaks his Ranks, and abandons his Colors, will be punished by Death by his own party, will take his chance against the enemy.
George Washington: Letter to the President of Congress, 9 February 1776

Cowards' funerals, when they come,
Are not wept so well at home.
A.E. Housman: A Shropshire Lad (The Day of Battle), 1896

Cowards are those who let their timidity get the better of their manhood.

George S. Patton, Jr.: Letter to Cadet George S. Patton, III, USMA, 6 June 1944

A coward's mother does not weep.
(Timidi mater non flet.)
Latin Proverb

(*See also* Cowardice, Fear, Timidity.)

Cowardice

To see the right and not to do it is cowardice.
Confucius, 551–478 B.C., Analects, ii, 24

As for those cowardly captains of yours, hang them up, for, by God! they deserve it.
Commodore M. du Casse: Letter to Admiral John Benbow, RN, mortally wounded after three days' action against du Casse, off Santa Marta, Colombia, 22 August 1702. Two of the "cowardly captains" were in fact shot.

Any Officer, or Soldier . . . who (upon the Approach, or Attack of the Enemy's Forces, by land or water) presumes to turn his back and flee, shall be instantly shot down, and all good officers are hereby authorized and required to see this done, that the brave and gallant part of the Army shall not fall a sacrifice to the base and cowardly part, or share their disgrace in a cowardly and unmannerly Retreat.
George Washington: Letter to the President of Congress, 20 September 1776

It is mutual cowardice that keeps us in peace. Were one-half of mankind brave, and one-half cowards, the brave would be always beating the cowards. Were all brave, they would lead a very uneasy life; all would be continually fighting: but being all cowards, we go on very well.
Samuel Johnson, 28 April 1778 (in Boswell's Life of Johnson)

I could not look on Death, which being known,
Men led me to him, blindfold and alone.
Rudyard Kipling: Epitaphs of the War (The Coward), 1919

Most of the time the word *realism* is a polite translation of the word *cowardice.*
Jean Dutourd: Taxis of the Marne, 1957

Nervous in the Service.
Enlisted men's phrase for cowardice

(*See also* Coward, Fear, Timidity.)

Crimean War (1854–1856)

The Crimean War is one of the bad jokes of history.
Philip Guedalla: The Two Marshals, 1943

Cromwell, Oliver (1599–1658)

As for Colonel Cromwell, he hath 2,000 brave men, well disciplined; no man swears but he pays his twelvepence; if he be drunk, he is set in the stocks, or worse; if one calls the other roundhead he is cashiered . . . How happy were it if all the forces were thus disciplined!
Puritan news letter, May 1643

I am a poor weak creature yet accepted to serve the Lord and His people. Indeed . . . ye know not me, my weakness, my immoderate passions, and everyway unfitness for my work; yet the Lord, who will have mercy on whom He will, does as you see.
Oliver Cromwell: Before Naseby, 14 June 1645

Cromwell, our chief of men, who through a cloud
Not of warr only, but detractions rude,
To peace and truth thy glorious way hast plough'd,
And on the neck of Fortune proud
Hast rear'd God's Trophies, and his work pursu'd.
Milton: Sonnet to the Lord Generall, May 1652

Crusade

In that enormous silence, tiny and unafraid,
Comes up along a winding road the noise of the Crusade.
Strong gongs groaning as the guns boom far,
Don John of Austria is going to the war;
Stiff flags straining in the night-blasts cold,
In the gloom black-purple, in the glint, old-gold,

Torchlight crimson on the copper kettle-
drums,
Then the tuckets, then the trumpets, then
the cannon, and he comes.
G.K. Chesterton, 1874–1936, Lepanto

Cryptanalysis

Stimson, as Secretary of State, was dealing as a gentleman with the gentleman sent as ambassadors and ministers from friendly nations, and, as he later said, "Gentlemen do not read each other's mail."
Henry L. Stimson and McGeorge Bundy: On Active Service in Peace and War, 1948 (on abolishing the U.S. "Black Chamber" in 1929)

When the fate of a nation and the lives of its soldiers are at stake, gentlemen do read each other's mail—if they can get their hands on it.
Allen Dulles: The Craft of Intelligence, 1963

D

Danger

Danger gleams like sunshine to a brave man's eyes.
Euripides: Iphigenia in Tauris, 412 B.C.

Constant exposure to dangers will breed contempt for them.
Seneca: De Providentia, 64 A.D.

Out of this nettle, danger, we pluck this flower, safety.
Shakespeare: I King Henry IV, ii, 1597

In meditation, all dangers should be seen; in execution, none, unless they are very formidable.
Attributed to Francis Bacon, 1561–1626

Danger, the spur of all great minds.
George Chapman: Bussy d'Ambois, v, 1613

To conquer without danger is to triumph without glory. (À vaincre sans péril, on triomphe sans gloire.)
Pierre Corneille: Le Cid, 1637

Dangers bring fears, and fears more dangers bring.
Richard Baxter, 1615–1691, Love Breathing Thanks

Dangers, by being despised, grow great; so they do by absurd provision against them.
Edmund Burke: Speech on the Petition of the Unitarians, 11 May 1792

If I had been censured every time I have run my ship, or fleets under my command, into great danger, I should long ago have been *out* of the Service, and never *in* the House of Peers.
Nelson: Letter to the Admiralty, March 1805

War is the province of danger.
Clausewitz: On War, 1832

(*See also* Risk.)

Daring

What man dare, I dare.
Shakespeare: Macbeth, iii, 1605

He who would greatly achieve must greatly dare. for brilliant victory is only achieved at the risk of disastrous defeat.
Washington Irving, 1783–1859 (of James Lawrence)

For great aims we must dare great things.
Clausewitz: Principles of War, 1812

(*See also* Audacity, Bravery, Gallantry, Heroism, Valor.)

Dead

Dead like the rest, for this is true: war never chooses an evil man, but the good.
Sophocles: Philoctetes, 408 B.C.

There be of them that have left a name behind them. And some there be which have no memorial . . . Their bodies are buried in peace; but their name liveth for evermore.
Ecclesiasticus, XLIV, 8–14

Sleep in peace, slain in your country's wars!
Shakespeare: Titus Andronicus, i, 1, 1593

When all those legs and arms and heads, chopped off in a battle, shall join together at the latter day, and cry all—we died at such a place.
Shakespeare: King Henry V, iv, 1, 1598

Those that leave their valiant bones in
France,
Dying like men, though buried in your
dunghills,
They shall be famed.
Shakespeare: King Henry V, iv, 3, 1598

Dead on the field of honor. (Mort au champs d'honneur.)
Theophile de la Tour d'Auvergne, whom Napoleon called "First Grenadier of France," was killed at Oberhausen 27 June 1800. Thereafter, at reveille roll-call each morning, the first sergeant answered his name with the above words. The same custom was later instituted in other French units in memory of other brave soldiers.

And they who for their country die
Shall fill an honored grave,

For glory lights the soldier's tomb,
And beauty weeps the brave.
Joseph Rodman Drake: To the Defenders of New Orleans, 1814

Then stand to your glasses steady!
We drink in our comrades' eyes:
One cup to the dead already—
Hurrah for the next that dies!
Bartholomew Dowling, 1823–1863, The Revel

On fame's eternal camping ground
Their silent tents are spread,
And glory guards with solemn round
The bivouac of the dead.
Theodore O'Hara: The Bivouac of the Dead, 1847 (written to commemorate the American dead at Buena Vista, 22 February 1847, and required, by a 19th century act of Congress, to be displayed in every National Cemetery)

He fell on the field:
His country mourned him,
And his father was resigned.
Bulwer-Lytton: The Caxtons, xviii, 1849

All quiet along the Potomac tonight,
No sound save the rush of the river,
While soft falls the dew on the face of the dead—
The picket's off duty forever.
Ethel Lynn Beers: The Picket Guard, 1861

We have fed our sea for a thousand years
And she calls us, still unfed,
Though there's never a wave of all her waves
But marks our English dead.
Rudyard Kipling: A Song of the English, 1893

Happy are those who have died in great battles,
Lying on the ground before the face of God.
Charles Péguy: Poem, c. 1910

They shall not grow old, as we that are left grow old:
Age shall not weary them nor the years condemn.
Laurence Binyon: For the Fallen, 21 September 1914

Blow out, you bugles, over the rich dead!
There's none of these so lonely and poor of old,
But, dying, has made us rarer gifts than gold.
Rupert Brooke: The Dead, 1914

And now these waiting dreams are satisfied;
From twilight to the halls of dawn he went;
His lance is broken; but he lies content
With that high hour, in which he lived and died.
And falling thus, he wants no recompense,
Who found his battle in the last resort;
Nor needs he any hearse to bear him hence,
Who goes to join the men of Agincourt.
Herbert Asquith, 1881–1947, The Volunteer.

O valiant hearts, who to your glory came
Through dust of conflict and through battle flame;
Tranquil you lie, your knightly virtue proved,
Your memory hallowed in the land you loved.
Proudly you gathered, rank on rank, to war,
As you had heard God's message from afar;
All you had hoped for, all you had, you gave
To save mankind—yourself you scorned to save.
John Stanhope Arkwright: O Valiant Hearts, 1919

Here dead lie we because we did not choose
To live and shame the land from which we sprung.
Life, to be sure, is nothing much to lose;
But young men think it is, and we were young.
A.E. Housman: Here Dead Lie We, 1922

... unknown victims of conflicting opinions.
Lieutenant Commander Arnold S. Lott, USN: Most Dangerous Sea, 1959

(*See also* Casualties, Death.)

Death

A glorious death is his
Who for his country falls.
Homer: Iliad, xv, c. 1000 B.C.

We count it death to falter, not to die.
Simonides of Cheos, 556–468 B.C., Epigram

U. S. Navy

GEORGE S. PATTON, JR.

1885–1945

"There is only one sort of discipline—perfect discipline."

It is sweet and fitting to die for one's country. (Dulce et decorum est pro patria mori.)
Horace, 65–8 B.C., Odes, iii, 2

A pale horse: and him that sat on him was Death.
Revelation, VI, 8

We who are about to die, salute you. (Morituri te salutamus.)
Traditional greeting of the gladiators to Roman Emperors

The wings of man's life are plumed with the feathers of death.
Sir Francis Drake, 1540–1596

Give me leave without offense always to live and die in this mind—that he is not worthy to live at all who in fear of danger or death shunneth his country's service and his own honor; seeing that death is inevitable and the fame of virtue immortal.
Sir Humphrey Gilbert: Discourse, 1576

Nothing in his life
Became him like the leaving of it; he died
As one that studied in his death
To throw away the dearest thing he owed
As 'twere a careless trifle.
Shakespeare: Macbeth, i, 4, 1605

Brave death outweighs bad life.
Shakespeare: Coriolanus, i, 6, 1607

O eloquent, just and mightie DEATH! whom none could advise, thou has perswaded; what none hath dared, thou hast done; and whom all the world hath flattered, thou only hast cast out of the world and despised; thou hast drawn together all the farre stretched greatnesse, all the pride, cruelltie, and ambition of men, and covered it all with these two narrow words, HIC JACET!
Sir Walter Raleigh: Historie of the World, 1615

I had rather die as a valiant Souldier, than be hanged as a Coward.
Reply by the Spanish Governor of Portobello to Sir Henry Morgan's demand that he surrender, 1668

. . . Have you seen the young,
How they walked steadfast, how they sprung
To meet deaths proud and memorable, yes,
But sure and sometimes cruel, none the less?
La Fontaine: Fables, 1668 (translation by Edward Marsh)

So he passed over, and all the trumpets sounded for him on the other side.
John Bunyan: Pilgrim's Progress (Death of Valiant for Truth), 1678

You die alone. (On se mourra seul.)
Blaise Pascal, 1623–1662

How sleep the brave, who sink to rest,
By all their country's wishes blest!
William Collins: Ode Written in the Year 1746

Life with disgrace is dreadful. A glorious death is to be envied.
Nelson: Letter to Lady Nelson, March 1795

What millions died—that Caesar might be great!
Thomas Campbell: The Pleasures of Hope, 1799

There is a victory in dying well.
Thomas Campbell, 1777–1844, Stanzas to the Memory of the Spanish Patriots

Death overtakes the coward, but never the brave until his hour is come.
Attributed to Napoleon I, 1769–1821

Fifty-seventh, die hard!
By a British colonel (Inglis) to the 57th Foot (West Middlesex Regiment) at Albuera, 16 May1811

I have looked death too often in the face to be afraid of him now.
Lieutenant John Cushing Aylwin, USN: Dying on board USS Constitution *after her victory over HMS* Java, *29 December 1812*

An the fellow died as he lived; but it is part of a sailor's life to die well.
Stephen Decatur, 1779–1820 (of James Lawrence)

She would not be comforted, saying that the camp was made up of the young men from the first and best families . . . and that they were proud, and would fight. I explained that young men of the best families did not like to be killed better than ordinary people.

W.T. Sherman: of a Confederate sympathizer, 1861

God has fixed the time for my death. I do not concern myself about that, but to be always ready, no matter when it may overtake me.—That is the way all men should live, and then all would be equally brave.
Stonewall Jackson: Letter, 1862

Major, tell my father I died with my face to the enemy.
Colonel I. E. Avery, CSA: At Gettysburg, 2 July 1863

Right in the van,
On the red ramparts' slippery swell,
With heart that beat a charge, he fell...
James Russell Lowell: Memoriae Positum, August 1863 (in memory of Colonel Robert Gould Shaw, 54th Massachusetts Infantry, killed in the assault on Fort Wagner, 18 July 1863)

Better like Hector in the field to die,
Than like a perfumed Paris turn and fly.
Henry Wadsworth Longfellow: Morituri Salutamus, 1875

Death is lighter than a feather; duty, heavy as a mountain.
Emperor Meiji of Japan, Imperial Rescript to Soldiers and Sailors, 4 January 1883

Legionnaires, you are here to die and I shall send you where one dies.
General de Négrier: To men of the Foreign Legion, Tonkin, 1884

Qui procul hinc, the legend's writ—
The frontier-grave is far away—
Qui ante diem periit:
Sed miles, sed pro patria.
Henry Newbolt, 1862–1938, Clifton Chapel (The Latin translates, "Who died far away, before his time, but as a soldier and for his country.")

To make war—to kill, without being killed, is an illusion.
General Mikhail Ivanovich Dragomirov, 1830–1905

To die with honor when one can no longer live with honor.
John Luther Long: Madame Butterfly (inscription on Samurai sword), 1897

Don't cheer, men; the poor devils are dying.
Captain John W. Philip, USN: As USS Texas *passed the burning hulk of the Spanish cruiser,* Vizcaya, *at Santiago, 3 July 1898*

We who must live salute you who have found strength to die.
Royal Cortissoz: Inscription in Memorial Hall, Yale University, 1915

I have a rendezvous with Death
At some disputed barricade.
Alan Seeger: I Have a Rendezvous with Death, 1916

Generals die in bed.
Soldier saying in France, World War I

I could not look on death, which being known,
Men led me to him, blindfold and alone.
Rudyard Kipling: Epitaphs of the War (The Coward), 1919

It is better to die on your feet than to live on your knees.
La Pasionaria (Dolores Ibarruri): Speech in Paris, 1936

Death stands at attention, obedient, expectant, ready to serve, ready to shear away the peoples *en masse*; ready, if called on, to pulverize, without hope of repair, what is left of civilization. He awaits only the word of command. He awaits it from a frail, bewildered being, long his victim, now—for one occasion only—his Master.
Winston Churchill: The Gathering Storm, iii, 1948

(*See also* Casualties, Dead.)

Deception

All warfare is based on deception.
Sun Tzu, 400–320 B. C., The Art of War, i

Though fraud in other activities be detestable, in the management of war it is laudable and glorious, and he who overcomes an enemy by fraud is as much to be praised as he who does so by force.
Niccolo Machiavelli: Discorsi, iii, 1531

Force, and fraud, are in war the two cardinal virtues.
Thomas Hobbes: Leviathan, i, 13, 1651

To achieve victory we must as far as possible make the enemy blind and deaf by sealing his eyes and ears, and drive his commanders to distraction by creating confusion in their minds.
Mao Tse-tung, 1893–,
On Protracted War

(*See also* Ruse de Guerre, Strategem.)

Decision

Quick decisions are unsafe decisions.
Sophocles, 495–406 B. C.,
Oedipus Tyrannus

The god of war hates those who hesitate.
Euripides: Heraclidae, c. 425 B. C.

The die is cast. (Alea iacta est.)
Julius Caesar: On crossing the Rubicon and commencing civil war, 49 B. C.

In distributing justice, you must always incline a little to the strongest side. Thus, if a dispute happens between a field officer and a subaltern, you must, if possible, give it in favor of the former. Force is indeed the ruling principle in military affairs; in conformity to which the French term their cannon, the *ultima ratio regum.*
Francis Grose: Advice to the Officers of the British Army, 1782

The wine is poured and we must drink it.
Marshal Ney: Order to advance, Jena, 14 October 1906

He who seizes on the moment, he is the right man.
Geothe: Faust, i, 1808

The decisions a general has to make would furnish a problem of mathematical calculations not unworthy of the powers of a Newton or an Euler.
Clausewitz: On War, 1832

You will usually find that the enemy has three courses open to him, and of these he will adopt the fourth.
Helmuth von Moltke ("The Elder"), 1800–1891

In all operations a moment arrives when brave decisions have to be made if an enterprise is to be carried through.
Sir Roger Keyes: Letter from the Dardanelles, 1915

When all is said and done the greatest quality required in a commander is "decision"; he must be able to issue clear orders and have the drive to get things done. Indecision and hesitation are fatal in any officer; in a C-in-C they are criminal.
Montgomery of Alamein: Memoirs, xxi, 1958

It was an example of inflexibility in the pursuit of previously conceived ideas that is, unfortunately, too frequent in modern warfare. Final decisions are made not at the front by those who are there, but many miles away by those who can but guess at the possibilities and potentialities.
Douglas MacArthur: Reminiscences, 1964

The rarest gift that God bestows on man is the capacity for decision.
Dean Acheson: Speech at Freedom House, New York City, 13 April 1965

Declaration of War

Cry "Havoc!" and let slip the dogs of war.
Shakespeare: Julius Caesar, iii, 1, 1599

My sentence is for open war.
Milton: Paradise Lost, ii, 1667

My voice is still for war.
Gods! can a Roman senate long debate
Which of the two to choose, slavery or death?
Joseph Addison: Cato, i, 4, 1713

It is wonderful with what coolness and indifference the greater part of mankind see war commenced.
Samuel Johnson: Thoughts Respecting Falkland's Islands, 1771

Hostilities must not be begun without previous and explicit warning in the form of a reasoned declaration of war, or of an ultimatum embracing a conditional declaration.
The Hague Convention, 1907

If the iron dice roll, may God help us.
Theobald von Bethmann-Hollweg: To the German Reichstag, as Foreign Minister, 1 August 1914

...the immense, delicate, and hazardous transition from peace to war.
Winston Churchill: The Gathering Storm, 1948

Decorations

Let all be present and expect the palm, the prize of victory.
Virgil: Aeneid, v, 21 B. C.

Those sweet rewards, which decorate
the brave,
'Tis folly to decline.
Samuel Johnson, 1709–1784

Show me a republic, ancient or modern, in which there have been no decorations. Some people call them baubles. Well, it is by such baubles that one leads men.
Napoleon I: Remark on establishing the Legion of Honor, 19 May 1802

Orders and decorations are necessary in order to dazzle the people.
Napoleon I: Political Aphorisms, 1848

The General got 'is decorations thick
(The men that backed 'is lies could not
complain),
The Staff 'ad DSO's till we was sick,
An' the soldier—'ad to do the work again!
Rudyard Kipling: Stellenbosch, 1903

(*See also* Awards, Medals.)

Defeat

There is one source, O Athenians, of all your defeats. It is that your citizens have ceased to be soldiers.
Demosthenes, 383–322 B. C.

Woe to the vanquished. (Vae victis.)
Livy, 59 B.C.–17 A.D., History, V, xviii

The painful warrior, famousèd for fight,
After a thousand victories, once foil'd,
Is from the books of honor razed quite,
And all the rest forgot for which he toil'd.
Shakespeare: Sonnet xxv, 1609

A battle lost is a battle one thinks one has lost.
Joseph de Maistre, 1754–1821

A defeated ruler should never be spared.
Stendhal (Henri Beyle): Life of Napoleon, 1818

Even the final decision of a war is not to be regarded as absolute. The conquered nation often sees it as only a passing evil, to be repaired in after times by political combinations.
Clausewitz: On War, 1832

Defeat is a school in which truth always grows strong.
Henry Ward Beecher, 1813–1887

We are not interested in the possibilities of defeat. They do not exist.
Victoria: During the Crimean War, 1854 (Winston Churchill kept this quotation at his place on the War Cabinet table, World War II.)

As to being prepared for defeat, I certainly am not. Any man who is prepared for defeat would be half defeated before he commenced. I hope for success; shall do all in my power to secure it and trust to God for the rest.
David G. Farragut: Letter to his wife, 1864

I had rather die than be whipped.
J. E. B. Stuart: To his troops while being carried mortally wounded from the field at Yellow Tavern, Virginia, 11 May 1864

Too much success is not wholly desirable; an occasional beating is good for men—and nations.
Mahan: Life of Nelson, ix, 1897

Errors and defeats are more obviously illustrative of principles than successes are...Defeat cries aloud for explanation; whereas success, like charity, covers a multitude of sins.
Mahan: Naval Strategy, 1911

They fought well. Let them keep their weapons.
Ferdinand Foch: Of the Imperial German Army, 1919

A beaten general is disgraced forever.
Ferdinand Foch: Precepts, 1919

Defeat brings prudence and concentration; it ennobles and fortifies.
Havelock Ellis, 1858–1939, The Task of Social Hygiene

The winner is asked no questions—the loser has to answer for everything.
Sir Ian Hamilton: Gallipoli Diary, 1920

France has lost a battle. But France has not lost the war.

Charles De Gaulle: Broadcast to the French people, 18 June 1940

We got a hell of a beating. We got run out of Burma, and it is humiliating as hell.

Joseph W. Stilwell: Statement after retreat from Burma, 25 May 1942

Man in war is not beaten, and cannot be beaten, until he owns himself beaten.

B. H. Liddell Hart: Thoughts on War, i, 1944

Fearful are the convulsions of defeat.

Winston Churchill: The Gathering Storm, 1948

A defeated man does not make a good philosopher.

Jean Dutourd: Taxis of the Marne, 1957

Defense

Brave men are a city's strongest tower of defense.

Alcaeus, fl. 600 B. C.

The true contempt of an invader is shown by deeds of valor in the field.

Hermocrates of Syracuse: Speech to the Syracusans, 415 BC

Those expert at preparing defenses consider it fundamental to rely on the strengths of such obstacles as mountains, rivers and foothills. They make it impossible for the enemy to know where to attack. They secretly conceal themselves as under the nine-layered ground.

Tu Yu, 735–812

What boots it at one gate to make defense,
And at another to let in the foe?

Milton: Samson Agonistes, 1671

Self-defense is nature's oldest law.

John Dryden: Absalom and Achitophel, i, 1681

The secret of success is to have a solid body so firm and impenetrable that wherever it is or wherever it may go, it shall bring the enemy to a stand like a mobile bastion, and shall be self-defensive.

Comte de Montecucculi: Principes de l'Art Militaire, 1712

Petty geniuses attempt to hold everything; wise men hold fast to the key points. They parry great blows and scorn little accidents. There is an ancient apothegm: he who would preserve everything, preserves nothing. Therefore, always sacrifice the bagatelle and pursue the essential.

Frederick the Great: Instructions for His Generals, vii, 1747

Stand your ground, men. Don't fire unless fired upon. But if they mean to have a war, let it begin here.

Captain Jonas Parker: To the Minute Men on Lexington Green as the redcoats deployed, 19 April 1775

Our business is to Defend the main chance; to Attack only by Detail; and when a precious advantage Offers.

Horatio Gates: Letter regarding operations of the Continental Army, 1776

Defense is of much more importance than opulence.

Adam Smith: An Inquiry into the Nature and Causes of the Wealth of Nations, 1776

A government without the power of defense is a solecism.

James Wilson: During debate on adoption of the Constitution, 1787

Safety from external danger is the most powerful director of national conduct. Even the most ardent love of liberty will, after a time, give way to its dictates.

Alexander Hamilton: The Federalist, 1787

War is honorable
In those who do their native rights
    maintain;
In those whose swords an iron barrier are
Between the lawless spoiler and the weak.

Joanna Baillie, 1762–1851

Millions for defense, but not a cent for tribute.

Robert Goodloe Harper: Toast at a dinner for John Marshall, 18 June 1798 (often attributed, despite his disavowal, to Charles Cotesworth Pinckney, when American Minister to France, 1797)

Never depend completely on the strength of the terrain and consequently never be enticed into passive defense by a strong terrain.

Clausewitz: Principles of War, 1812

A swift and vigorous transition to attack—the flashing sword of vengeance—is the most brilliant point of the defensive.

Clausewitz: On War, 1832

Self-defense is a virtue,
Sole bulwark of all right.

Byron, 1788–1824, Sardanapalus, ii, 1

The passive defense is always pernicious.

Jomini: Precis de l'Art de la Guerre, 1838

This place should be defended with the spirit which actuated the defenders of Thermopylae, and if left to myself such is my determination.

Stonewall Jackson: Letter to R. E. Lee concerning Maryland Heights, June 1861

The best protection against the enemy's fire is a well directed fire from our own guns.

David G. Farragut: General Order for the attack on Port Hudson, 14 March 1863

There is no better way of defending a long line than by moving into the enemy's territory.

R. E. Lee: Letter to Brigadier General John R. Jones, CSA, 21 March 1863

A clever military leader will succeed in many cases in choosing defensive positions of such an offensive nature from the strategic point of view that the enemy is compelled to attack us in them.

Helmuth Von Moltke ("The Elder"), 1800–1891

In war, the defensive exists mainly that the offensive may act more freely.

Mahan: Naval Strategy, 1911

Every position must be held to the last man; there must be no retirement. With our backs to the wall, and believing in the justice of our cause, each one of us must fight on to the end.

Sir Douglas Haig: Order to the British Army in France, 12 April 1918

Every development or improvement in firearms favors the defensive.

Giulio Douhet: The Command of the Air, 1921

The influence of growing firepower on tactical defense is evident...The defensive is able more than before to carry out its original mission, which is to break the strength of the attacker, to parry his blows, to weaken him, to bleed him, so as to reverse the relation of forces and lead finally to the offensive, which is the only decisive form of warfare.

Field Marshal General Ritter Wilhelm Von Leeb: Defense, 1938

We shall defend every village, every town and every city. The vast mass of London itself, fought street by street, could easily devour an entire hostile army; and we would rather see London laid in ruins and ashes than that it should be tamely and abjectly enslaved.

Winston Churchill: Broadcast to the British people, 14 July 1940

You can squeeze a bee in your hand until it suffocates, but it will not suffocate without having stung you. You may say that is a small matter, and, indeed, it is a small matter. But if the bee had not stung you, bees would have long ago ceased to exist.

Jean Paulhan, 1884–

Cet animal est très méchant; quand on l'attaque, il se defend. (This is a very vicious animal: when attacked, he defends himself.)

Notice said to have been posted in the Paris Zoo.

(*See also* Last Stand, Resistance, Scorched Earth.)

Defense Organization

The separation of the military and naval professions is at once the effect and the cause of modern improvements in the science of navigation and maritime war.

Edward Gibbon: Decline and Fall of the Roman Empire, xii, 1782

Each nation that is organized democratically is not organized militarily; it is, as opposed to the other, in a state of inferiority for war.

Ardant Du Picq, 1821–1870, Battle Studies

Whatever the system adopted, it must aim above all at perfect efficiency in military action; and the nearer it approaches to

this ideal the better it is. It would seem that this is too obvious for mention. It may be for mention; but not for reiteration.
Mahan: Naval Administration and Warfare, 1903

There are great dangers in giving absolute priority to any department.
Winston Churchill: Memorandum for Prime Minister, 18 September 1939

The politician in search of military efficiency arbitrarily simplifies the society with which he has to deal. But whereas the scientist simplifies by a process of analysis and isolation, the politician can only simplify by compulsion, by a Procrustean process of chopping and stretching designed to make the living social organism conform to a certain easily understood and readily manipulated mechanical pattern.
Aldous Huxley: Grey Eminence, x, 1941

In this age of specialization it is manifestly impossible for a single man to master the intricacies of both the Army and the Navy. As far as the Air Force is concerned, it will always remain an auxiliary arm. The Army and Navy are different. Each is decisive. Each can carry through a decisive operation. The Air Force cannot.
General Lesley James McNair, USA, 1883–1944

Good will can make any organization work; conversely, the best organization chart in the world is unsound if the men who have to make it work don't believe in it.
James Forrestal: Regarding unification proposals, 1946

Just as the Industrial Revolution eclipsed one-man direction in economic and governmental management, it likewise rendered obsolete one-man military direction of a nation's armed forces.
Brigadier General J. D. Hittle, USMC, 1915–

I suggest that great care be exercised lest the Office of the Secretary of Defense, instead of being a small and efficient unit which determines the policies of the Military Establishment and controls and directs the Departments, feeding on its own growth, becomes a separate empire.
Ferdinand Eberstadt: To the Senate Armed Services Committee, 1949

Fighting spirit is not primarily the result of a neat organization chart nor of a logical organizational setup. The former should never be sacrificed to the latter.
Ferdinand Eberstadt: Testimony on the Marine Corps Bill, Senate Armed Forces Committee, 1951

(*See also* Pentagon, Unification.)

Defiance

Come the three corners of the world in arms,
And we shall shock them.
Shakespeare: King John, v, 7, 1596

What though the field be lost?
All is not lost; th'unconquerable will,
And study of revenge, immortal hate,
And courage never to submit or yield.
Milton: Paradise Lost, i, 1667

Don't tread on me.
Motto on the American colonists' Rattlesnake Flag, 1776

To defy Power that seems omnipotent...
Never to change, nor falter, nor repent.
Percy Bysshe Shelley, 1792–1822

We give up the fort when there's not a man left to defend it.
Captain George Croghan, USA: At Fort Stephenson, to the British General Proctor, Lower Sandusky, 1 August 1813

Nuts!
Major General Anthony C. McAuliffe, USA: In reply to a German demand that he surrender his beleaguered force at Bastogne, 23 December 1944

(*See also* Resistance, Resolution, Will to Fight.)

Delay

Delay breeds dangers.
John Lyly: Euphues, 1578

Defer no time, delays have dangerous ends.
Shakespeare: I King Henry VI, iii, 2, 1591

All delays are dangerous in war.
John Dryden: Tyrannic Love, i, 1669

Demobilization

Scatter thou the people that delight in war.
Psalm LXVIII, 30

Deliver to the army this news of peace; let them have pay, and part.
Shakespeare: II King Henry IV, iv, 2, 1597

Resolved, That the commanding officer be and he is hereby directed to discharge the troops now in the service of the United States, except twenty-five privates, to guard the stores at Fort Pitt, and fifty-five to guard the stores at West Point and other magazines, with a proportionate number of officers; no officer to remain in service above the rank of a captain.
Resolution of the Continental Congress disbanding the Continental Army, 2 June 1784

Get the boys out of the trenches and back to their homes by Christmas.
Henry Ford: On chartering the "Peace Ship," Oscar II, *December 1915*

America fought the war like a football game, after which the winner leaves the field and celebrates.
General Albert C. Wedemeyer, USA, 1896-, on the demobilization after World War II.

It was no demobilization, it was a rout.
George C. Marshall, 1880–1959

Dependents

Absent from me
So let him be,
If fame and glory come
With him triumphant home,
Bear and forebear,
Make my heart strong,
Through bitter care,
Days sad and long,
All this I can endure, all and yet more,
If I may hear him hailed at last victor in war.
That prize enough for me.
Plautus: Amphitryon, 3rd century, B. C.

He that hath wife and children hath given hostages to fortune, for they are impediments to great enterprises, either or virtue or of mischief.
Francis Bacon, 1561–1626, Essays

A wife is a useless piece of furniture for a soldier.
Madame de Pompadour, 1721–1764, letter to the Duchess d'Estrees

Tell me not, sweet, I am unkind,
That from the nunnery
Of thy chaste breast, and quiet mind,
To war and arms I fly.
Richard Lovelace, 1618–1658

If any of the soldiers' wives or children happen to be taken ill, never give them any assistance. You receive no pence from them, and you know *ex nihilo nihil sit.*
Francis Grose: Advice to the Officers of the British Army, 1782

The duty of the captain of a battleship in wartime is incessant, requiring the most constant and unremitting attention, and the best first lieutenant in the world cannot be a sufficient substitute for him, in that his whole body and soul should be in it, which, with his wife and family near him, never has been, or can be; and, unless the husband and wife can make up their minds to separation *during war*, it would be unwise and unjust to the Service, and to me, for him to delay one day his intention of retiring.
Lord St. Vincent: Letter to a captain's wife, 1800

Tell my sister not to weep for me, and sob with drooping head,
When the troops come marching home again, with glad and gallant tread,
But to look upon them proudly, with a calm and steadfast eye,
For her brother was a soldier, too, and not afraid to die.
Caroline Elizabeth Norton: 1808–1877, Soldier of the Rhine

Thy voice is heard thro' rolling drums,
That beat to battle where he stands;
Thy face across his fancy comes,
And gives the battle to his hands:
A moment, while the trumpets blow,
He sees his brood about thy knee;
The next, like fire he meets the foe,
And strikes him dead for thine and thee.
Tennyson: The Princess, 1847

Marriage is good for nothing in the military profession.
Napoleon I: Political Aphorisms, 1848

It were better to be a soldier's widow than a coward's wife.

Thomas Bailey Aldrich, 1836–1907, Mercedes, ii, 2

For the colonel's lady an' Judy O'Grady
Are sisters under their skins.

Rudyard Kipling: The Ladies, 1896

Man has two supreme loyalties—to country and to family...So long as their families are safe, they will defend their country, believing that by their sacrifice they are safeguarding their families also. But even the bonds of patriotism, discipline, and comradeship are loosened when the family itself is threatened.

B. H. Liddell Hart: Sherman, 1927

Sons of heroes are a plague.

Greek Proverb

(*See also* Love and War, Officers' Sons, Women.)

De Ruyter, Michiel Adrianzoon (1607–1676)

I never saw him other than even-tempered, and when victory was assured, saying always that it was the good God that gives it to us. Amid the disorders of the Fleet and the appearance of loss, he seemed only to be moved by the misfortune to his country...The day after victory I found him sweeping his own cabin and feeding his chickens.

Comte de Guiche: After the Four-Days' Battle with the English, 1666

Desert Warfare

My skin is black upon me, and my bones are burned with heat.

Job, XXX, 30

The panting thirst that scorches in the breath
Of those that die the soldier's fiery death.

Byron: Lara, 1814

Desertion

An adversary is more hurt by desertion than by slaughter.

Vegetius: De Re Militari, iii, 378

Let soldiers marry; they will no longer desert. Bound to their families, they are bound to their country.

Voltaire, 1694–1778, Satirical Dictionary

Desperation

Follow me—You have the advantage of necessity, that last and most powerful of weapons.

Vettius Messius of Volscia: To his troops before breaking out of a Roman encirclement, 5th century, B. C.

No persons will behave so desperately in action as those who are tired of their lives.

Francis Grose: Advice to the Officers of the British Army, 1782

Destroyers

There must be developed in the men that handle [destroyers] that mixture of skill and daring which can only be attained if the boats are habitually used under circumstances that imply the risk of accident.

William S. Sims, 1858–1936, endorsement (pre-World War I) on correspondence regarding destroyers

Thirty-one knot Burke, get athwart the Buka-Rabaul evacuation line...If enemy contacted, you know what to do.

William F. Halsey: Message to Captain Arleigh Burke, USN, 24 November 1943 (This was the origin of the nickname, "31-Knot" Burke, based on the fact that, though his destroyer squadron could in theory only make 30 knots, his speed of advance toward the enemy on this occasion was 31 knots.)

Of all the tools the Navy will employ to control the seas in any future war...the destroyer will be sure to be there.

C. W. Nimitz, 1962

Detachments

A defensive war is apt to betray us into too frequent detachments. Those generals who are better acquainted with their profession, having only the capital end in view, guard against a decisive blow, and acquiesce in smaller misfortunes to avoid a greater.

Frederick the Great: Instructions for His Generals, 1747

Detente

...enough reduction of tension between enemies to encourage fights between friends.

Hervé Alphand, 1963 (M. Alphand, French Ambassador in Washington, was commenting on Franco-American relations)

Determination

Let us do or die.

Robert Burns, 1759–1796, Scots Wha Hae

With God's blessing, let us make thorough work of it.

Stonewall Jackson: At the outset of the Valley Campaign, 1862

Every attack, once undertaken, must be fought to a finish; every defense, once begun, must be carried on with the utmost energy.

Ferdinand Foch: Precepts, 1919

Determination will carry the soldier only as far as the place where he dies, and few soldiers care to go quite that far.

A. J. Liebling: in the New Yorker, 19 October 1963

(*See also* Resolution, Will to Fight, Tenacity.)

Deterrence

When there is mutual fear, men think twice before they make aggression upon one another.

Hermocrates of Syracuse: To the Sicilian envoys at Gela, 424 B. C.

I deplore a tendency to accept "mutual deterrence" as a blessing.

General Thomas D. White, USAF: in Newsweek, 24 February 1964

The central concept of modern strategy is deterrence.

Abba Eban: During televised interview Washington, 7 March 1965

In the military operations off Cuba President Kennedy did not look for military victory, he sought to change Mr Khrushchev's mind, and he succeeded.

Vice Admiral Sir Peter Gretton: Lecture, Royal United Service Institution, London, 7 April 1965

Dewey, George (1837–1917)

O Dewey was the morning
Upon the first of May,
And Dewey was the Admiral
Down in Manila Bay...

Eugene Fitch Ware: Verse in Topeka Daily Capital, 3 May 1898

Difficulties

I am not come forth to find difficulties, but to remove them.

Horatio Nelson, 1758–1805

The friction inherent in the tremendous war-machine of an armed power is so great in itself that it should not be increased unnecessarily.

Clausewitz: Principles of War, 1812

Everything is very simple in war, but the simplest thing is difficult. These difficulties accumulate and produce a friction which no man can imagine exactly who has not seen war.

Clausewitz: On War, 1832

It is the property of ordinary men, in times of danger, to see difficulties more clearly than advantages, and to shrink from steps which involve risk.

Mahan: Life of Nelson, x, 1897

Never, never, never believe any war will be smooth and easy.

Winston Churchill: A Roving Commission, 1930

I commend these ideas to your study, hoping that the intention will be to solve the difficulties.

Winston Churchill: Minute to Director of Naval Construction, 12 September 1939

I wish I could persuade you to try to overcome the difficulties instead of merely intrenching yourself behind them.

Winston Churchill: Memorandum for Minister of Agriculture, 28 February 1943

Campaigns and battles are nothing but a long series of difficulties to be overcome.
George C. Marshall: To the First Officer Candidate Class, Fort Benning, Georgia, 18 September 1941

"Difficulties" is the name given to things which it is our business to overcome.
E. J. King: Address to the graduating class, U. S. Naval Academy, 19 June 1942

War is of its nature digressive, exasperating, and easy to conduct only in retrospect.
A. J. Liebling: in the New Yorker, 19 October 1963

Diplomacy

He had rather spend £10,000 on Embassies to keep or procure peace with dishonor, than £10,000 on an army that would have forced peace with honor.
Sir Anthony Weldon: The Court and Character of King James, 1650

The more powerful the prince, the more suave his diplomacy should be. Since power of that kind is likely to waken jealousy...the diplomat should let it speak for itself.
Francois De Callières: The Art of Negotiating with Princes, 1733

Diplomacy without arms is music without instruments.
Frederick the Great, 1712–1786

When my profession fails, yours has to come to the rescue.
Talleyrand, 1754–1838, to Marshal Ney.

...Mentir et démentir. (To lie and deny.)
Attributed to Talleyrand, 1754–1838, as his definition of diplomacy

A fleet of British ships of war are the best negotiators in Europe.
Nelson: Letter to Lady Hamilton, March 1801

Foreign policy demands scarcely any of those qualities which are peculiar to a democracy; on the contrary it calls for the perfect use of almost all those qualities in which a democracy is deficient. Democracy...cannot combine its measures with secrecy or await their consequences with patience. These are qualities which are more characteristic of an individual or an aristocracy.
Alexis de Tocqueville, 1805–1859

Mr. Prime Minister, I don't resent your having a card up your sleeve, but I do resent your thinking that God put it there.
Ascribed to a French diplomat in conversation with Gladstone, c. 1885

Diplomacy is utterly useless where there is no force behind it.
Theodore Roosevelt: Address to the Naval War College, Newport, Rhode Island, 2 June 1897

He lied; I knew he lied and he knew I knew he lied. That was diplomacy.
Rear Admiral William Wirt Kimball, USN, of negotiations with the Mexican Minister to Nicaragua, c. 1909

Open covenants of peace, openly arrived at, after which there shall be no private international understandings of any kind but diplomacy shall proceed always frankly and in the public view.
Woodrow Wilson: To Congress, 8 January 1918 (Point I of the Fourteen Points)

Guns are left to do what words
Might have done earlier, rightly used
John Waller, 1917–, in Beirut

Influence is founded on seven specific diplomatic virtues, namely: truthfulness, precision, calm, good temper, patience, modesty, and loyalty.
Sir Harold Nicolson, 1886–, Diplomacy

A diplomat's words must have no relation to actions—otherwise what kind of diplomacy is it? Words are one thing, actions another. Good words are a concealment of bad deeds. Sincere diplomacy is no more possible than dry water or iron wood.
Attributed to J. V. Stalin, 1879–1951

Diplomacy has rarely been able to gain at the conference table what cannot be gained or held on the battlefield.
General Walter Bedell Smith, USA: on his return from the Geneva Conference on Indo-China, 1954

Diplomacy is the expression of national strength in terms of gentlemanly discourse.
Robert McClintock: Lecture,

U. S. Naval War College, 17 September 1964

Diplomacy is strategy's twin.
Anthony Eden: The Reckoning, 1965

Successful diplomacy, like successful marriage, is not much publicized.
John Paton Davies: In New York Times Magazine Section, 23 May 1965

When a man fights it means that a fool has lost his argument.
Chinese Proverb.

(*See also* Alliances, Ambassador, Diplomat, Gunboat Diplomacy, Negotiations, Politico-Military Affairs.)

Diplomat

One knows too well how these diplomats are, with what kind of ravenous hunger this sort of people reaches for negotiations, and how reluctantly, having once entered upon them, they give them up.
Gneisenau: Letter, 1814

The diplomat is the servant, not the master of the soldier.
Theodore Roosevelt: Address to the U. S. Naval War College, Newport, Rhode Island, 2 June 1897

A BEAR. How do you know I am diplomat?
CHINESE WOMAN. Why, by the skilful way you hide your claws.
Edmond Rostand: L'Aiglon, iv, 1900

The ability to get to the verge without getting into the war is the necessary art. If you cannot master it, you inevitably get into war. If you try to run away from it, if you are scared to go to the brink, you are lost.
John Foster Dulles: In Life Magazine, January 1956

The basic quality of the diplomat is not intelligence but loyalty.
C. Northcote Parkinson: Comment on "democratization" in recruitment of foreign service and military officers, New York Times, 20 August 1961

(*See also* Ambassador, Diplomacy.)

Dipping Colors

When any of His Majesty's ships shall meet with any ship or ships belonging to any foreign Prince within His Majesty's seas (which extend to Cape Finisterre) it is expected that the said foreign ships do strike their topsail and take in their flag, in acknowledgement of His Majesty's Sovereignty in those seas.
British order in Council, 18th century (the origin of the custom of dipping colors, however, goes back to a provision in the Treaty of Westminster, 5 April, 1654, between England and the Netherlands.)

Disarmament

He maketh wars to cease in all the world: he breaketh the bow, and knappeth the spear in sunder, and burneth the chariots in the fire.
Psalm XLVI

How are the weapons of war perished.
II Samuel, I, 27

They shall beat their swords into plowshares, and their spears into pruning-hooks: nation shall not lift up sword against nation, neither shall they learn war any more.
Isaiah, II, 4

Among other evils which being unarmed brings you, it causes you to be despised.
Niccolo Machiavelli: The Prince, vi, 1513

Were half the power, that fills the world with terror,
Were half the wealth, bestowed on camps and courts,
Given to redeem the human mind from error,
There were no need for arsenals and forts.
Henry Wadsworth Longfellow: The Arsenal at Springfield, 1841

An end to these bloated armaments.
Benjamin Disraeli: To the House of Commons, 1862

Adequate guarantees given and taken that national armaments will be reduced to the lowest point consistent with domestic safety.
Woodrow Wilson: To Congress, 8 January 1918 (Point IV of the Fourteen Points)

Our government should secure a joint agreement with all nations for world disarmament and also for a referendum on war, except in case of actual or threatened attack. Those who furnish the blood and bear the burdens imposed by war should, whenever possible, be consulted before this supreme sacrifice is required of them.
Democratic National Platform, 1924

I am quite prepared to disarm—provided that the others disarm first.
Benito Mussolini, 1883–1945

It is the greatest possible mistake to mix up disarmament with peace. When you have peace you will have disarmament.
Winston Churchill: To the House of Commons, 13 July 1934

When the animals had gathered, the lion looked at the eagle and said gravely, "We must abolish talons." The tiger looked at the elephant and said, "We must abolish tusks." The elephant looked back at the tiger and said, "We must abolish claws and jaws." Thus each animal in turn proposed the abolition of the weapons he did not have, until at last the bear rose up and said in tones of sweet reasonableness: "Comrades, let us abolish everything—everything but the great universal embrace."
Attributed to Winston Churchill, 1874–1965

From time immemorial, the idea of disarmament has been one of the most favored forms of diplomatic dissimulation of the true motives and plans of those governments which have been seized by sudden "love of peace." This phenomenon is very understandable. Any proposal for the reduction of armaments could invariably count upon broad popularity and support from public opinion.
E. V. Tarlé, 1862–1917, History of Diplomacy (Tarlé was a Russian historian.)

It never has made and never will make any sense trying to abolish any particular weapon of war. What we have to abolish is war.
Sir John Slessor: Strategy for the West, 1954

Arms control and military strategy have been regarded as opposites; the first directed towards the removal of the second, the second toward avoiding the shackles of the first.
Hedley Bull, The Control of the Arms Race, 1961

Disarmament, to be successful, must be a symptom of underlying friendliness.
Samuel Eliot Morison: History of the American People, xxv, 1965

(*See also* Armament, Pacifism.)

Discharge

There is no discharge in that war.
Ecclesiastes, VIII, 8

Discipline

Maintain discipline and caution above all things, and be on the alert to obey the word of command. It is both the noblest and the safest thing for a great army to be visibly animated by one spirit.
Archidamus of Sparta: To the Lacaedaemonian expeditionary force departing for Athens, 431 B. C.

In the moment of action remember the value of silence and order.
Phormio of Athens: To the Athenian seamen and Marines before action in the Crisaean Gulf, 429 B.C.

A large army is always disorderly.
Euripides: Hecuba, c. 426 B.C.

When every man is his own master in battle, he will readily find a decent excuse for saving himself.
Brasidas of Sparta: To the Lacadaemonian Army, 423 B. C.

The strength of an army lies in strict discipline and undeviating obedience to its officers.
Thucydides: History of the Peloponnesian Wars, c. 404 B.C.

Self-control is the chief element in self-respect, and self-respect is the chief element in courage..
Thucydides: History of the Peloponnesian Wars, c. 404 B.C.

The orders of the superiors is the source whence discipline is born.
Wu Ch'i, 430–381 B. C., Art of War, iii, 2

It is discipline that makes one feel safe, while lack of discipline has destroyed many people before now.

Xenophon: Speech to the Greek officers after the defeat of Cyrus at Cunaxa, 401 B. C.

If troops are punished before their loyalty is secured they will be disobedient. If not obedient, it is difficult to employ them. If troops are loyal, but punishments are not enforced, you cannot employ them. Thus, command them with civility and imbue them uniformly with martial ardor and it may be said that victory is certain . . . When orders are consistently trustworthy and observed, the relationship of a commander with his troops is satisfactory.

Sun Tzu, 400–320 B. C., The Art of War, ix

Pardon one offense and you encourage the commission of many.

Publilius Syrus: Sententiae, c. 42 B.C.

No-one can rule except one that can be ruled.

Seneca, 4 B.C.–65 A.D., De Ira, ii

He shall rule them with a rod of iron.

Revelation, II, 27

No state can be either happy or secure that is remiss and negligent in the discipline of its troops.

Vegetius: De Re Militari, i, 378

The ancients, taught by experience, preferred discipline to numbers.

Vegetius: De Re Militari, iii, 378

Good order makes men bold, and confusion, cowards.

Niccolo Machiavelli: Arte della Guerra, 1520

As for Colonel Cromwell, he hath 2,000 brave men, well disciplined; no man swears but he pays his twelvepence; if he be drunk, he is set in the stocks, or worse; if one calls the other roundhead he is cashiered . . . How happy were it if all the forces were thus disciplined!

Puritan news letter, May 1643

He that enrolleth himself a soldier taketh away the excuse of a timorous nature, and is obliged not only to go into battle, but also not to run from it without his captain's leave.

Thomas Hobbes: Leviathan, xxi, 1651

The fleet was in such a condition as to discipline, as if the Devil had commanded it.

Samuel Pepys: Diary entry, 20 October 1666

Self-discipline is that which, next to virtue, truly and essentially raises one man above another.

Joseph Addison, 1672–1719

The Romans conquered all peoples by their discipline. In the measure that it became corrupted their success decreased. When the Emperor Gratian permitted the legions to give up their cuirasses and helmets because the soldiers complained that they were too heavy, all was lost. The barbarians whom they had defeated for so many centuries vanquished them in their turn.

Maurice de Saxe: Mes Rêveries, x, 1732

After the organization of troops, military discipline is the first matter that presents itself. It is the soul of armies. If it is not established with wisdom and maintained with unshakable resolution you will have no soldiers. Regiments and armies will be only contemptible, armed mobs, more dangerous to their own country than to the enemy . . . It has always been noted that it is with those armies in which the severest discipline is enforced that the greatest deeds are performed.

Maurice de Saxe: Mes Rêveries, xviii, 1732

The Troops must pay exact obedience to all Orders & they may be assured of the most impartial Justice; It is recommended to them to live in great Friendship, and Harmony amongst each other, and to carry on the publick Business, as becomes Men who mean to do honor to themselves, & to their Country by their behavior.

Lord Jeffrey Amherst: Order before landing at Louisbourg, 1758

Discipline is simply the art of inspiring more fear in the soldiers of their officers than of the enemy.

Helvetius: de l'Esprit, 1758

Discipline is the soul of an army. It makes small numbers formidable; procures success to the weak, and esteem to all.

George Washington: Letter of Instruction to the Captains of the Virginia Regiment, 29 July 1759

Men accustomed to unbounded freedom, and no control, cannot brook the Restraint which is indispensably necessary to the good order and government of an Army.

George Washington: Letter to the President of Congress, 1776

There are two systems which, generally speaking, divide the disciplinarians, the one is that of *training men like spaniels, by the stick;* the other . . . of *substituting the point of honor in place of severity.* The followers of the first are for reducing the nature of man as low as it will bear . . . The admirers of the latter are for exalting rationality, and they are commonly deceived in their expectations . . . I apprehend a just medium between the two extremes to be the surest means to bring English soldiers to perfection.

Major General John Burgoyne: Code of Instructions for the 15th Dragoons, 1762

It was an inflexible maxim of Roman discipline that a good soldier should dread his own officers far more than the enemy.

Edward Gibbon: The Decline and Fall of the Roman Empire, i, 1776 (cf. Helvetius, ante.)

Since Discipline's the strongest cord
That ties the martial Ranks,
Attention to the soldier's word,
To win his country's thanks.
Each cannot be a General,
Nor lead the glorious van:
To be a hero, stand, or fall,
Depend upon the man.
Let all then in their station stand;
Each point of duty weigh;
Rememb'ring those can best command,
Who best know to obey.

Charles Dibdin, 1745–1814, Discipline

Without discipline is well planned and strictly supported, a military corps or a ship's crew are no better than a disorderly mob; it is a well-formed discipline that gives force, preserves order, obedience, and cleanliness, and causes alertness and despatch in execution of business.

Admiral Richard Kempenfelt, RN: Letter to Sir Charles Middleton, 28 December 1779

Mobs will never do to govern states or command armies.

John Adams: Letter to Benjamin Hichborn, 27 January 1787

A free people ought not only to be armed, but disciplined.

George Washington: To Congress 8 January 1790

Discipline begins in the Wardroom. I dread not the seamen. It is the *indiscreet* conversations of the officers and their *presumptuous* discussions of the orders they receive that produce all our ills.

Lord St. Vincent, 1735–1823

Discipline is summed up in the one word obedience.

Lord St. Vincent, 1735–1823

The officers of companies must attend to the men in their quarters as well as on the march, or the army will soon be no better than a banditti.

Wellington: General Order of 19 May 1800

Popularity, however desirable it may be to individuals, will not form, or feed, or pay an army; will not enable it to march, and fight; will not keep it in a state of efficiency for long and arduous service.

Wellington: Letter from Portugal, 8 April 1811

We have in the service the scum of the earth as common soldiers; and of late years we have been doing everything in our power, both by law and by publications, to relax the discipline by which alone such men can be kept in order. The officers of the lower ranks will not perform the duty required from them for the purpose of keeping their soldiers in order; and it is next to impossible to punish any officer for neglects of this description. As to the non-commissioned officers, as I have repeatedly stated, they are as bad as the men, and too near them, in point of pay and situation, by the regulations of late years, for us to expect them to do anything to keep the men in order. It is really a disgrace to have anything to say to such men as some of our soldiers are.

Wellington: After Vitoria, 21 June 1813

In the Service, Mr. Simple, one is obliged to appear angry without indulging the sentiment.

Frederick Marryat: Peter Simple, 1834

They [the soldiers] should be made to understand that discipline contributes no less to their safety than to their efficiency . . . Let officers and men be made to feel that they will most effectively secure their safety by remaining steadily at their posts, preserving order, and fighting with coolness and vigor.

R.E. Lee: Circular to Troops, Army of Northern Virginia, 1865

Discipline is not made to order, cannot be created offhand; it is a matter of the institution of tradition. The Commander must have absolute confidence in his right to command, must have the habit of command, pride in commanding.

Ardant du Picq, 1821–1870, Battle Studies

Discipline and blind obedience are things which can be produced and given permanence only by long familiarity.

Wilhelm I of Prussia, 1797–1888

The discipline that makes the soldiers of a free country reliable in battle is not to be gained by harsh or tyrannical treatment. On the contrary, such treatment is far more likely to destroy than to make an army. It is possible to impart instruction and give commands in such a manner and in such a tone of voice as to inspire in the soldier no feeling but an intense desire to obey, while the opposite manner and tone of voice cannot fail to excite strong resentment and a desire to disobey.

J.M. Schofield: Address to the Corps of Cadets, West Point, and now inscribed in the Sally Port

As the severity of military operations increases, so also must the sternness of the discipline. The zeal of the soldiers, their warlike instincts, and the interests and excitements of war may ensure obedience of orders and the cheerful endurance of perils and hardships during a short and prosperous campaign. But when fortune is dubious or adverse; when retreats as well as advances are necessary; when supplies fail, arrangements miscarry, and disasters impend, and when the struggle is protracted, men can only be persuaded to accept evil things by the lively realization of the fact that greater terrors await their refusal.

Winston Churchill: The River War, ii, 1899

To be disciplined . . . means that one frankly adopts the thoughts and views of the superior in command, and that one uses all humanly practicable means in order to give him satisfaction.

Ferdinand Foch: Precepts, 1919

We was rotten 'fore we started—we was
never disciplined;
We made it out a favor if an order was
obeyed.
Yes, every little drummer 'ad 'is rights an'
wrongs to mind,
So we had to pay for teachin'—an' we
paid!

Rudyard Kipling: That Day, 1895

Discipline, in the sense in which it is restrictive, [is] submergent of individuality, the Lowest Common Denominator of man. In regular armies in peace it means the limits of energy attainable by everybody present: it is the hunt not of an average, but of an absolute . . . The aim is to render the unit a unit, and the man a type, in order that their effort shall be calculable, their collective output even in grain and in bulk. The deeper the discipline, the lower the individual efficiency, and the more sure the performance. It is a deliberate sacrifice of capacity in order to reduce the uncertain element, the bionomic factor, in enlisted humanity.

T.E. Lawrence, 1888–1935, The Science of Guerrilla Warfare

Be kindly and just in your dealings with your men. Never play favorites. Make them feel that justice tempered with mercy may always be counted on. This does not mean a slackening of discipline. Obedience to orders and regulations must always be insisted upon, and good conduct on the part of the men exacted. Especially should this be done with reference to the civilian inhabitants of foreign countries in which Marines are serving.

John A. Lejeune: Letter No. 1, to all officers, U.S. Marine Corps, 1920

True military discipline stems not from knowledge but from habit.

Hans von Seeckt, 1866–1936

Discipline is willing obedience to attain the greatest good by the greatest number. It means laying aside, for the time being, of ordinary everyday go-as-you-please and do-what-you-like. It means one for all and all for one—teamwork. It means a machine—not of inert metal, but one of

living men—an integrated human machine in which each does his part and contributes his full share.
E.J. King: 1878–1956

I don't mind being called tough, since I find in this racket it's the tough guys who lead the survivors.
Colonel Curtis Le May, USA: To Lieutenant General Ira Eaker, USA, in England, 1943

There is only one sort of discipline—*perfect discipline.* If you do not enforce and maintain discipline, you are potential murderers.
George S. Patton, Jr.: Instructions to Third Army corps and division commanders, 1944

No sane man is unafraid in battle, but discipline produces in him a form of vicarious courage.
George S. Patton, Jr.: War as I Knew It, 1947

If you can't get them to salute when they should salute and wear the clothes you tell them to wear, how are you going to get them to die for their country?
George S. Patton, Jr., 1885–1945

Discipline is teaching which makes a man do something which he would not, unless he had learnt that it was the right, the proper, and the expedient thing to do. At its best, it is instilled and maintained by pride in oneself, in one's unit, in one's profession; only at its worst by a fear of punishment.
Sir A. P. Wavell: Soldiers and Soldiering, 1953

Discipline is as necessary to the soldier as the air he breathes. It is not only the source of his strength, it is the source of his contentment.
Jean Dutourd: Taxis of the Marne, 1957

Fighting men can never feed on syllogisms. Their only law is the law of averages. Only discipline can master this. And discipline yoked to the love of comrades is beyond defeat.
Robert Leckie: The March to Glory, xviii, 1960

Soldiers deserve to be well paid and well hanged.
German Proverb

. . . Doing the right thing when there's no one to tell you right from wrong.
Navy saying

A taut ship's a happy one.
Navy saying

(*See also* Command, Leadership, Obedience.)

Dispositions

Anciently the skilful warriors first made themselves invincible and awaited the enemy's moment of vulnerability. Invincibility depends on one's self; the enemy's vulnerability on him. It follows that those skilled in war can make themselves invincible but cannot cause an enemy to be certainly vulnerable.
Sun Tzu, 400–320 B.C., The Art of War, iv

Diversion

Every diversion brings war into a district into which it would not have penetrated; for that reason it will always be the means, more or less, of calling forth military forces which would otherwise have continued in abeyance.
Clausewitz: On War, vii, 1832

A diversion is only a subordinate part of the drama of war. It is either a deceit, the success of which depends rather upon the incapacity of the opponent than upon its own merits; or it is an indirect use of forces which, from their character or position, cannot be made to conduce directly to the main effort of the enterprise in hand.
Mahan: The Influence of Sea Power upon the French Revolution and Empire, i, 1892

Division

The Pope! How many divisions has *he* got?
J. V. Stalin: To Pierre Laval during conversations in Moscow, May 1935

. . . the smallest formation that is a complete orchestra of war and the largest in which every man can know you.
Sir William Slim: Defeat into Victory, 1956

A division is something that comes into being very laboriously and it is created out of many things—out of blood, sacrifice, and the effort and a good part of the productive life of tens or hundreds of thousands of individuals who have passed through its ranks, A division is something that is very easy to strike off on a troop-list tally sheet. But when you need a division, when war begins, you can't bring it into being overnight. It is easy to destroy. It is extremely difficult to re-create.

Brigadier General J. D. Hittle USMC, 1962

Division of Forces

How frequently an army is separated on its march by bad roads, rivers, difficult passes? And how many such situations will enable you to surprise some part of it? How often do opportunities present themselves of separating it, so as to be able, although inferior, to attack one part with advantage and at the same time, by the proper placement of a small number of troops prevent its being relieved by the other?

Maurice de Saxe: Mes Rêveries, xiv, 1732

It may be laid down as a principle that any movement is dangerous which is so extended as to give the enemy an opportunity, while it is taking place, of beating the remainder of the army in position.

Jomini: Précis de l'Art de la Guerre, 1838

I can only advise the party on the defensive not to divide his forces too much by attempting to cover every point.

Jomini: Précis de l'Art de la Guerre, 1838

It is not enough to pierce and divide the enemy's army; the advantage thus gained must be followed up, or a subsequent reunion of the parts may nullify all the previous operations . . . That is to say, the first stroke must be followed up by successive blows on one or both sides, which shall, at once, keep the enemy asunder, and destroy his force.

Sir Edward Hamley: Operations of War, 1866

Divide and conquer.

Ancient maxim

Doctrine

Doctrine is nothing but the skin of truth set up and stuffed.

Henry Ward Beecher, 1813–1887, Life Thoughts

Doctrines are the most frightful tyrants to which men ever are subject.

William Graham Sumner, 1840–1910, Essays

A doctrine of war consists first in a common way of objectively approaching the subject; second, in a common way of handling it.

Ferdinand Foch: Precepts, 1919

Official manuals, by the nature of their compilation, are merely registers of prevailing practice, not the log-books of a scientific study of war.

B. H. Liddell Hart: Thoughts on War, xii, 1944

Generals and admirals stress the central importance of "doctrine." Military doctrine is the "logic" of their professional behavior. As such, it is a synthesis of scientific knowledge and expertise on the one hand, and of traditions and political assumptions on the other.

Morris Janowitz: The Professional Soldier, xii, 1960

Doctrine is indispensable to an army . . . Doctrine provides a military organization with a common philosophy, a common language, a common purpose, and a unity of effort.

General George H. Decker, USA: Address, Command and General Staff College, Fort Leavenworth, Kansas, 16 December 1960

Wars in the main start with textbooks.

Alan Moorehead: The Blue Nile, vi, 1962

Doctrine is codified common sense . . . It is what tells a commander or a soldier what to do when specific directions are lacking.

Captain C. H. Amme, USN: In U.S. Naval Institute Proceedings, March 1964

Our forefathers defeated our enemies without ever suspecting that such a thing as military doctrine existed.

Colonel-General S. Shtemenko: In Nedelya, 31 January 1965

(*See also* Principles of War.)

Dogs of War

Cry "Havoc!" and let slip the dogs of war.
Shakespeare: Julius Caesar, iii, 1, 1599

Douhet, Giulio (1869–1930)

Let us take care not to treat lightly, as a Utopian dreamer, a man who may later be regarded as a Prophet.
Marshal Petain: Introduction to La Doctrine de Guerre du General Douhet, 7 June 1934

Drake, Francis (1540–1596)

Drake he's in his hammock an' a
thousand miles away,
(Capten, art tha sleepin' there below?)
Slung atween the round shot in Nombre
Dios Bay,
An' dreaming arl the time o' Plymouth
Hoe.
Henry Newbolt, 1862–1938, Drake's Drum

Dreadnought

Fear God and dread nought.
Motto selected by Sir John Fisher on entering the peerage (Admiral Fisher developed the first dreadnought battleship, HMS Dreadnought.)

For each and all, as for the Royal Navy, the watchword should be, "Carry on, and dread nought."
Winston Churchill: To the House of Commons, 6 December 1939

Dress Uniform

Even good men like to make the public stare.
Byron: Don Juan, iii, 1823

Gold lace has a charm for the fair.
W. S. Gilbert, Patience, i, 1881

(*See also* Uniform.)

Drill

What is necessary to be performed in the heat of action should constantly be practiced in the leisure of peace.
Vegetius: De Re Militari, 378

Troops who march in an irregular and disorderly manner are always in great danger of being defeated.
Vegetius: De Re Militari, 378

The [soldiers] must learn to keep their ranks, to obey words of command, and signals by drum and trumpet, and to observe good order, whether they halt, advance, retreat, are upon a march, or engaged with an enemy.
Niccolo Machiavelli: Arte della Guerra, 1520

Drill is necessary to make the soldier steady and skillful, although it does not warrant exclusive attention.
Maurice de Saxe: Mes Rêveries, v, 1732

When you have nothing more necessary for the men to do, let them be exercised at small-arms; it makes the men straight, gives them an easy and graceful motion of the limbs, shakes off the awkward clown, and gives that military air which shows a man to advantage.
Admiral Richard Kempenfelt, RN: Letter to Sir Charles Middleton, 28 December 1779

In exercising the regiment, call out frequently to some of the most attentive men and officers to dress, cover, or something of that nature: the less they are reprehensible, the greater will your discernment appear to the bystanders, in finding out a fault invisible to them.
Francis Grose: Advice to the Officers of the British Army, 1782

. . . pirouetting up and down a barracks yard.
Carnot, 1753–1823

The exterior splendor, the regularity of movements, the adroitness and at the same time firmness of the mass—all this gives the individual soldier the safe and calming conviction that nothing can withstand his particular regiment or battalion.
Colmar von der Goltz, 1843–1916, Rossbach und Jena

(*See also* Ceremonies, Training.)

Drill Instructor

The finest edge is made with the blunt whetstone.
John Lyly: Euphues, 1579

But now, my boys, leave off, and list to me,
That mean to teach you rudiments of war.
I'll have you learn to sleep upon the ground,
March in your armor thorough watery fens,
Sustain the scorching heat and freezing cold,
Hunger and thirst, right adjuncts of the war.
Christopher Marlowe: Tamburlaine the Great, iii, 2, 1587

Be copy now to men of grosser blood,
And teach them how to war.
Shakespeare: King Henry V, iii, 2, 1598

We but teach bloody instructions . . .
Shakespeare: Macbeth, i, 7, 1605

. . . a man in khaki kit who could handle men a bit.
Rudyard Kipling: Pharaoh and the Sergeant, 1900

(*See also* Recruit, Recruit Training, Training.)

Drink

Drinking is the soldier's pleasure.
John Dryden: Alexander's Feast, 1697

Whenever an invasion of enemy territory is planned, all the beer and brandy possible to furnish to the army should be brewed on the frontier. And within the enemy territory all the breweries that are found near the encampment are put to work.
Frederick The Great: Instructions for His Generals, 1747

Here sleeps in peace a Hampshire grenadier,
Who caught his death by drinking cold small beer.
Soldiers! Take heed from his untimely fall,
And when you're hot, drink strong or none at all.
Epitaph on Thomas Thetcher, Winchester, England, 1764

The British soldiers are fellows who have all enlisted for drink—that is the plain fact—they have all enlisted for drink.
Wellington: Letter from Portugal, 1811

If the facilities for washing were as great as those for drink, our Indian Army would be the cleanest body of men in the World.
Florence Nightingale: Observations on Evidence Contained in Stational Reports Submitted to the Royal Commission on the Sanitary State of the Army in India, 1863

Red wine for children, champagne for men, and brandy for soldiers.
Otto von Bismarck, 1815–1898

(*See also* Drunkenness.)

Drum

Strike up the drum and march courageously.
Christopher Marlowe: Tamburlaine the Great, ii, 2, 1587

Strike up the drums, and let the tongue of war
Plead for our interest.
Shakespeare: King John, v, 2, 1596

. . . with boisterous, untuned drums.
Shakespeare: King Richard II, i, 3, 1595

We scorn the vulgar Ways to bid you come,
Whole Europe now obeys the Call of Drum.
George Farquhar: The Recruiting Officer, 1706

A drum and its appurtenances may, in the hands of a clever fellow, answer many good purposes besides that of being beaten. Should a flock of ducks or geese obstruct your line of march, two or three may be safely and secretly lodged in it; and the drum case will hold peas, beans, apples and potatoes, when the haversack is full.
Francis Grose: Advice to the Officers of the British Army, 1782

Hark! I hear the tramp of thousands,
And of armed men the hum;
Lo! A nation's hosts have gathered
Round the quick alarming drum—
Saying, "Come . . ."
Bret Harte, 1839–1902, The Reveille

Drum Major

In marching by the commanding officer, when you beat the short troop, look as stern as possible, and appear as if you could eat him up at a mouthful.

Francis Grose: Advice to the Officers of the British Army, 1782

(*See also* Bands, Martial Music.)

Drunkenness

That which hath made them drunk hath made me bold.

Shakespeare: Macbeth, ii, 1, 1605

A drunken quarrel is very bad, and is always to be lamented, but probably the less it is enquired into the better.

Wellington: Letter from India, 1802

The military profession is not designed for debauchees.

The Military Mentor: Being a Series of Letters Recently Written by a General Officer to His Son, 1804

Every soldier or ranger who shall be found drunk or sensibly intoxicated shall be compelled, as soon as his strength will permit, to dig a grave at a suitable burying place large enough for his own reception, as such graves cannot fail soon to be wanted for the drunken man himself or some drunken companion.

General Order attributed to Winfield Scott, 1786–1866

He is a respectable old man and has no other failing than that which but too often attends an old soldier.

Colonel Archibald Henderson USMC: Letter to Lieutenant Colonel Wainwright, 1 July 1837 (speaking of a Sergeant Triguet)

"Drunk and resisting the Guard!"
Mad drunk and resisting the Guard—
'Strewth, but I socked it them hard!
So it's pack-drill for me and a fortnight's
C.B.
For "Drunk and resisting the Guard."

Rudyard Kipling: Cells, 1892

(*See also* Drink.)

Du Picq, Ardant (1821–1870)

Never has a man of action, of brutal action, and in the eyes of universal prejudice, more spendidly glorified the spirituality of war.

Barbey d'Aurevilly, 1808–1889

Duty

In doing what we ought, we deserve no praise, because it is our duty.

St. Augustine, 354–430

I hold my duty as I hold my soul.

Shakespeare: Hamlet, ii, 2, 1600

Do your duty, and leave the rest to heaven.

Pierre Corneille: Horace, 1639

He trespasses against his duty who sleeps upon his watch, as well as he that goes over to the enemy.

Edmund Burke: Thoughts on the Cause of the Present Discontentments, 1770

Three things prompt men to a regular discharge of their Duty in time of Action: natural bravery, hope of reward, and fear of punishment. The first two are common to the untutor'd, and the Disciplin'd Soldiers; but the latter, most obviously distinguishes the one from the other. A Coward, when taught to believe, that if he breaks his Ranks, and abandons his Colours, he will be punished by Death by his own party, will take his chance against the Enemy.

George Washington: Letter to the President of Congress, 9 February 1776

When ordered for duty, always grumble and question the roster. This will procure you the character of one that will not be imposed on.

Francis Grose: Advice to the Officers of the British Army, 1782

We all did our duty, which, in the patriot's, and gentleman's language, is a very comprehensive word, of great honor, meaning, and import.

Rudolph Erich Raspe: Travels of Baron Munchausen, v, 1785

Perish discretion when it interferes with duty.

Hannah More, 1745–1833

A most minute Attention to Duty is the making of a good Officer.

Thomas Truxtun: Letter, 1798

Duty is the great business of a sea-officer; all private considerations must give way to it, however painful it may be.
Nelson: Letter to Frances Nisbet, 1786

Say to the fleet, England expects that every man will do his duty.
Nelson: To his flag-lieutenant, aboard HMS Victory, *before action at Trafalgar, 21 October 1805*

Thank God I have done my duty.
Nelson: While lying mortally wounded in the cockpit, HMS Victory, *Trafalgar, 21 October 1805*

Duty—the sublimest word in the English language.
Attributed to Horatio Nelson, 1758–1805

Stern daughter of the Voice
Of God! Oh Duty . . .
William Wordsworth: Ode to Duty, 1805

The brave man, inattentive to his duty, is worth little more to his Country, than the coward who deserts her in the hour of danger.
Andrew Jackson: Address to the American troops, New Orleans, 8 January 1815

Whatever happens, you and I will do our duty.
Wellington: Remark to Lord Uxbridge on the eve of Waterloo, 17 June 1815

First let mankind be concerned for its duties, and then only for its rights!
Gneisenau, 1760–1831

Duty before decency!
Frederick Marryat: Mr. Midshipman Easy, 1836

Where duty calls, or danger, be never wanting there.
George Duffield, Jr.: Stand Up, Stand Up, for Jesus, 1858

Duty is ours; consequences are God's.
Stonewall Jackson, 1862

Don't flinch from the fire, boys! There's a hotter fire than that waiting for those who don't do their duty! Give that rascally tug a shot!
David G. Farragut: To gun-crews, USS Hartford, *as a tug pushed a Confederate fire-ship alongside during the passage of the forts below New Orleans, 24 April 1863*

Do your duty in all things. You cannot do more. You should never wish to do less.
R. E. Lee, 1807–1870

No one will consider the day as ended, until the duties it brings have been discharged.
Major General Joseph Hooker, USA: General Order on assuming command of the Department of the Northwest, 1865

A soldier's vow to his country is that he will die for the guardianship of her domestic virtue, of her righteous laws, and her anyway challenged or endangered honor. A state without virtue, without laws, and without honor, he is bound not to defend.
John Ruskin: The Crown of Wild Olive, iii, 1866

When Duty whispers low, *Thou must,*
The youth replies, *I can.*
R. W. Emerson: Voluntaries, iii, 1867

He seen his duty, a dead-sure thing—
And went for it, thar and then;
And Christ ain't a-going to be too hard
On a man that died for men.
John Hay: Jim Bludso, 1871

Knowledge is a steep which few may climb,
While Duty is a path which all may tread.
William Morris, 1834–1896, The Epic of Hades (Héré)

Duty, duty must be done;
The rule applies to everyone.
W. S. Gilbert: Ruddigore, i, 1887

Death is lighter than a feather; duty, heavy as a mountain.
Emperor Meiji of Japan: Imperial Rescript to Soldiers and Sailors, 4 January 1883

The moril of this story, it is plainly to be seen:
You 'avn't got no families when servin' of the Queen—
You 'avn't got no brothers, fathers, sisters, wives, or sons—

If you want to win your battles take an'
work your bloomin' guns!
Rudyard Kipling: Snarleyow, 1890

On he went up the great, bare staircase of his duty, uncheered and undepressed.
R. L. Stevenson: Weir of Hermiston, 1896

War is only a sort of dramatic representation, a sort of dramatic symbol of a thousand forms of duty.
Woodrow Wilson: Speech at Brooklyn, 11 May 1914

Duty, oh Duty,
Why art thou not a
Sweetie or a cutie?
Ogden Nash, 1935

We gave thanks to God for the noblest of all His blessings, the sense that we had done our duty.
Winston Churchill, 1874–1965

If I do my full duty, the rest will take care of itself.
George S. Patton, Jr.: Diary entry before the North African landings, 8 November 1942

I pray daily to do my duty, retain my self confidence and accomplish my destiny.
George S. Patton, Jr.: Diary entry, 20 June 1943

Any commander who fails to attain his objective, and who is not dead or severely wounded, has not done his full duty.
George S. Patton, Jr.: War as I Knew It, 1947

Duty is never simple, never easy, and rarely obvious.
Jean Dutourd: Taxis of the Marne, 1957

Duty, honor, country.
Motto of the U. S. Military Academy, West Point.

The path of Duty is the way to Glory.
Graven in stone over a gateway at Wellington School, Somerset, England.

Duty-Roster

Never suffer your roster to be questioned, and though it should be wrong, never condescend to alter it. The roster is the adjutant's log-book, which he will manage as will be most conducive to his own private views.
Francis Grose: Advice to the Officers of the British Army, 1782

Economy of Force

I could lick those fellows any day, but it would cost me 10,000 men, and, as this is the last army England has, we must take care of it.

Wellington: Of Massena's army in Portugal, October 1810

A war should only be undertaken with forces proportioned to the obstacles to be overcome.

Napoleon I: Maxims of War, 1831

The principle of economy of force is the art of pouring out *all* one's resources at a given moment on one spot.

Ferdinand Foch: Precepts, 1919

. . . The art of making the weight of all one's forces successively bear on the resistances one may meet.

Ferdinand Foch: Principles of War, 1920

To me, an unnecessary action, or shot, or casualty, was not only waste but sin.

T. E. Lawrence, 1888–1935

We used the smallest force, in the quickest time, in the farthest place.

T. E. Lawrence, 1888–1935

Efficiency

Favoritism is the secret of efficiency.

Sir John Fisher, 1841–1920

Egypt

From these pyramids, forty centuries gaze down upon you. (Soldats! du haut de ces monuments, quarante siècles vous regardèrent.)

Napoleon I: General Order to the Army of the Nile, 21 July 1798

Elite Troops

A small army consisting of chosen troops is far better than a vast body chiefly composed of rabble.

The Hitopadesa, iii, c. 500

Chosen men who were warriors.

I Kings, xii, 21

It is better to have a small number of well-kept and well disciplined troops than to have a great number who are neglected in these matters. It is not big armies that win battles; it is the good ones.

Maurice de Saxe: Mes Rêveries, iv, 1732

(*See also* Shock Troops.)

Embarkation

The citizens came to take farewell,
one of an acquaintance, another
of a kinsman, another of a son, and as they passed along were full of hope and full of tears.

Thucydides: Peloponnesian Wars, vi (on the embarkation of the Syracusan Expedition from Athens, 415 B. C.)

Stuck in 'eavy marchin'-order, sopped and wringin'—
Sick, before our time, to watch 'er 'eave an' fall,
'Ere's your 'appy 'ome at last, an' stop your singin!
'Alt! Fall in along the troop-deck! Silence all!

Rudyard Kipling: "Birds of Prey" March (Troops for Foreign Service), 1895

To Members of the United States Marine Corps Expeditionary Forces:
You have embarked for distant places where the war is being fought. Upon the outcome depends the freedom of your lives: the freedom of the lives of those you love—your fellow citizens—your people . . . We who stay at home have our duties to perform—duties owed in many parts to you. You will be supported by the whole force and power of this Nation . . . You bear with you the hope, the confidence, the gratitude and the prayers of your family, your fellow-citizens, and your President—

Franklin Delano Roosevelt: Special message sent to Marines embarking for the South Pacific, February 1942

(*See also* Expeditionary Service.)

Emergencies

New emergencies must be met in war as they come.

Winston Churchill: Telegram to Anthony Eden, 3 November 1940

Encouragement

No need to write long messages to a good soldier. Fight on. I am on my way. Maurice.

Maurice de Saxe: Field message to Marquis de Courtivron, besieged with 600 troops by 4,000 Croats, 1742

Endurance

Even the bravest cannot fight beyond his strength.

Homer: The Iliad, xiii, c. 1000 B. C.

The strong did what they could, and the weak suffered what they must.

Thucydides: History of the Peloponnesian Wars, 404 B. C.

He that shall endure unto the end, the same shall be saved.

Matthew, XXIV, 13

He conquers who endures.

Persius, 34–62 A. D.

If Job had been a General in my situation, his memory had not been so famous for patience.

General Philip John Schuyler: Letter to Washington, 26 September 1775 (of the difficulties in supplying Montgomery's expedition against Quebec)

To endure is greater than to dare; to tire out hostile fortune; to be daunted by no difficulty; to keep heart when all have lost it . . .

W. M. Thackeray: The Virginians, 1857

It's dogged as does it.

Anthony Trollope: Last Chronicles of Barset, lxi, 1867

(*See also* Resolution, Tenacity.)

Enemy

It is right to be taught, even by an enemy. (Fas est et ab hoste doceri.)

Ovid: Metamorphoses, iv, c. 8 A. D.

A dead enemy always smells good. (Optime olere hostem occisum.)

Alus Vitellius: On the battlefield of Beariacum, 69 A.D. (also ascribed to Charles IX of France after the St. Bartholomew's Night massacres of 1571)

You must not fight too often with one enemy, or you will teach him all your art of war.

Napoleon I, 1769–1821

The wish of a possible enemy is the beacon which suggests the shoal.

Mahan: Some Neglected Aspects of War, 1907

The enemy never sleeps. (L'Ennemi ne s'endort.)

French Proverb

Though thine enemy seem like a mouse, watch him like a lion.

Italian Proverb.

Enemy Capabilities

'Tis best to weigh
The enemy more mighty than he seems.

Shakespeare: King Henry V, ii, 4, 1598

A general in all of his projects should not think so much about what he wishes to do as what his enemy will do; that he should never underestimate this enemy, but he should put himself in his place to appreciate difficulties and hindrances the enemy could interpose; that his plans will be deranged at the slightest event if he has not foreseen everything and if he has not devised means with which to surmount the obstacles.

Frederick the Great: Instructions for His Generals, iii, 1747

However absorbed a commander may be in the elaboration of his own thoughts, it is sometimes necessary to take the enemy into account.

Attributed to Winston Churchill, 1874–1965

Experienced military men are familiar with the tendency that always has to be watched in staff work, to see all our own difficulties but to credit the enemy with the ability to do things we should not dream of attempting.

Sir John Slessor: Strategy for the West, 1954

(*See also* Intelligence.)

U. S. Navy

DAVID G. FARRAGUT

1801–1870

"Damn the torpedoes! Captain Drayton, go ahead!"

Engagement

It was thought that to engage the enemy to fight was our business.
Oliver Cromwell: At Preston, 17 August 1648

I engage, and after that I see what to do. (Je m'engage, et après ça, je voie.)
Napoleon I: Remark during the Italian Campaign, 1796

Going into action today reminds one of a struggle between two blind men, between two adversaries who perpetually seek each other but cannot see.
Ferdinand Foch: Precepts, 1919

(*See also* Contact.)

Engineer

The rough ways shall be made smooth.
Luke, III, 5

Which of you, intending to build a tower, sitteth not down first, and counteth the cost, whether he have sufficient to finish it?
Luke, XIV, 28

For 'tis the sport to have the engineer
Hois'd with his own petard.
Shakespeare: Hamlet, iii, 4, 1600

Essayons. (Let's try!)
Motto of the Corps of Engineers

Enlisted Men

I have a lovely company; you would respect them, did you know them. They are no Anabaptists, they are honest sober Christians: they expect to be used as men!
Oliver Cromwell: Letter to Oliver St. John, 11 September 1643

As a private soldier, you should consider all your officers as your natural enemies, with whom you are in a perpetual state of warfare: you should reflect that they are constantly endeavoring to withhold from you all your just dues, and to impose on you every unnecessary hardship; and this for the mere satisfaction of doing you an injury.
Francis Grose: Advice to the Officers of the British Army, 1782

I don't know what effect these men will have on the enemy, but, by God, they frighten *me*.
Wellington: On a draft of troops sent him during the Peninsular Campaign, 1809

We have in the service the scum of the earth as common soldiers.
Wellington: Letter to Earl Bathurst after Vitoria, 2 July 1813

The humblest bred man who stands in the ranks of an army, is as susceptible of the sentiment of glory, and honor, and shame, as the proudest captain that ever carried a plume.
Brigadier General William Duane, USA: A Handbook for Infantry, 1814 (Duane was Adjutant-General of the Army)

I have never been able to join in the popular cry about the recklessness, sensuality, and helplessness of the soldiers. On the contrary I should say...that I have never seen so teachable and helpful a class as the Army generally. Give them opportunity promptly and securely to send money home and they will use it. Give them schools and lectures and they will come to use them. Give them books and games and amusements and they will leave off drinking. Give them suffering and they will bear it. Give them work and they will do it.
Florence Nightingale: Letter to her sister, March 1856

An' if sometimes our conduck isn't all
your fancy paints,
Why, single men in barricks don't grow
into plaster saints.
Rudyard Kipling: Tommy, 1890

Service fellows . . . enlist mainly for a refuge against the pain of making a living.
T. E. Lawrence: Letter to James Hanley, 2 July 1931

The ordinary soldier has a surprisingly good nose for what is true and what false.
Erwin Rommel: The Rommel Papers, ix, 1953

You can't snow the troops.
U. S. Marine Corps saying ("snow" = hoodwink.)

(*See also* Soldier.)

Ensign

Ay, tear her tattered ensign down!
Long has it waved on high,
And many an eye has danced to see
That banner in the sky.
Oliver Wendell Holmes, Sr.: Old Ironsides, 1830

And there, while thread will hang to thread,
Oh, let that ensign fly!
The noblest constellation set
Against the Northern sky.
George Henry Boker, The Cumberland, 1862 (USS Cumberland, *wooden sloop-of-war, was sunk with all guns in action and colors flying, by the Confederate ironclad,* Virginia, *8 March 1862.)*

(*See also* Colors, Flag.)

Entrenchments

When I hear talk of lines, I always think I am hearing talk of the walls of China. The good ones are those that nature has made, and the good entrenchments are good dispositions and brave soldiers.
Maurice de Saxe: Mes Rêveries, xxv, 1732

To bury an army in entrenchments, where it may be outflanked and surrounded or forced in front even if secure from a flank attack, is manifest folly.
Jomini: Précis de l'Art de la Guerre, 1838

Entrench, entrench, entrench...
Sir Henry Lawrence: Dying instructions, 4 July 1857, during the defense of Lucknow

(*See also* Fortifications, Foxholes.)

Envelopment

The deep envelopment based on surprise, which severs the enemy's supply lines, is and always has been the most decisive maneuver of war. A short envelopment which fails to envelop and leaves the enemy's supply system intact, merely divides your own forces and can lead to heavy loss and even jeopardy.
Douglas MacArthur: Conference before Inchon, 23 August 1950

(*See also* Flank.)

Equipment

The fate of an army may depend on a buckle.
Major General George H. Thomas, 1816–1870

(*See also* Maintenance, Logistics.)

Escape

To a surrounded enemy you must leave a way of escape.
Sun Tzu, 400–320 B. C., The Art of War

Espionage

See the land, what it is; and the people that dwelleth therein, whether they be strong or weak, few or many.
Numbers, XIII, 18

And Joshua, the son of Nun, sent out of Shittim two men to spy secretly, saying Go view the land, even Jericho. And they went, and came into a harlot's house, named Rahab, and lodged there.
Joshua, II, i

Every kind of service, necessary to the public good, becomes honorable by being necessary.
Nathan Hale: Letter to Captain William Hull, 10 September, 1775 (Hull had objected to Hale's entering the British lines as an American spy.)

It is one of the dangerous characteristics of the sort of information supplied by secret agents that it becomes rare and less explicit as the peril increases and the need for information becomes greater.
Alexis de Tocqueville, 1805–1859

(*See also* Intelligence, Spies.)

Esprit de Corps

To insure victory the troops must have confidence in themselves as well as in their commanders.
Niccolo Machiavelli: Discorsi, xxxiii, 1531

We would not die in that man's company
That fears his fellowship to die with us.
Shakespeare: King Henry V, iv, 3, 1598

We few, we happy few, we band of brothers . . .

Shakespeare: King Henry V, iv, 3, 1599

That silly, sanguine notion, which is firmly entertained here, that one Englishman can beat three Frenchmen, encourages, and has sometimes enabled, one Englishman, in reality, to beat two.

Lord Chesterfield: Letters, 7 February 1749

All that can be done with the soldier is to give him *esprit de corps*—i. e., a higher opinion of his own regiment than all the other troops in the country.

Frederick the Great: Military Testament, 1768

Consider your corps as your family; your commander as your father; your comrade as your brother; your inferior as a young relative. Then all will be happy and friendly and easy. Don't think of yourself, think of your comrades; they will think of you. Perish yourself, but save your comrades.

General Mikhail Ivanovich Dragomirov: Notes for Soldiers, 1890

Personal bravery of a single individual alone is not decisive on the day of battle, but rather bravery of the corps, and the latter rests on the good opinion and the confidence that each individual places in the corps to which be belongs.

Colmar von der Goltz, 1843–1916, Rossbach und Jena

[Esprit de corps] is the soul of the Navy and ours from the junior cadet to the senior admiral, the charge to "save that soul alive."

Rear Admiral C. F. Goodrich, USN 1845–1925

Now is the time to popularize with the troops by giving to all regiments and units the little badges and distinctions they like so much . . . All regimental distinctions should be encouraged.

Winston Churchill: To Secretary of State for War July 1940

. . . that regimental *esprit de corps* upon which all armies worthy of the name are founded.

Winston Churchill: Note for Secretary of State for War, 21 November 1942

Living in an atmosphere of soldierly duty and *esprit de corps* permeates the soul, where drill merely attunes the muscles.

B. H. Liddell Hart: Thoughts on War, v, 1944

He was as loyal to his caste as the good soldier is to his regiment, the good undergraduate to his school, the good Communist or Nazi to his party. Loyalty to organization A always entails some degree of suspicion, contempt, or downright loathing of organizations B, C, D, and all the rest.

Aldous Huxley: The Devils of Loudun, 1952

Esprit de corps thrives not only on success, but on hardships and adversity shared with courage and fortitude.

Major General Orlando Ward, USA: Letter, May 1965

(*See also* Morale.)

Estimate of the Situation

Employment of troops must be in accord with determination of the enemy's strong and weak points, after which you speedily attack his critical positions.

Wu Ch'i, 430–381 B. C., Art of War, ii

With many calculations, one can win; with few one cannot. How much less chance of victory has one who makes none at all! By this means I examine the situation and the outcome will be clearly apparent.

Sun Tzu, 400-320 B. C., The Art of War, i

The wise general in his deliberations must consider both favorable and unfavorable factors. By taking into account the favorable factors, he makes his plan feasible; by taking into account the unfavorable, he may resolve the difficulties.

Sun Tzu, 400–320 B. C., The Art of War, viii

Weigh the situation, then move.

Sun Tzu, 400–320 B. C., The Art of War, vii

A general is not easily overcome who can form a true judgment of his own and the enemy's forces.

Vegetius: De Re Militari, 378

The ability of a commander to comprehend a situation and act promptly is the talent which great men have of conceiving in a moment all the advantages of the terrain and the use that they can make of in with their army.
Frederick the Great: Instructions for His Generals, 1747

The one who is to draw up a plan of operations must possess a minute knowledge of the power of his adversary and of the help the latter may expect from his allies. He must compare the forces of the enemy with his own numbers and those of his allies so that he can judge which kind of war he is able to lead or to undertake.
Frederick the Great: Letter, 1748

De quoi s'agit-il? (What is the problem?)
Ferdinand Foch: Favorite question asked by Foch of himself and his staff when confronted by a problem or decision (Foch once attributed the question to General Verdy du Vernois on arrival on the battlefield of Nachod.)

(*See also* Enemy Capabilities, Intelligence, Plans.)

Evacuations

Wars are not won by evacuations.
Winston Churchill: To the House of Commons after Dunkirk, June 1940

(*See also* Retreat, Withdrawal.)

Example

A brave captain is as a root, out of which, as branches, the courage of his soldiers doth spring.
Sir Philip Sidney, 1554–1586

One must never excuse oneself by pointing to the soldiers: . . . I have never seen faults happen through them that had not been done by the captain.
Blaise Montluc: Commentaires, 1592

Who can look for modestie and sobrietie in the souldiers, where the captaine is given to wine, or women, and spendeth his time in riot and excesse?
Mathew Sutcliffe: The Practice, Proceedings, and Lawes of Armes, 1593

No man should possess him with any appearance of fear, lest he, by shewing it, should dishearten his army.
Shakespeare: King Henry V, iv, 1, 1598

If you choose godly, honest men to be captains of Horse, honest men will follow them.
Oliver Cromwell: To Sir William Springe, September 1643

As I would deserve and keep the kindness of this army, I must let them see that when I expose them, I would not exempt myself.
Duke of Marlborough, 1650–1722

Let your character be above reproach, for that is the way to earn men's obedience.
Mathias von Schulenburg: To Marshal Saxe when a young officer, 1709

Example is the best General Order.
Major General George Crook, USA, 1828–1890

I am the very model of a modern major-general.
W. S. Gilbert: The Pirates of Penzance, i, 1880

There is an inheritance of heroic example which is necessary to a nation's life... National character is a more sacred thing than human life.
H. W. Wilson: Ironclads in Action, 1895

With two thousand years of examples behind us, we have no excuse when fighting, for not fighting well.
T. E. Lawrence, 1888–1935

Be an example to your men, both in your duty and in private life. Never spare yourself, and let the troops see that you don't, in your endurance of fatigue and privation. Always be tactful and well mannered and teach your subordinates to be the same. Avoid excessive sharpness or harshness of voice, which usually indicates the man who has shortcomings of his own to hide.
Erwin Rommel: Remarks to graduating cadets, Wiener Neustadt Military School, 1938

In moments of panic, fatigue, or disorganization, or when something out of the ordinary has to be demanded from [his troops], the personal example of the commander works wonders, especially if he has had the wit to create some sort of legend round himself.

Erwin Rommel: The Rommel Papers, ix, 1953

The first leader under whom a young officer serves, especially if he be admired, can often fix a character beyond recall.

Fletcher Pratt: Preble's Boys, ix, 1950

Excuses

And, oftentimes, excusing of a fault
Doth make the fault the worse by the excuse.

Shakespeare: King John, iv, 2, 1596

How vulnerable are those who explain.

Attributed to Dean Acheson: In Time magazine, 26 June 1964

Qui s'excuse, s'accuse. (He who makes excuses, accuses himself.)

French Proverb

Expedition

There has never been a protracted campaign from which a country has benefitted.

Sun Tzu, 400–320 B. C., Art of War

In war, expedition often proves of more consequence than courage.

Vegetius: De Re Militari, 378

"Make the war short and sharp," as the French say.

Niccolo Machiavelli: Discorsi, vi, 1531

The advantage of time and place in all martial actions is half a victory, which being lost is irrecoverable.

Sir Francis Drake: To Queen Elizabeth, 1588

Then fiery expedition by my wing . . .

Shakespeare: King Richard III, iv, 3, 1592

(*See also* Promptness, Rapidity, Speed.)

Expeditionary Service

And bold and hard adventures t'undertake,
Leaving his country for his country's sake.

Charles Fitzgeffry: Life and Death of Sir Francis Drake, 1596

We go to gain a little plot of ground,
That hath in it no profit but the name.

Shakespeare: Hamlet, iv, 1600

Our conjunct expeditions go forth freighted with good wishes, blessing and huzzas. These they soon disburthen and have too often come home loaded with reproaches, sorrow, and disappointment.

Thomas More Molyneux: Conjunct Expeditions, 1759

You are ordered abroad as a soldier of the King to help your French comrades against the invasion of a common enemy. You have to perform a task which will need your courage, your energy, your patience. Remember that the honor of the British Army depends on your individual conduct. It will be your duty not only to set an example of discipline and perfect steadiness under fire but also to maintain the most friendly relations with those whom you are helping in this struggle . . . Do your duty bravely. Fear God. Honor the King.

Lord Kitchener: To the soldiers of the British Expeditionary Force, on embarking for France, August 1914

(*See also* Embarkation.)

Experience

Experience is by industry achieved.

Shakespeare: Two Gentlemen of Verona, i, 3, 1594

War, like most things, is a science to be acquired and perfected by diligence, by perseverance, by time, and by practice.

Alexander Hamilton: The Federalist, No. xxv, 1787

It was not until after Gettysburg and Vicksburg that the war professionally began. Then our men had learned in the dearest school on earth the simple lessons of war. Then we had brigades, divisions and corps which could be handled professionally, and it was then that we as professional soldiers could rightly be held to a just responsibility.

W. T. Sherman, 1820–1891

Military men who spend their lives in the uniform of their country acquire experience in preparing for war and waging it. No theoretical studies, no intellectual attainments on the part of the layman can be a substitute for the experience of having lived and delivered under the stress of war.

Maxwell D. Taylor: Graduation address at West Point, June 1963

(*See also* Combat Experience.)

Exploitation

When we have incurred the risk of a battle, we should know how to profit by the victory, and not merely content ourselves, according to custom, with possession of the field.

Maurice de Saxe: Mes Rêveries, 1732

Faithful

Be thou faithful unto death.
Revelations, II, 10

The latest breath that gave the sound of words
Was deep-sworn faith.
Shakespeare: King John, iii, 1, 1596

Among the faithless, faithful only he.
Milton: Paradise Lost, 1667 (of Abdiel)

Ever faithful. (Semper Fidelis)
Motto of the U. S. Marine Corps

(*See also* Loyalty.)

Famine

Famine makes greater havoc in an army that the enemy, and is more terrible than the sword.
Vegetius: De Re Militari, iii, 378

Yet famine,
Ere it clean o'erthrow nature, makes it valiant.
Shakespeare: Cymbeline, iii, 6, 1609

(*See also* Hunger.)

Farragut, David Glasgow (1801–1870)

To Farragut all glory!
The Sea King's worthy peer,
Columbia's greatest seaman,
Without reproach or fear.
Verse in a New York newspaper, 13 December 1864, on arrival of Farragut and USS Hartford *in New York.*

Fatigue

Would God that night or Blücher would come.
Attributed to Wellington: On the afternoon of Waterloo, 18 June 1815

After a certain period, even the victor becomes tired of war; and the more civilized a people is, the more quickly will this weakness become apparent.
Colmar von der Goltz: The Conduct of War, 1883

Fatigue makes cowards of us all.
George S. Patton, Jr.: War As I Knew It, 1947

(*See also* Combat Fatigue.)

Fear

Fear makes men forget, and skill which cannot fight is useless.
Phormio of Athens: To the Athenian seamen and Marines before action in the Crisaean Gulf, 429 B. C.

In battle those who are most afraid are always in most danger.
Catiline: To his troops in the field, Pistoria, 63 B. C.

The man that is roused neither by glory nor danger it is vain to exhort; terror closes the ears of the mind.
Sallust, 86–34 B. C., Catilina

Even the bravest are frightened by sudden terrors.
Tacitus: Annals, xv, c. 70 A.D.

Fear, the worst prophet in misfortune, anticipates many evils.
Publius Statius; Thebaid, iii, c. 90 A. D.

Fear makes men ready to believe the worst.
(Ad deteriora credenda proni metu.)
Quintus Curtius Rufus: De Rebus Gesti Alexandri Magni, iv, c. 2d century A. D.

Nothing is to be feared but fear.
Francis Bacon, 1521–1566

Of all base passions, fear is the most accursed.
Shakespeare: King Henry VI, Pt I, v, 2, 1590

To fear the foe, since fear oppresseth strength,
Gives in your weakness strength unto your foe.
And so your follies fight unto yourself.
Fear and be slain; no worse can come to fight:
And fight and die is death destroying death;
Where fearing dying pays death servile breath.
Shakespeare: King Richard II, iii, 2, 1595

Fie, my lord, fie! a soldier, and afear'd?
Shakespeare: Macbeth, v, 1, 1605

Fear has many eyes.
Cervantes: Don Quixote, iii, 1605

A coward's fear can make a coward valiant.
Owen Feltham, 1602–1688, Resolves (Of Cowardice)

Early and provident fear is the mother of safety.
Edmund Burke: Speech on the petition of the Unitarians, 11 May 1792

Nothing so much to be feared as fear.
Henry David Thoreau, 1817–1862, Journal

Never take counsel of your fears.
Stonewall Jackson: To Major Hotchkiss, 18 June 1862

The direct foe of courage is the fear itself, not the object of it; and the man who can overcome his own terror is a hero and more.
George MacDonald, 1824–1905, Sir Gibbie

Keep your fears to yourself but share your courage.
Robert Louis Stevenson, 1850–1894

The man who conquers in war is the man who is least afraid of death.
Alexei Kuropatkin, 1848–1925

The one permanent emotion of the inferior man is fear—fear of the unknown, the complex, the inexplicable. What he wants beyond everything else is safety.
H. L. Mencken: Prejudices: Second Series, 1920

All men are frightened. The more intelligent they are, the more they are frightened. The courageous man is the man who forces himself, in spite of his fear, to carry on. Discipline, pride, self-respect, self confidence, and love of glory are attributes which will make a man courageous even when he is afraid.
George S. Patton, Jr.: War As I Knew It, 1947

No sane man is unafraid in battle.
George S. Patton, Jr.: War As I Knew It, 1947

(*See also* Cowardice, Timidity.)

Field Equipment

There are five things the soldier should never be without—his musket, his ammunition, his rations (for at least four days), and his entrenching tool. The knapsack may be reduced to the smallest size possible . . but the soldier should always have it with him.
Napoleon I: Maxims of War, 1831

Field Fortification

Field fortification, when well conceived, is always useful and never harmful.
Napoleon I: Maxims of War, 1831

The principles of field fortification need to be perfected. This part of the art of war is susceptible of making great progress.
Napoleon I: From St. Helena, 1816

(*See also* Entrenchments.)

Field Hospital

Keep two lancets; a blunt one for the soldiers, and a sharp one for the officers: this will be making a proper distinction between them.
Francis Grose: Advice to the Officers of the British Army, 1782

I have seen the bodies of the dead, stores for the living, munitions of war, sick men staggering from weakness, wounded men helpless on stretchers, and invalid orderlies . . .
Newspaper account of a British field hospital in the Crimea, 1854

(*See also* Health, Surgeon.)

Field Music

By your profession you are evidently destined to make a noise in the world: and your party-colored coat and drum carriage, like the zone of Venus, or halter about the neck of a felon, makes you appear a pretty fellow in the eyes of the ladies. So you may always, if not over-modest (which I must own is not often the failing of gentlemen of your calling), be sure of bringing off a girl from every quarter.

Francis Grose: Advice to the Officers of the British Army, 1782 (of field musics)

(*See also* Martial Music.)

Field Service

Henceforth, in fields of conquest, the tents shall be your home.
E. W. Shurtleff: Lead on, O King Eternal, 1887

I was lying on the ground, with the soft side of a stone for my pillow.
John S. Mosby: War Reminiscences, ii, 1887

In war a soldier must expect short commons, short sleep, and sore feet. Even an old soldier finds it difficult, and for a green one it is hard. But if it's hard for you it isn't easier for the enemy; perhaps harder still. Only you see your own hardships, but don't see the enemy's.
General Mikhail Ivanovich Dragomirov: Notes for Soldiers, c. 1890

Officers should live under the same conditions as their men, for that is the only way in which they can gain from their men the admiration and confidence so vital in war. It is incorrect to hold a theory of equality in all things, but there must be equality of existence in accepting the hardships and dangers of war.
Mao Tse-tung: On Guerrilla Warfare, 1937

Fifth Column

We have four columns advancing upon Madrid. The fifth column will rise at the proper time.
Emilio Mola: Radio broadcast, October 1936 (Ernest Hemingway to some extent popularized the phrase in the title of his 1938 play, The Fifth Column.)

When I wage war . . . in the midst of peace troops will suddenly appear, let us say, in Paris. They will wear French uniforms. They will march through the streets in broad daylight. No one will stop them. Everything has been thought out, prepared to the last detail. They will march to the headquarters of the General Staff. They will occupy the ministries, the Chamber of Deputies . . . Peace will be negotiated before the war has begun. We shall have enough volunteers, men like our S. A., trustworthy and ready for any sacrifice. We shall send them across the border in peacetime. Gradually. The place of artillery preparation for frontal attack by the infantry will in future be taken by revolutionary propaganda, to break down the enemy psychologically before the armies begin to function at all.
Adolph Hitler, 1889–1945

(*See also* Quisling, Treason.)

Fight

Blessed be the Lord my strength, which teacheth my hands to war, and my fingers to fight.
Psalm CXXXXIV

Fight on, my men, Sir Andrew says,
A little I'm hurt, but not yet slain;
I'll lie me down and bleed awhile,
And then I'll rise and fight again.
Ballad, Sir Andrew Barton, c. 1550

Fight to the last gasp.
Shakespeare: I King Henry VI, i, 1, 1591

I'll fight, till from my bones my flesh be hacked.
Shakespeare: Macbeth, v, 3, 1605

They now to fight are gone,
Armour on armour shone,
Drum now to drum did groan,
To hear was wonder...
Michael Drayton: To the Cambro-Britans, Agincourt, 1627

There is a time to pray and a time to fight. This is the time to fight.
Pastor Muhlenberg: Sermon at Woodstock, Virginia, 1775

I have not yet begun to fight.
John Paul Jones: In reply to the hail, "Have you struck?" from Captain Richard Pearson, RN, commanding HM late ship Serapis, *off Flamborough Head, 23 September 1775*

Fight as they fathers fought,
Fall as thy fathers fell!
Thy task is taught, thy shroud is wrought;—
So—forward—and farewell!
W. M. Praed, 1802–1839

U. S. Navy

ERNEST J. KING

1878–1956

"Difficulties is the name given to things it is our business to overcome."

Fifty-four forty or fight.
William Allen: To the U. S. Senate, during debate on the Oregon boundary, 1844 (reference is to the parallel of latitude, 54–40 N.)

One fight more, the best and the last!
Robert Browning: Prospice, 1861

Let your watchword be, fight, fight, fight. If you cannot cut off from [the enemy's] columns large slices, the general desires that you will not fail to take small ones.
Major General Joseph Hooker USA: To General Stoneman, 1863

Fight the good fight with all thy might.
J. S. B. Monsell: Fight the Good Fight, 1863

I propose to fight it out on this line, if it takes all summer.
U. S. Grant: Despatch to General H. W. Halleck from Spottsylvania Court House, 11 May 1864

Fight anything that comes.
W. T. Sherman: Order to General J. B. McPherson before Atlanta, 11 May 1864

If we must be enemies, let us be men, and fight it out as we propose to do,
and not deal in hypocritical appeals to God and humanity.
W. T. Sherman: Letter to General John B. Hood, CSA, Atlanta, 10 September 1864

Thrice armed is he that hath his quarrel
just;
And four times he who gets his fist in fust.
Artemus Ward (Charles Farrar Browne), 1834–1867, Shakespeare Up-to-Date

Do not get into a fight if you can possibly avoid it. If you get in, see it through. Don't hit if it is honorably possible to avoid hitting, but never hit soft. Don't hit at all if you can help it; don't hit a man if you can possibly avoid it; but if you do hit him, put him to sleep.
Theodore Roosevelt: Speech in Washington, 24 January 1918

I would fight without a break. I would fight in front of Amiens. I would fight in Amiens. I would fight behind Amiens.
I would never surrender.
Ferdinand Foch: To Sir Douglas Haig during the German offensive, 26 March 1918

I shall fight before Paris, I shall fight in Paris, I shall fight behind Paris. (Je me bats devant Paris, je me bats à Paris, je me bats derrière Paris.)
George Clemenceau: To the Chanber of Deputies, 4 June 1918 (cf. Foch, ante, 26 March 1918)

He is a fighter . . . a fighter . . . a fighter!
John J. Pershing: Favorite expression of praise while commanding the AEF, 1918

We shall fight on for ever and ever and ever.
Winston Churchill: Message to Paul Reynaud, fall of France, 1940

We will fight the enemy where we now stand; there will be no withdrawal and no surrender.
Montgomery of Alamein: To the Eighth Army after assuming command, 13 August 1942

Still, if you will not fight for the right when you can easily win without bloodshed; if you will not fight when your victory will be sure and not too costly; you may come to the moment when you will have to fight with all the odds against you and only a precarious chance of survival. There may even be a worse case. You may have to fight when there is no hope of victory, because it is better to perish than live as slaves.
Winston Churchill: The Gathering Storm, xx, 1948

It's an ill fight where he that wins has the warst o't.
Scotch Proverb

(*See also* Battle, Combat, Fighting.)

Fighting

'Tis no festival unless there be some fighting.
Thomas Hall: Funebria Floria, 1660

Most sorts of diversion in men, children, and other animals, are in swift imitation of fighting.
Jonathan Swift: Thoughts on Various Subjects, 1706

He rushed into the field, and, foremost fighting, fell.

Byron: Childe Harold's Pilgrimage, iii (Eve of Waterloo), 1816

War in its literal meaning is fighting, for fighting alone is the efficient principle in the manifold activity which in a wide sense is called war.

Clausewitz: On War, 1832

Fighting is to war what cash payment is to trade, for however rarely it may be necessary for it actually to occur, everything is directed toward it.

Friedrich Engels: Letter to Karl Marx, 25 September 1857

Go in anywhere, Colonel! You'll find lovely fighting along the whole line.

Major General Philip Kearny USA: To the colonel of a reinforcing regiment at Seven Pines, 31 May 1862

War means fighting . . . The business of the soldier is to fight . . . to find the enemy and strike him; to invade his country, and do him all possible damage in the shortest possible time.

Stonewall Jackson, 1824–1863

War means fighting, and fighting means killing.

Attributed to Nathan Bedford Forrest, 1821–1877

For the Lord abideth back of me to guide my fighting arm.

Rudyard Kipling: Mulholland's Contract, 1895

His zeal for close fighting was more than calculated risk required, surely not a bad fault.

Commander Frederick L. Sawyer, USN: Sons of Gunboats, 1946

(*See also* Battle, Combat, Fight.)

Fighting in Towns

What is the position about London? I have a very clear view that we should fight every inch of it, and that it would *devour* quite a large invading army.

Winston Churchill: Memorandum to General Ismay, 2 July 1940

Finance

They [the Athenians] have abundance of gold and silver, and these make war, like other things, go smoothly.

Hermocrates of Syracuse: Speech to the Syracusans, 415 B. C.

War is not so much a matter of weapons as of money, for money furnishes the material for war. And this is especially true when a land power fights those whose strength is on the sea.

Thucydides: The Peloponnesian Wars, 404 B. C.

Fight with silver spears and you will conquer everywhere.

Reply by the Oracle at Delphi to a question by Philip of Macedon, c. 385 B.C.

The sinews of war are infinite money.

Cicero: Philippics, v, c. 60 B. C.

To carry on war, three things are necessary: money, money, and yet more money.

Gian Jacopo Trivulzio: To Louis XII of France, 1499

Coin is the sinews of war.

Francois Rabelais, 1495–1533, Works, i, 46

Money is not the sinews of war, although it is generally so considered . . . It is not gold but good soldiers that insure success in war.

Niccolo Machiavelli: Discorsi, ii, 1531

Trade is the source of finance, and finance is the vital nerve of war.

Jean Baptiste Colbert, 1619–1683

Now the whole art of war is in a manner reduced to money; and nowadays that prince who can best find money to feed, clothe, and pay his army, not he that hath the most valiant troops, is surest to success and conquest.

Charles Davenant: Essay on Ways and Means of Supplying the War, 1695

Victory always falls to the side having the last crown.

John Law: As French Finance Minister, 1718

Have money and a good army; they ensure the glory and safety of a prince.

Frederick Wilhelm I of Prussia: To his son (later, Frederick the Great), 1724

Without a military chest it is next to impossible to employ an army with effect.
Attributed to Nathanael Greene, 1724–1786

They all present me with that Gorgon Head, an empty treasury.
Anthony Wayne: Reporting on his mission to the Continental Congress to plead for funds for Washington's army, 1781

Was it Bonaparte who said that he found the vices very good patriots?—"he got five millions from the love of brandy, and he should be glad to know which of the virtues would pay him as much." Tobacco and opium have broad backs, and will cheerfully carry the load of armies.
R. W. Emerson, 1802–1883, Society and Solitude

Money, credit, is the life of war; lessen it, and vigor flags; destroy it, and resistance dies.
Mahan: Sea Power in Its Relations to the War of 1812, i, 1905

. . . the financial potency that determines the issues of war.
Mahan: Sea Power in Its Relations to the War of 1812, i, 1905

Fire

The best protection against the enemy's fire is a well directed fire from our own guns.
David G. Farragut: General Order for the attack on Port Hudson, 14 March 1863

You may fire when you are ready, Gridley.
George Dewey: To Captain C. V. Gridley USN, Admiral Dewey's flag captain at Manila Bay, 1 May 1898

Keep up the fire.
Colonel E. H. Liscum, USA: Dying words after being mortally wounded at the head of the 9th Infantry at Tientsin, 13 July 1900

Fire has become the decisive argument.
Ferdinand Foch: Principles of War, 1920

Fire and Sword

Thou hadst fire and sword on thy side.
Shakespeare: I King Henry IV, ii, 4, 1597

I want no prisoners. I wish you to burn and kill; the more you burn and kill, the better it will please me.
Brigadier General Jacob H. Smith, USA: Order to Major L.W.T. Waller for the pacification of Samar, October 1901

Fire Discipline

Fire low.
Oliver Cromwell: Order to his troops at the battle of Preston, 17 August 1648

By push of bayonets—no firing till you see the whites of their eyes.
Frederick The Great: At Prague, 6 May 1757

Silent till you see the whites of their eyes.
Prince Charles of Prussia: At Jägerndorf, 23 May 1745

Don't fire 'til you see the whites of their eyes.
William Prescott: Order at Bunker Hill, 17 June 1775

Men, you are all marksmen—don't one of you fire until you see the white of their eyes.
Israel Putnam: Order at Bunker Hill, relaying Prescott's order using the same phrase (cf. Prescott ante), 17 June 1775

Boys, aim at their waistbands.
John Stark: Order at Bunker Hill, 17 June 1775

Firepower

Imagine all the earthquakes in the world, and all the thunder and lightnings together in a space of two miles, all going off at once.
Description by unknown U.S. Army officer of night engagement when Farragut ran Forts Jackson and St. Philip, 24 April 1862

There is but one means to extenuate the effects of enemy fire: it is to develop a more violent fire oneself.
Ferdinand Foch: Precepts, 1919

All movements on the battlefield have but one end in view, the development of fire in greater volume and more effectively than that of the opposing force.

T. Miller Maguire: The Development of Tactics, 1904

Firepower kills.

Henri-Philippe Petain, 1856–1951

It is firepower, and firepower that arrives at the right time and place, that counts in modern war.

B. H. Liddell Hart: Thoughts on War, iv, 1944

The tendency towards under-rating fire-power . . . has marked every peace interval in modern military history.

B. H. Liddell Hart: Thoughts on War, 1944

The greater the visible effect of fire on the attacking infantry, the firmer grows the defenders' morale, whilst a conviction of the impregnability of the defense rapidly intensifies in the mind of the attacking infantryman.

B. H. Liddell Hart: Thoughts on War, xiv, 1944

Battles are won by fire and by movement. The purpose of the movement is to get the fire in a more adventageous place to play on the enemy. This is from the rear or flank.

George S. Patton, Jr.: War As I Knew It, 1947

(*See also* Fire Discipline, Firepower, Fire Superiority, Fire Support, Musketry.)

Fire Superiority

Battles are won by superiority of fire.

Frederick The Great: Military Testament, 1768 (this is probably the first recorded appearance of the phrase, "fire superiority")

Whatever happens, we have got
The Maxim Gun, and they have not.

Hilaire Belloc, 1870–1953, The Modern Traveler

A superiority of fire, and therefore a superiority in directing and delivering fire and in making use of fire, will become the main factors upon which the efficiency of a force will depend.

Ferdinand Foch: Precepts, 1919

Fire Support

I am persuaded that unless troops are properly supported in action, they will be defeated.

Maurice de Saxe: Mes Rêveries, 1732

The better the infantry, the more it should be economized and supported by good batteries. Good infantry is without doubt the sinews of an army; but if it has to fight a long time against very superior artillery, it will become demoralized and will be destroyed.

Napoleon I: Maxims of War, 1831

It is a fact that infantry attacks always halt and fail at those points where preparation has been insufficient; once more we see that the power of organization is greater than the bravery of the troops.

Ferdinand Foch: After the Artois and Champagne battles, 1917

As a fortifier of morale, an immediate and visible form of support is infinitely more efficacious than a distant and unseen one.

B. H. Liddell Hart: Great Captains Unveiled, 1927

First Sergeant

The first sergeant must be the first sergeant of the unit in fact as well as in name.

U. S. Marine Corps Manual, 1920

(*See also* Noncommissioned Officer, Sergeant.)

Fisher, John Arbuthnot (1841–1920)

Fear God and Dread Nought.

Motto chosen by Admiral Fisher when appointed to the peerage (As First Sea Lord, Fisher conceived and build the first dreadnought battleship, HMS Dreadnought.*)*

He was a mixture of Machiavelli and a child, which must have been extraordinarily baffling to politicians and men of the world.

Esther Meynell: A Woman Talking, 1940

[Fisher's] great claim to fame is that he succeeded in making us *think*.

Admiral Sir Sydney Fremantle, RN: Letter to A. J. Marder, 1946

Fitness Report

All his faults observed,
Set in a notebook, learn'd and conn'd by rote.
Shakespeare: Julius Caesar, iv, 3, 1599

Personally, I would not breed from this officer.
Remark on a Cavalry officer's efficiency report, c. 1900

This officer has the manners of an organ-grinder and the morals of his monkey. I am unable to report on his work, as he has done none . . .
Said to be excerpted from a British Army fitness report

Flag

There it is—Old Glory!
William Driver: As the Colors were first broken out on board a new Salem merchantman of which he had been appointed master, 10 August 1831 (probably the origin of the name, "Old Glory.")

Have I not myself known five hundred living soldiers sabred into crows' meat for a piece of glazed cotton which they call their flag; which, had you sold it in any market-cross, would not have brought above three groschen?
Thomas Carlyle: Sartor Resartus, iii, 1836

If anyone attempts to haul down the American flag, shoot him on the spot.
John A. Dix: Order given, as Collector of the Port of New Orleans, to the local Treasury officers, 29 January 1861

"Shoot, if you must, this old grey head,
But spare your country's flag," she said.
John Greenleaf Whittier: Barbara Frietchie, 1863

Yes, we'll rally round the flag, we'll rally once again,
Shouting the battle-cry of Freedom,
We will rally from the hillside, we'll gather from the plain,
Shouting the battle-cry of Freedom.
George Frederick Root: Battle-Cry of Freedom, 1963

Lay me down, and save the Flag!
Colonel James A. Mulligan, USA: Last command to his regiment, at Kernstown, Virginia, while, being carried to the rear, mortally wounded, he saw the regimental color in danger of being taken by the enemy, 23 March 1862

A moth-eaten rag on a worm-eaten pole,
It does not look likely to stir a man's soul,
'Tis the deeds that were done 'neath the moth-eaten rag,
When the pole was a staff, and the rag was a flag.
Sir Edward B. Hamley, 1824–1893

I have seen the glories of art and architecture, and mountains and rivers; I have seen the sunset on the Jungfrau, and the full moon rise over Mont Blanc. But the fairest vision on which these eyes ever looked was the flag of my country in a foreign land.
George Frisbie Hoar, 1826–1924

My good mother,
Be proud: I carry the flag.—
Rainer Maria Rilke: The Lay of the Love and Death of Cornet Christopher Rilke, 1904

Stiff flags straining in the night-blasts cold,
In the gloom black-purple, in the glint, old-gold.
G. K. Chesterton, 1874–1936, Lepanto

Hats off!
Along the street there comes
A blare of bugles, a ruffle of drums,
A flash of color beneath the sky:
Hats off!
The flag is passing by.
Henry Holcomb Bennett, 1863–1924, The Flag Goes By

Where American citizens go, that flag goes with them to protect them.
Lyndon B. Johnson: Speech to AFL/CIO Convention, Washington, 3 May 1965 (after Marine landings in Santo Domingo).

The flag is a jealous mistress.
Soldiers' proverb

All for the flag!
Military saying

(*See also* Colors.)

Flagpole

I see that the old flagpole still stands. Have your troops hoist the Colors to its peak, and let no enemy ever haul them down.

Douglas MacArthur: To Colonel George M. Jones, USA, on the reoccupation of Corregidor, 2 March 1945

When you raise hell, do it at least a mile from the flagpole.

Service saying

Flamethrower

They sawed in two and hollowed out a great beam, which they joined together very exactly, like a flute, and suspended a vessel by chains at the end of the beam; the iron mouth of a bellows directed downward into the vessel was attached to the beam, of which a great part was itself overlaid with iron. This machine they brought up from a distance on carts to various points of the rampart where vine stems and wood had been most extensively used, and when it was quite near the wall they applied a large bellows to their own end of the beam and blew through it. The blast, prevented from escaping, passed into the vessel which contained burning coals and sulphur and pitch; these made a huge flame, and set fire to the rampart, so that none could remain upon it.

Thucydides: The Peloponnesian Wars, Siege of Delium, 424 B. C.

Flank

Flank attack is the essence of the whole history of war.

Alfred von Schlieffen, 1933–1913

If I had worried about flanks, I could never have fought the war.

George S. Patton, Jr.: War As I Knew It, 1947

(*See also* Envelopment.)

Fleet

A city on the inconstant billows dancing;
For so appears this fleet majestical . . .

Shakespeare: King Henry V, iii, prologue, 1598

I am sure I need not point out to you the immense advantage it will be to us to have a formidable fleet in readiness.

Lord Sandwich: Letter to Lord North, 10 September 1772

With Ships the sea was sprinkled far and nigh.

William Wordsworth, 1770–1850, With Ships the Sea Was Sprinkled

We must all put our shoulders to the wheel, and make the great machine of the Fleet intrusted to our charge go on smoothly.

Nelson: Letter, 30 September 1805

Not alone is the strength of the Fleet measured by the number of its fighting units, but by its efficiency, by its ability to proceed promptly where it is needed and to engage and overcome an enemy.

Admiral Richard Wainwright, USN: Letter, 1911

The moral effect of an omnipresent fleet is very great, but it cannot be weighed against a main fleet known to be ready to strike and able to strike hard.

Sir John Fisher: To Lord Stamfordham, 25 June 1912

The functions of the Fleet are (1) to cut the communications of the enemy nation; (2) to keep open the communications of its own nation. The royal road to success in these endeavors is to defeat the enemy's grand fleet.

Sir Ian Hamilton: Soul and Body of an Army, iv, 1921

You seldom hear of the fleets except when there's trouble, and then you hear a lot.

Vice Admiral J. S. McCain, Jr., USN: In Norfolk Star Ledger, 4 August 1964

(*See also* Naval Operations, Naval Warfare, Navy.)

Fleet in Being

Most men were in fear that the French would invade; but I was always of another opinion, for I always said that whilst we had a fleet in being, they would not dare to make an attempt.

Admiral Sir Arthur Torrington, RN: Report, 1690, defending his unwillingness, with inferior forces, to engage a stronger French fleet (the origin of the "fleet in being" phrase and concept).

The probable value of a "fleet in being" has, in the opinion of the writer, been much overstated; for, even at the best, the game of evasion, which this is, if persisted in, can have but one issue. The superior force will in the end run the inferior to earth.

Mahan: Lessons of the War with Spain, 1899

Fleet Train

In any event the stores and provisions for the supply of the fleet should be kept afloat, and to provide for other exigencies a few large armed transports would afford great resource.

Lord St. Vincent: Memorandum to First Lord of the Admiralty, 1795

Flight

He rode upon the cherub, and did fly; yea, he did fly upon the wings of the wind.

Psalm XVIII

They shall mount up with wings as eagles.

Isaiah, XL, 31

There shall be wings.

Leonardo Da Vinci, 1452–1519

The fated sky
Gives us free scope.

Shakespeare, 1514–1616

What can you conceive more silly and extravagant than to suppose a man racking his brains, and studying night and day how to fly?

William Law: A Serious Call to a Devout and Holy Life, xi, 1728

Bishop Wilkins prophesied that the time would come when gentlemen, when they were to go on a journey, would call for their wings as regularly as they call for their boots.

Maria Edgworth: Essay on Irish Bulls, ii, 1802

Are there no foolish projects in Great Britain? Did not good Bishop Wilkins project a scheme to fly?

Timothy Dwight: Remarks on the Review of Inchiquin's Letters, 1815

Thou born to match the gale (thou art all wings,)
To cope with heaven and earth and sea and hurricane . . .
At dusk thou look'st on Senegal, at morn America.

Walt Whitman, 1819–1892, Man-of-War Bird

The example of the bird does not prove that man can fly. Imagine the proud possessor of the aeroplane darting through the air at a speed of several hundred feet per second. It is the speed alone that sustains him. How is he ever going to stop?

Simon Newcomb: In The Independent, 22 October 1903

(*See also* Aviation.)

Flodden Field, (9 September 1513)

Still from the sire the son shall hear
Of the stern strife, and carnage drear
Of Flodden's fatal field,
Where shiver'd was fair Scotland's spear,
And broken was her shield!

Walter Scott: Marmion, xxxiv, 1808

Foch, Ferdinand (1851–1929)

This officer, during his professorship at the Ecole de Guerre, taught metaphysics, and metaphysics so abstruse that it made idiots of a number of his pupils.

Report rendered by the Securité Nationale to Clemenceau when he was considering Foch to command the Ecole Superieure de Guerre, 1908

. . . only a frantic pair of moustaches.

T. E. Lawrence: Of Foch to Liddell Hart, 23 April 1932

Foe

That stern joy which warriors feel
In foemen worthy of their steel.

Walter Scott: The Lady of the Lake, 1810

. . . whispering with white lips "The foe! they come! they come!"

Byron: Childe Harold's Pilgrimage, iii, 1816

(*See also* Enemy.)

Fog of War

The *coup d'oeuil* is a gift of God and cannot be acquired; but if professional knowledge does not perfect it, one only sees things imperfectly and in a fog.

Chevalier Folard: Nouvelles Decouvertes sur la Guerre, 1724 (this is the earliest reference so far encountered to "the fog of war.")

He who wars walks in a mist through which the keenest eye cannot always discern the right path.

Sir William Napier: History of the War in the Peninsula, 1840

Friends, how goes the fight?

T. B. Macaulay: The Battle of Lake Regillus, 1842

Fontenoy, (11 May 1745)

Gentlemen of the French Guard, fire first! (Messieurs les gardes francaises, tirez!)

Lord Charles Hay: To the French at Fontenoy (but he is actually reported to have said "Gentlemen of the French Guard, I hope you will stand and fight us today and not escape by swimming the Scheldt, as you did at Dettingen.")

And, standing on his charger
the brave King Louis spoke,
"Send on my Irish cavalry,"
the headlong Irish broke,
At Fontenoy, At Fontenoy!
"Remember Limerick,
Dash down the Sassenach!"

Thomas Davis, 1814–1845, The Battle of Fontenoy.

Foote, Andrew Hull (1806–1863)

He prays like a saint and fights like the devil.

Rear Admiral Francis H. Gregory, USN: d. 1866

Foraging

If you wish to remain in a camp and the enemy is in your vicinity, your foraging between his army and yours amounts to living at his expense.

Frederick The Great: Instructions for His Generals, 1747

Force

"Aye," quoth Sancho Panza, "you rake-helly fellows have a saying which is pat to your purpose: 'Never cringe nor creep for what you by force may reap.' "

Cervantes: Don Quixote, 1604

Force is indeed the ruling principle in military affairs; in conformity to which the French term their cannon, the *ultima ratio regum.*

Francis Grose: Advice to the Officers of the British Army, 1782

There never was a government without force.

James Madison: During debate on adoption of the Constitution, 1787

Force is the vital principle and immediate parent of despotism.

Thomas Jefferson: Inaugural address, 4 March 1801

Where I would take a penknife Lord St. Vincent takes a hatchet.

Horatio Nelson, 1758–1805

The moral is to the material as three to one.

Napoleon I, 1769–1821

Force is only justifiable in extremes; when we have the upper hand, justice is preferable.

Napoleon I: Political Aphorisms, 1848

Force does not exist for mobility but mobility for force. It is of no use to get there first unless, when the enemy arrives, you have also the most men—the greater force.

Mahan: Lessons of the War with Spain, 1899

An efficient military body depends for its effect in war—and in peace—less upon its position than upon its concentrated force.

Mahan: Naval Administration and Warfare, 1906

Between two groups that want to make inconsistent kinds of worlds, I see no remedy except force.

Oliver Wendell Holmes, Jr., 1841–1935, letter to Sir Frederick Pollock

Superior force is a powerful persuader.

Winston Churchill: Note to the First Sea Lord, 15 October 1942

I have always been ready to use force in order to defy tyranny or ward off ruin.
Winston Churchill: The Gathering Storm, iii, 1948

The more I reflect on the experience of history the more I come to see the instability of solutions achieved by force, and to suspect even those instances where force has had the appearance of resolving difficulties.
B.H. Liddell Hart: Thoughts on War, i, 1944

State power is armed force and without armed force there can be no state power. What is independence? Independence is armed force and without armed force there can be no independence . . . State power, independence, freedom and equality are won by armed force and armed force alone.
People's Daily, Peking, 24 June 1964

The distinction between private violence and public force is the central principle of a civilized society.
Walter Lippmann: in Washington Post, 21 July 1964

. . . An iron hand in a velvet glove.
Italian Proverb

(*See also* Strength, Superiority.)

Foreign Legion

There will be formed a Legion composed of foreigners. This Legion will be known as the Foreign Legion. It will be stationed in Africa.
Louis-Phillippe of France: Decree of 9 March 1831, establishing the Legion

A soldier of the Legion lay dying in Algiers . . .
Caroline Elizabeth Norton, 1808–1877, Bingen on the Rhine

Honor and Fidelity (Honneur et Fidélité).
Foreign Legion motto

The Legion is our fatherland. (Legio patria nostra.)
Foreign Legion motto

We're the blood sausage . . .
Tiens, Voila du Boudin, Foreign Legion marching song, 19th century

March or die. (Marche ou crêve.)
Traditional saying of the Foreign Legion

Fort

A ship's a fool to fight a fort.
Attributed to Horatio Nelson, 1758–1805

We give up the fort when there's not a man left to defend it.
Captain George Croghan USA: To the British General Proctor at Fort Stephenson, Lower Sandusky 1 August 1813

Were half the power, that fills the world
with terror,
Were half the wealth, bestowed on camps
and courts,
Given to redeem the human mind from
error,
There were no need for arsenals and forts.
Henry Wadsworth Longfellow: The Arsenal at Springfield, 1841

A single shot will sink a ship, while a hundred rounds cannot silence a fort.
John Ericsson: Regarding the bombardment of Charleston, 1863

Hold the fort; I am coming.
W. T. Sherman: Signal to Brigadier General J. M. Corse, at Allatoona, Georgia, 5 October 1864

Fortifications

Happy they whose walls already rise.
Virgil, 70–19 B.C., Aeneid, i

It [the Siegfried Line] is a monument to human stupidity. When natural obstacles—oceans and mountains—can be so readily overcome, anything that man makes, man can overcome.
George S. Patton, Jr.: To the Third Army staff, November 1944

Fortress

[Fortresses] are not needed by those . . . that have good armies . . . but fortresses without good armies are incompetent for defense.
Niccolo Machiavelli: Discorsi, xxiv, 1531

The fate of a nation may depend sometimes upon the position of a fortress.
Napoleon I: Political Aphorisms, 1848 (referring to the defense of Acre, 1799)

It is bad policy to cover a frontier with fortresses very close together . . . To bury an army in entrenchments where it may be outflanked and surrounded, or forced in front even if secure from a flank attack, is manifest folly.
Jomini: Précis de l'Art de la Guerre, 1838

Fortresses are the tombs of armies.
Military maxim

(*See also* Fort, Fortifications.)

Fortunes of War

Fortune has heretofore been a fickle Goddess to us—and like some other females changed for the first new face she saw.
Anthony Wayne: Letter to his wife, 1776

I am called to sup, but where to breakfast, either within the enemy's lines in triumph, or in the other world!
Anthony Wayne: Letter prior to night assault on Stony Point, 15 July 1779

The fortunes of war flow this way and that, and no prudent fighter holds his enemy in contempt.
Goethe: Iphigenie auf Tauris, v, 1787

A stout heart breaks bad luck.
Spanish Proverb

(*See also* Chance, Luck.)

Foxhole

There are no atheists in the foxholes.
Chaplain W. T. Cummings, USA: Sermon on Bataan, March 1942

Goddamn it, you'll never get the Purple Heart hiding in a foxhole! Follow me!
Lieutenant Colonel Henry P. Crowe USMC: Guadalcanal, 13 January 1943

Franco-Prussian War (1870–1871)

Think of it ever; speak of it never. (N'en parlez jamais; pensez-y toujours.)
Leon Gambetta, 1838–1882 (specifically of the French defeat at Sedan)

Frederick the Great (1712–1786)

He fiddles and fights as well as any man in Christendom.
Voltaire: Letter to Sir Everard Fawkenden, June 1742

The King of Prussia is a mischievous rascal, a base friend, a bad ally, and bad relation and a bad neighbor: in fact, the most dangerous and evil-disposed Prince in Europe.
William Pitt: Letter, May 1756

He was above all great in the most critical moments.
Napoleon I: Correspondance de Napoleon, xxxii, 1816

. . . a mixture of Puck and Machiavelli welded together on the anvil of Vulcan by the hammer of Thor.
J.F.C. Fuller: Decisive Battles, xvi, 1939

Frigate

Frigates are the eyes of a fleet.
Nelson: Letter, 1804

. . . A handful of fir-built frigates under a bit of striped bunting, manned by bastards and outlaws.
The London Times, 1812 (of the U. S. Navy)

The public will learn with sentiments which we shall not presume to anticipate, that a third British frigate has struck her flag to an American.
The Pilot, English shipping journal, 1813 (of the capture of HMS Java *by USS* Constitution)

No names can be too fine for a pretty girl or a good frigate.
Frederick Marryat: Peter Simple, 1834

Scarce one tall frigate walks the sea
Or skirts the safer shores,
Of all that bore to victory

Our stout old Commodores;
Hull, Bainbridge, Porter—where are they?
The answering billows roll,
Still bright in memory's sunset ray—
God rest each gallant soul!

Oliver Wendell Holmes, Sr., 1809–1894, A Toast to the Vice Admiral

Front

The army report confined itself to a single sentence: All quiet on the Western Front.

Erich Maria Remarque: In Westen Nichts Neues, 1929

Look to the front.

Motto of the Rifle Brigade, British Army

Frontal Attack

Close combat, man to man, is plainly to be regarded as the real basis of combat.

Clausewitz: On War, 1832

Are there not other alternatives than sending our armies to chew barbed wire in Flanders?

Winston Churchill: Memorandum for the Prime Minister, 24 December 1914

Gallant fellows, these soldiers; they always go for the thickest place in the fence.

Admiral Sir John de Robeck: Watching initial landings at Gallipoli, 25 April 1915

Fuel

Excuses for failure attributed to shortness of coal will be closely scrutinized; and justly.

Mahan: Naval Strategy, 1911

G

Gallantry

Nothing in his life
Became him like the leaving of it; he died
As one that had studied in his death
To throw away the dearest thing he owed
As 'twere a careless trifle.
Shakespeare: Macbeth, i, 4, 1605

Nothing on earth is so stupid as a gallant officer.
Wellington: After the battle of Fuentes de Oñoro, 3 May 1811

But here are men who fought in gallant actions
As gallantly as ever heroes fought . . .
Byron, 1788–1824

(*See also* Audacity, Bravery, Courage, Daring, Heroism.)

Gallipoli (16 February 1915 – 9 January 1916)

Damn the Dardanelles. They will be our grave.
Sir John Fisher: To the Dardanelles Committee, 1915

The Dardanelles operations hang like a millstone about our necks, and have brought upon us the most vast disaster that has happened in the course of the war.
Sir Edward Carson: To the House of Commons, October 1916

Garrison Duty

Now we suffer the ills of a long peace.
Juvenal, 60–130, Satires

Soldiers in peace are like chimneys in summer.
Lord Burghley, 1520–1598, Ten Precepts

I would advise you to get the command of a camp or district in old England; there you may enjoy all the pomp and parade of war, and, at the same time, be tolerably secure from those hard knocks which your necessities impelled you to risk in your younger days.
Francis Grose: Advice to the Officers of the British Army, 1782

. . . the treadmill of the garrison.
Stonewall Jackson: Letter, 1861

General Staff

The ideal General Staff should, in peace time, do nothing! They deal in an intangible stuff called thought. Their main business consists in thinking out what an enemy may do and what their Commanding Generals ought to do, and the less they clank their spurs the better.
Sir Ian Hamilton: The Soul and Body of an Army, iii, 1921

Doesn't it make you surer you were right, to see all the General Staff opposing you?
T.E. Lawrence: Letter to Ernest Thurtle, 2 May 1930

The General Staff was truly . . . an all-powerful military priesthood, linked by ties of intellectual and professional comradeship. A corps of directors, a society within a society, they were to the German Army what the Jesuits at their political zenith were to the Church of Rome.
B.H. Liddell Hart: Thoughts on War, iv, 1944

(*See also* Staff.)

Generals

The general must be first in the toils and fatigues of the army. In the heat of summer he does not spread his parasol nor in the cold of winter don thick clothing. In dangerous places he must dismount and walk. He waits until the army's wells have been dug and only then drinks; until the army's food is cooked before he eats; until the army's fortifications have been completed, to shelter himself.
The Ping Fa (Chinese military code), c. 5th century, B.C.

Ah! the generals! they are numerous but not good for much!
Aristophanes, c. 450–388 B.C.

The general who in advancing does not seek personal fame, and in withdrawing is not concerned with avoiding punishment, but whose only purpose is to protect the people and promote the best interests of his sovereign, is the precious jewel of the state. Because such a general regards his men as infants they will march with him into the deepest valleys. He treats them as his own beloved sons and they will die with him.
Sun Tzu, 400–320 B.C., The Art of War, x

Nothing is more useful, or of greater importance, in the conduct of a general, than to examine with the nicest care, into the character and natural disposition of the opposite commander. A general in the field should endeavor to discover in the chief that is against him, whether there be any weakness in his mind and character, through which he may be attacked with some advantage.

Polybious, History, c. 150 B.C.

Comrades, you have lost a good captain to make a bad general.

Saturninus, d. 100 B.C.

Adversity reveals the genius of a general; good fortune conceals it.

Horace, Satires, ii, c. 25 B.C.

Our gen'rals now, retired to their estates,
Hang their old trophies o'er the garden gates.

Horace: Epistles, i, c. 20 B.C.
(translation by Pope, 1735)

A general must be continent, sober, frugal, hard-working, middle-aged, eloquent, father of a family, and member of an illustrious house. Soldiers do not like being under the command of one who is not of good birth. In addition, a general should be polite, affable, easy of approach, and cool-headed.

Onosander, fl. 1st century A.D.

What is essential in the temperament of a general is steadiness.

Wang Hsi, fl. 100 A.D.

A good general should know the characters of the generals opposed to him—if they are prudent or rash, if they fight according to rule or haphazard.

Vegetius: De Re Militari, 378

The best general is that one who is able to bring about peace from war.

Belisarius, 505–565

The general is the Minister of Death, who is not responsible to the heavens above, to the earth beneath, to the enemy to his front, or to the sovereign in his rear.

Tu Mu, 803–852, Wei Liao Tzu

A general who is stupid and courageous is a calamity.

Tu Mu, 803–852, Wei Liao Tzu

A general should be a very model of temperance, expecially with regard to female captives.

Leo VI of Byzantium, 886–911

The Romans were not only less ungrateful than other republics but were also more lenient and considerate in the punishment of the generals of their armies.

Machiavelli: Discorsi, xxxi, 1531

The principall care that a Prince or State that entreth into warres is to have, if that there be, choyse made of a sufficient Generall. If Princes looke for good successe in their warres, let them without affection and partialitie, make choyse of a sufficient Generall, religious, skilfull, couragious, and adorned with such vertues, both for warre and peace, as the importance of the matters which he manageth requireth . . . After knowledge and judgment in matters of warre, the next vertue required in a Generall is courage and speed to execute that which is wisely determined. There are also other vertues required in a Generall, which although they be not so necessarie as the former, yet for the execution of matters are very requisite and profitable, as namely justice, liberalitie, courtesie, temperance, and loyaltie.

Mathew Sutcliffe: The Practice, Proceedings, and Lawes of Armes, 1593

'Tis fit a general
Should not endanger his own person oft . . .

John Webster: The White Devil, ii, c. 1608

There is only one general in a century.

Baltasar Gracian: The Art of Worldly Wisdom, cciii, 1647

I have formed a picture of a general commanding, which is not chimerical. I have seen such men. The first of all qualities is *Courage.* Without this the others are of little value, since they cannot be used. The second is *Intelligence,* which must be strong and fertile in expedients. The third is *Health.*

Maurice de Saxe: Mes Rêveries, xxxi, 1732

The greatest Generals ow'd their high Renown not so much to Atchievements perform'd in War as to their intimate acquaintance with the Muses.

The Cadet (author unknown), 1756

. . . a man who would sacrifice everything, the Navy in particular, to military aggrandizement.

Commodore Sir John Duckworth, RN: Of General Stuart, landing force commander at Minorca, October, 1798

Really when I reflect upon the character and the attainments of some of the General officers of this army, and consider that these are the persons on whom I am to rely to lead columns against the French Generals, and who are to carry my instructions into execution, I tremble; and, as Lord Chesterfield said of the Generals of his day, "I only hope that when the enemy reads the list of their names he trembles as I do."

Wellington: Letter from Portugal, 1810

Many of the greatest military commanders owe their exaltation and celebrity to the Art of letter writing.

Colonel J.G.D. Tucker: Advice to Young Officers, 1826

The best generals are those who have served in the artillery.

Napoleon I: To General Gaspard Gourgaud, St. Helena, 1817

It is exceptional and difficult to find in one man all the qualities necessary for a great general. What is most desirable, and which instantly sets a man apart, is that his intelligence or talent, are balanced by his character or courage.

Napoleon I: Maxims of War, 1831

The Gauls were not conquered by the Roman legions, but by Caesar. It was not before the Carthaginian soldiers that Rome was made to tremble, but before Hannibal. It was not the Macedonian phalanx which reached India, but Alexander. It was not the French army that reached the Weser and the Inn, it was Turenne. Prussia was not defended for seven years against the three most formidable European powers by the Prussian soldiers but by Frederick the Great.

Attributed to Napoleon I, 1769–1821 (quoted by Foch in Precepts)

Generals have never risen from the very learned or really erudite class of officers, but have been mostly men who, from the circumstances of their position, could not have attained to any great amount of knowledge.

Clausewitz: On War, 1832

People are rather inclined to look upon a subordinate general grown grey in the service, and in whom constant discharge of routine duties has produced a decided poverty of mind, as a man of failing intellect, and, with all respect for his bravery, to laugh at his simplicity.

Clausewitz: On War, 1832

The most essential qualities of a general will always be: *first,* a high moral courage, capable of great resolution; *second,* a physical courage which takes no account of danger. His scientific or military acquirements are secondary to these.

Jomini: Précis de l'Art de la Guerre, 1838

Hail, ye indomitable heroes, hail!
Despite of all your generals, ye prevail.

Walter Savage Landor: The Crimean Heroes, 1856

Let no man be so rash as to suppose that, in donning a general's uniform, he is forth-with competent to perform a general's functions.

Dennis Hart Mahan, 1802–1871 (Professor Mahan was the father of Alfred Thayer Mahan, the naval strategist.)

I am the very model of a modern major-general.
I've information vegetable, animal and mineral;
I know the kings of England, and I quote the fights historical,
From Marathon to Waterloo, in order categorical.

W.S. Gilbert: The Pirates of Penzance, i, 1880

The prize of the general is not a bigger tent, but command.

Oliver Wendell Holmes, Jr.: Speech in New York, 15 February 1913

Generals die in bed.

Soldier saying on the Western Front, World War I

"Good morning! Good morning!" the General said
When we met him last week on the way to the Line.

Now the soldiers he smiled at are most of
'em dead,
And we're cursing his staff for incompetent swine.
"He's a cheery old card," grunted Harry
to Jack
As they slogged up to Arras with rifle and
pack.
But he did for them both with his plan of
attack.

Siegfried Sassoon: The General, 1917

Battles are lost or won by generals, not by the rank and file.

Ferdinand Foch: Principles of War, 1920

It takes 15,000 casualties to train a major general.

Attributed to Ferdinand Foch, 1851–1929

The perfect general would know everything in heaven and earth.

T.E. Lawrence: Letter to Liddell Hart, 26 June 1933

We must have "character", which simply means that he knows what he wants and has the courage and determination to get it. He should have a genuine interest in, and a real knowledge of, humanity, the raw materials of his trade, and, most vital of all, he must have what we call the fighting spirit, the will to win.

Sir A.P. Wavell: Lecture at Trinity College, Cambridge University, 1939

The weakness of democracies is that once a general has been built up in public opinion it becomes impossible to remove him.

Edouard Daladier: Quoted in The Grave-Diggers of Europe, 1944

No dead general has ever been criticized, so you have that way out, always.

George S. Patton, Jr.: Letter to Cadet George S. Patton, III, USMA, 6 June 1944

. . . the trades-union of Generals.

Ernest Bevin, 1881–1951

No general in his right mind wants to be downgraded to a secondary role.

S.L.A. Marshall: in Saturday Review, 12 December 1958

Nothing raises morale better than a dead general.

John Masters: The Road Past Mandalay, 1961

A general is just as good or as bad as the troops under his command make him.

Douglas MacArthur: To Congress, 20 August 1962

Soldiers ought to fear their general more than their enemy.

Military maxim

One bad general is better than two good ones.

French Proverb

(*See also* Commander, Generalship, High Command, Leadership.)

Generalship

Now there are five matters to which a general must pay strict heed. The first of these is administration; the second, preparedness; the third determination; the fourth, prudence; and the fifth, economy. Administration means to control many as he controls few. Preparedness means that when he marches forth from the gates he acts as if he perceives the enemy. Resolution means that when he approaches the enemy he does not worry about life. Prudence means that although he has conquered, he acts as if he were just beginning to fight. Economy means being sparing.

Wu Ch'I, 430–381 B.C., Art of War, iv

The general must know how to get his men their rations and every other kind of stores needed in war. He must have imagination to originate plans, practical sense and energy to carry them through. He must be observant, untiring, shrewd, kindly and cruel, simple and crafty, a watchman and a robber, lavish and miserly, generous and stingy, rash and conservative. All these and many other qualities, natural and acquired, must he have. He should also, as a matter of course, know his tactics; for a disorderly mob is no more an army than a heap of building materials is a house.

Socrates, 470–399 B.C. (Quoted in Memorabilia, iii, Xenophon)

When capable, feign incapacity; when active, inactivity. When near, make it appear that you are far away; when far

U. S. Navy

HORATIO NELSON

1758–1805

"Thank God I have done my duty."

away, that you are near. Offer the enemy a bait to lure him; feign disorder and strike him. When he concentrates, prepare against him; where he is strong, avoid him. Anger his general and confuse him. Pretend inferiority and encourage his arrogance. Keep him under strain and wear him down. When he is united, divide him. Attack where he is unprepared; sally out when he does not expect you. These are the strategist's keys to victory.
Sun Tzu, 400–320 B.C., The Art of War, i

He who knows when he can fight and when he cannot will be victorious.
Sun Tzu: 400–320 B.C., The Art of War, iii

To capture the enemy's army is better than to destroy it; to take intact a battalion, a company, a five-man squad, is better than to destroy them. For to win a hundred victories in a hundred battles is not the acme of skill. To subdue the enemy without fighting is the acme of skill.
Sun Tzu, 400–320 B.C., The Art of War,iii

When campaigning, be swift as the wind; in leisurely march, majestic as the forest; in raiding and plundering, like fire; in standing, firm as the mountains. As unfathomable as the clouds, move like a thunderbolt.
Sun Tzu, 400–320 B.C., The Art of War, vii

A good general not only sees the way to victory; he also knows when victory is impossible.
Polybius: Histories, i, c. 125 B.C.

Who asks whether bravery or cunning beat the enemy?
Virgil: Aeneid, ii, 19 B.C.

The proper arts of a general are judgment and prudence.
Tacitus: Annals, iii, c. 110

A general unable to estimate his capabilities or comprehend the arts of expediency and flexibility when faced with the opportunity to engage the enemy will advance in a stumbling and hesitant manner, looking anxiously first to his right and then his left, and be unable to produce a plan. Credulous, he will place confidence in unreliable reports, believing at one moment this and at another that. As timorous as a fox in advancing and retiring, his groups will be scattered about. What is the difference between this and driving innocent people into boiling water or fire? Is this not exactly like driving cows and sheep to feed wolves or tigers?
Tu Mu: 803–852, Wai Liao Tzu

A general cannot avoid a battle when the enemy is resolved upon it at all hazards.
Niccolo Machiavelli: Arte della Guerra, 1520

The object of a good general is not to fight, but to win. He has fought enough if he gains a victory.
Duke of Alba, c. 1560

The first quality of a general in chief is a great knowledge of the art of war. This is not intuitive, but the result of experience. A man is not born a commander. He must become one. Not to be anxious; to be always cool; to avoid confusion in his commands; never to change countenance; to give his orders in the midst of battle with as much composure as if he were perfectly at ease. These are the proofs of valor in a general.
Count de Montecucculi: Commentarii Bellici, 1740

We should try to make war without leaving anything to chance. In this lies the talent of a general.
Maurice de Saxe: Mes Rêveries, 1732

When the Duke of Cumberland has weakened his army sufficiently, I shall teach him that a general's first duty is to provide for its welfare.
Maurice de Saxe: Before Laufeldt, 1747

The principal task of the general is mental, involving large projects and major arrangements.
Frederick The Great: Instructions for His Generals, 1747

Success is only to be obtained by simultaneous efforts, directed upon a given point, sustained with constancy, and executed with decision.
Archduke Charles of Austria, 1771–1841

A great captain can only be formed by long experience and intense study.
Archduke Charles of Austria, 1771–1841

The real reason why I succeeded in my own campaigns is because I was always on the spot.

Wellington: Letter, 1805

The creator has not thought proper to mark those in the forehead who are of the stuff to make good generals. We are first, therefore, to seek them blindfold, and let them learn the trade at the expense of great losses.

Thomas Jefferson: Letter to General Bailey, February 1813

I engage and after that I see what to do. (Je m'engage, et après ça, je voie.)

Napoleon I: Remark during the Italian campaign, 1796

The mind of a general ought to resemble and be as clear as the field-glass of a telescope.

Napoleon I: To Barry E. O'Meara, St. Helena, 20 September 1817

The only true wisdom in a general is determined courage.

Napoleon I: Maxims of War, 1831

The first qualification in a general is a cool head—that is, a head which receives accurate impressions, and estimates things and objects at their real value. He must not allow himself to be elated by good news, or depressed by bad.

Napoleon I: Maxims of War, 1831

Every general is culpable who undertakes the execution of a plan which he considers faulty. It is his duty to represent his reasons, to insist upon a change of plan—in short to give in his resignation rather than allow himself to be made the instrument of his army's ruin.

Napoleon I: Maxims of War, 1831

In war the general alone can judge of certain arrangements. It depends on him alone to conquer difficulties by his own superior talents and resolution.

Napoleon I: Maxims of War, 1831

Generals-in-chief must be guided by their own experience, or their genius. Tactics, evolutions, the duties and knowledge of an engineer or artillery officer, may be learned in treatises, but the science of strategy is only to be acquired by experience and by studying the campaigns of all the great captains. Gustavus Adolphus, Turenne, and Frederick, as well as Alexander, Hannibal, and Caesar, have all acted upon the same principles. These have been—to keep their forces united; to leave no weak part unguarded; to seize with rapidity on important points. Such are the principles which lead to victory, and which, by inspiring terror at the reputation of your arms, will at once maintain fidelity and secure reputation.

Napoleon I: Maxims of War, 1831

Correct theories, founded upon right principles, sustained by actual events of wars, and added to accurate military history, will form a true school of instruction for generals.

Jomini: Précis de l'Art de la Guerre, 1838

How different is almost every military problem, except in the bare mechanism of tactics. In almost every case the data on which a solution depends are lacking . . . Too often the general has only conjectures to go on, and these based on false premises . . . What is true now, at the next moment may have no existence, or exist in a contrary sense . . . These considerations explain why history produces so few great captains.

Dennis Hart Mahan: Outpost, 1847

No man can justly be called a great captain who does not know how to organize and form the character of an army, as well as to lead it when formed.

Sir William Napier, 1785–1860

To move swiftly, strike vigorously, and secure all the fruits of the victory is the secret of successful war.

Stonewall Jackson, 1820–1863

I think and work with all my power to bring the troops to the right place at the right time; then I have done my duty. As soon as I order them forward into battle, I leave my army in the hands of God.

R.E. Lee, 1807–1870

The chief duties of a general to his command may be classified—the enforcement of discipline—tactical instruction—care of the health of his men—and they are important because tending to efficiency, the measure of which is the exact measure of his own efficiency.

Brigadier General C. F. Smith, USA: To Colonel Lew Wallace, September 1861

Beware of rashness, but with energy and sleepless vigilance go forward and give us victories.

Abraham Lincoln: Letter to Major General Joseph Hooker, 26 January 1863

The art of war is simple enough. Find out where your enemy is. Get at him as soon as you can. Strike at him as hard as you can, and keep moving on.

U.S. Grant, 1822–1885

The modern commander-in-chief is no Napoleon who stands with his brilliant suite upon a hill . . . The commander is farther to the rear in a house with roomy offices, where telegraph and wireless, telephone and signalling instruments are at hand, while a fleet of automobiles and motorcycles, ready for the longest trips, wait for orders. Here, in a comfortable chair before a large table, the modern Alexander overlooks the entire battlefield on a map. From here he telephones inspiring words, and here he receives the reports from army and corps commanders and from balloons and dirigibles which observe the enemy's movements and detect his positions.

Alfred von Schlieffen: Cannae, 1913

No victory is possible unless the commander be energetic, eager for responsibilities and bold undertakings; unless he possess and can impart to all the resolute will of seeing the thing through; unless he be capable of exerting a personal action composed of will, judgment, and freedom of mind in the midst of danger.

Ferdinand Foch: Precepts, 1919

A general has much to bear and needs strong nerves. The civilian is too inclined to think that war is only like the working out of an arithmetical problem with given numbers. It is anything but that. On both sides it is a case of wrestling with powerful, interwoven physical and psychological forces, a struggle which inferiority in numbers makes all the more difficult . . . The only quality which is known and constant is the will of the leader.

Erich Ludendorff: My War Memories, 1919

Tactics are based on weapon-power . . . strategy is based on movement . . . movement depends on supply.

J.F.C. Fuller: The Generalship of Ulysses S. Grant, 1929

Success has a soporiferous influence on generalship.

J.F.C. Fuller, Decisive Battles, xx, 1939

In battle, two things are usually required of the Commander-in-Chief: to make a good plan for his army and, secondly, to keep a strong reserve . . . But in order to make his plan, the General must not only reconnoiter the battle-ground, he must also study the achievements of the great Captains of the past.

Winston Churchill: Painting as a Pastime, 1932

Generalship, at least in my case, came not by instinct, unsought, but by understanding, hard study and brain concentration. Had it come easy to me, I should not have done it as well.

T.E. Lawrence, 1888–1935

The ever-victorious General is rare and there have been very few of these in history, but what is necessary is that our generals should have studied the art of war and paid attention to its rules; it is then that, with this wisdom tempered by courage, our military leaders will have better chances of success.

Mao Tse-tung: On the Study of War, 1936

Efficiency in a general, his soldiers have a right to expect; geniality they are usually right to suspect.

Sir Archibald Wavell: Generals and Generalship, 1939

An important difference between a military operation and a surgical operation is that the patient is not tied down. But it is a common fault of generalship to assume that he is.

B.H. Liddell Hart: Thoughts on War, xii, 1944

Two fundamental lessons of war experience are—never to check momentum; never to resume mere pushing.

B.H. Liddell Hart: Thoughts on War, xvi, 1944

Use steamroller strategy; that is, make up your mind on course and direction of action, and stick to it. But in tactics, do not

steamroller. Attack weakness. Hold them by the nose and kick them in the pants.
George S. Patton, Jr.: War As I Knew It, 1947

The modern army commander must free himself from routine methods and show a comprehensive grip of technical matters, for he must be in a position continually to adapt his ideas of warfare to the facts and possibilities of the moment.
Erwin Rommel: The Rommel Papers, ix, 1953

I have always believed that a motto for generals must be "No regrets", no crying over spilt milk.
Sir William Slim: Defeat Into Victory, 1956

The acid test of an officer who aspires to high command is his ability to be able to grasp quickly the essentials of a military problem, to decide rapidly what he will do, to make it quite clear to all concerned what he intends to achieve and how he will do it, and then to see that his subordinate commanders get on with the job.
Montgomery of Alamein; Memoirs, xxi, 1958

I am not a bit anxious about my battles.
If I am anxious I don't fight them.
I wait until I am ready.
Attributed to Montgomery of Alamein, 1887–

The capacity to understand the workings of the other man's mind is an essential element in generalship.
John Connell: Wavell, Soldier and Scholar, v, 1964

(*See also* Command, Coup d'Qeuil, General, High Command, Leadership, Strategy, Tactics.)

Geopolitics

Antwerp is a pistol pointed at the heart of England.
Attributed to Napoleon I, c. 1810

Essentially the great question remains: who will hold Constantinople?
Attributed to Napoleon I, 1769–1821

Italy is merely a geographical expression.
Prince Metternich, 1773–1859

The heart of France lies between Brussels and Paris.
Clausewitz, 1780–1831

Circumstances have caused the Mediterranean Sea to play a greater part in the history of the world, both in a commercial and a military point of view, than any other sheet of water of the same size. Nation after nation has striven to control it, and the strife still goes on.
Mahan: The Influence of Sea Power upon History, 1890

One thing is sure: the Caribbean Sea is the strategic key to the two great oceans, the Atlantic and Pacific, our chief maritime frontier.
Mahan: Naval Strategy, 1911

Cuba is as surely the key to the Gulf of Mexico as Gibraltar is to the Mediterranean.
Mahan, 1840–1914

Who rules East Europe commands the Heartland: Who rules the Heartland commands the World-Island: Who rules the World-Island commands the World.
Halford J. Mackinder: Democratic Ideals and Reality, 1919

When you think about the defense of England you no longer think of the chalk cliffs of Dover. You think of the Rhine. That is where our frontier lies today.
Stanley Baldwin: To the House of Commons, 30 July 1934

. . . the soft under-belly of the Axis.
Winston Churchill: To the House of Commons, 11 November 1942

Who controls the Rimland rules Eurasia; who rules Eurasia controls the destinies of the world.
Nicholas Spykman: America's Strategy and World Politics, 1942

Manchuria, with its industrial complex, coal, and iron ore, is the Ruhr of China.
Lieutenant General James M. Gavin, USA: In Harper's magazine, February 1966

Gettysburg (1–3 July 1863)

Never mind, General, all this has been my fault, it is I that have lost this fight, and

you must help me out of it the best way you can.
R.E. Lee: To General C.M. Wilcox after Pickett's Charge, 3 July 1863

We cannot hallow this ground. The brave men living and dead who struggled here have consecrated it far above our poor power to add or detract.
Abraham Lincoln: Gettysburg Address, 19 November 1863

Then at the brief command of Lee
Moved out that matchless infantry,
With Pickett grandly leading down,
To rush against the roaring crown
Of those dread heights of destiny.
Will Henry Thompson: The High Tide At Gettysburg, 1888

Gibraltar

Ay this is the famed rock which Hercules
And Goth and Moor bequeath'd us. At this door
England stands sentry. God! to hear the shrill
Sweet treble of her fifes upon the breeze,
And at the summons of the rock gun's roar
To see her red coats marching from the hill!
Wilfred Scawen Blunt, 1840–1922, Gibraltar

Glorious First of June (1 June 1794)

Our wives are now about going to church, but we will ring about these Frenchmen's ears a peal which will drown their bells.
Cuthbert Collingwood: Aboard HMS Barfleur, *going into action 1 June 1794*

Glory

It is the brave man's part to live with glory, or with glory die.
Sophocles, 495–406 B.C., Ajax

Glory is a mighty spur.
Ovid, 43 BC–18A.D., Epistulae ex Ponto, iv

Each of us must his end abide
In the ways of the world; so win who may
Glory ere death! When his days are told
That is the warrior's worthiest doom.
Beowulf, c. 1000

There must be a beginning of any great matter, but the continuing unto the end until it be thoroughly finished yields the true glory.
Sir Francis Drake: To Lord Walsingham, 17 May 1587

O, the fierce wretchedness that glory brings us!
Shakespeare: Timon of Athens, iv, 2, 1607

There is no road of flowers leading to glory.
La Fontaine, 1621–1695, Fables, x

To overcome in battle and subdue
Nations, and bring home spoils with infinite
Manslaughter, shall be the highest pitch
Of human glory.
Milton: Paradise Lost, xi, 1667

Glory is the true and honorable recompense of gallant actions.
Le Sage: Gil Blas, xii, 1735

Glory is the fair child of peril.
Tobias Smollet, 1721–1771

Glory is the sodger's prize;
The sodger's wealth is honor.
Robert Burns, 1759–1796, The Sodger's Return

Laurels grow in the bay of Biscay—I hope a bed of them may be found in the Mediterranean.
Nelson: Letter to Sir George Elliott, 4 August 1794

I am envious only of glory; for if it be a sin to covet glory, I am the most offending soul alive.
Nelson: Letter to Lady Hamilton, 18 February 1800

The combat deepens. On, ye brave,
Who rush to glory, or the grave!
Thomas Campbell, 1777–1844, Hohenlinden

The love of glory can only create a great hero; the contempt of it creates a great man.
Talleyrand, 1754–1838

We carved not a line, and we raised not a stone
But we left him alone with his glory.
Charles Wolfe: Burial of Sir John Moore at Corunna, 1817

Military glory—that attractive rainbow that rises in showers of blood, that serpent's eye that charms to destroy.

Abraham Lincoln: Speech against the Mexican War, House of Representatives, 12 January 1848

Military honor and glory have declined steadily since the destruction of the feudal system, which assured pre-eminence to men-at-arms.

Napoleon I: Political Aphorisms, 1848

His food
Was glory, which was poison to his mind
And peril to his body.

Henry Taylor, 1800–1886, Philip van Artevelde

Through the red gap of glory, they marched to the grave.

The Bold Fenian Men, Irish song, 19th century

Glory is that bright tragic thing
That for an instant
Means Dominion.

Emily Dickinson, 1830–1886, The Single Hound, v

The love of glory, the ardent desire for honorable distinction by honorable deeds, is among the most potent and elevating of military motives.

Mahan: Life of Nelson, viii, 1897

Common men endure these horrors and overcome them, along with the insistent yearnings of the belly and the reasonable promptings of fear; and in this, I think, is glory.

John W. Thomason, Jr.: Fix Bayonets! 1926

I buried death in the shroud of glory.

Jean-Paul Sartre: The Making of a Writer, 1964

(*See also* Awards, Decorations, Medals)

God of Battles

O God of battles! steel my soldiers' hearts . . .

Shakespeare: King Henry V, iv, 1, 1598

The Lord Himself hath a controversy with your enemies; even with that Romish Babylon of which the Spaniard is the great underpropper. In that respect we fight the Lord's battles.

Oliver Cromwell: Instructions to his admirals in the West Indies, 1655

God with us. (Gott mit uns.)

Motto on the badges of the Imperial German Army, World War I

(*See also* Christian Soldier, Religious War.)

Gordon, Charles George (1833–1885)

I am quite happy, thank God, and, like Lawrence, I have *tried* to do my duty.

Charles George Gordon: Postscript to his last letter from Khartoum, 29 December 1884 (Sir Henry Lawrence, hero of Lucknow, had as his epitaph, "Here lies Henry Lawrence, who tried to do his duty.")

Grant, Ulysses Simpson (1822–1885)

I can't spare this man; he fights.

Abraham Lincoln: To A.K. McClure, April 1862

When Grant once gets possession of a place, he holds on to it as if he had inherited it.

Abraham Lincoln: To Benjamin F. Butler, 22 June 1864

You just tell me the brand of whiskey Grant drinks—I would like to send a barrel of it to my other generals.

Attributed to Abraham Lincoln (originally reported in the New York Herald, 26 November 1863) as his reply to advisors who complained of Grant's drinking habits

Since Vicksburg they have not a word to say against Grant's habits. He has the disagreeable habit of not retreating before irresistible veterans.

Mary Chesnut: Diary, 1 January 1864

Each epoch creates its own agents, and General Grant more nearly than any other man impersonated the American character of 1861–5. He will stand, therefore, as the typical hero of the Great Civil War.

W.T. Sherman, 1820–1891

Grant stood by me when I was crazy, and I stood by him when he was drunk, and now we stand by each other.
Attributed to W.T. Sherman, c. 1870

I am more of a farmer than a soldier. I take little or no interest in military affairs.
U.S. Grant: To Bismarck at a review, Potsdam, 1879

You ask me what state he comes from. My answer shall be, he hails from Appomattox and its famous apple tree.
Roscoe Conkling: Nominating Grant for the Presidency, 1880

Grape

A whiff of grapeshot.
Thomas Carlyle: Frederick the Great, v, 1858

A little more grape, Captain Bragg.
Zachary Taylor: Order to Captain Braxton Bragg, of the Flying Artillery, at Buena Vista, 22 February 1847 (one account says Taylor's actual order was, "Double-shot your guns and give 'em hell!")

Grave

The fields of Flanders are his sepulcher
And the moon shot with blood, his epitaph.
Francisco De Quevedo, 1580–1645

The snow shall be their winding-sheet,
And every turf beneath their feet
Shall be a soldier's sepulcher.
Thomas Campbell, 1777-1844, Hohenlinden

There may they dig each other's graves,
And call the sad work glory.
Percy Bysshe Shelley: Queen Mab, vi, 1813

Seek out—less often sought than found—
A soldier's grave, for thee the best . . .
Byron: On My Thirty-Sixth Year, 1824

If I should die, think only this of me:
That there's some corner of a foreign field
That is forever England.
Rupert Brooke: The Soldier, 1914

Grenades

These balles, after they are fired and well kindeled, and having blowen a little, must be quickly throwen, lest they hurte such as hurtle them.
Peter Whitehorne: Certayne Wayes for the Ordering of Soldiours in Battelray, 1573

Grenadiers

I have seen sieges where the companies of grenadiers had to be replaced several times. This is easily explained: grenadiers are wanted everywhere. If there are four cats to chase, it is grenadiers who are called for, and usually they are killed without any necessity.
Maurice de Saxe: Mes Rêveries, viii, 1732

Is this a fit time, said my father to himself, to talk of Pensions and Grenadiers?
Laurence Sterne: Tristram Shandy, iv, 5, 1759

Some talk of Alexander, and some of Hercules,
Of Hector and Lysander, and such great names as these;
But of all the world's brave heroes, there's none that can compare
With a tow-row-row-row-row-row-row,
For a British Grenadier!
Song, The British Grenadiers, attributed to Charles Dibdin and first performed 17 January 1780 in celebration of the attack and capture of Savannah (but a minority view dates words and music from the 17th century).

. . . The first grenadier of France.
Napoleon I: Of Théophile de la Tour d'Auvergne, killed at Oberhausen, 27 June 1800. Thereafter, at reveille roll-call each morning, the first sergeant answered his name with, "Mort au champ d'honneur." ("Dead on the field of honor.")

Grenville, Sir Richard (1542–1591)

Heere die I, Richard Grenville, with a joyfull and quiet mind, for that I haue ended my life as a true soldier ought to do, yat hath fought for his countrey, Queene, religion, and honour, whereby my soule most joyfully departeth out of this bodie, and shall alwaies leaue behinde it an euerlasting fame of a valiant and true soldier that hath done his dutie, as he was bound to doe.

Sir Richard Grenville: Dying words, on board the Spanish flagship, 14 September 1591, off the Azores, after the singlehanded fight of HMS Revenge *against the Spanish fleet.*

Grog

For grog is our starboard, our larboard,
Our mainmast, our mizzen, our log—
At sea, or ashore, or when harbour'd,
The Mariner's compass is grog.

U.S. Navy song, early 19th century

Sailors will never be convinced that rum is a dangerous thing, by taking it away from them, and giving it to the officers; nor that temperance is their friend, which takes from them what they have always had, and gives them nothing in place of it.

Richard Henry Dana, Jr.: Two Years Before the Mast, xxxi, 1840

Jack's happy days will soon be gone,
To return again, ah, never!
For they've raised his pay by five cents a day,
But stopped his grog forever.
All hands to splice the mainbrace call,
But splice it now in sorrow;
For the spirit-room key must be laid away,
Forever on tomorrow.

Paymaster Caspar Schenck, USN: Farewell to Grog (written on board USS Portsmouth, *on abolition of the Navy's grog ration, 1862)*

Guadalcanal (7 August 1942–9 February 1943)

And when he gets to Heaven,
To St. Peter he will tell:
"One more Marine reporting, Sir—
"I've served my time in Hell."

Epitaph on the grave of a Marine, Guadalcanal, 1942

Guadalcanal is no longer merely a name of an island. It is the name of the graveyard of the Japanese Army.

Major General Kiyotake Kawaguchi: After Guadalcanal, c. 1943

. . . Long may the tale be told in the great Republic.

Winston Churchill: Of the conquest of Guadalcanal, Closing the Ring, i, 1951

There is no question that Japan's doom was sealed with the closing of the struggle for Guadalcanal.

Vice Admiral Raizo Tanaka, IJN: Struggle for Guadalcanal, 1956

Guard Duty

For this relief, much thanks; 'tis bitter cold.

Shakespeare: Hamlet, i, 1, 1600

Life is a tour of guard duty; you must mount guard properly and be relieved without reproach.

Charlet, 1650–1720

He trespasses against his duty who sleeps upon his watch, as well as he that goes over to the enemy.

Edmund Burke: Thoughts on the Present Discontents, 1770

Put none but Americans on guard tonight.

Attributed to George Washington, 30 April 1777

If you are sentinel at the tent of one of the field-officers, you need not challenge in the fore-part of the evening, for fear of disturbing his honour, who perhaps may be reading, writing, or entertaining company. But as soon as he is gone to bed, roar out every ten minutes at least, *Who comes there?* though nobody is passing. This will give him a favorable idea of your alertness; and though his slumbers may be broken, yet will they be the more pleasing, when he finds out that he reposes in perfect security. When the hour of relief approaches, keep constantly crying out, *Relief, relief!* it will prevent the guard from forgetting you, and prove that you are not asleep.

Francis Grose: Advice to the Officers of the British Army, 1782

All quiet along the Potomac tonight.

Ethel Lynn Beers: The Picket Guard, 1861

A picket frozen on duty . . .

W.H. Carruth, 1859–1924, Each in His Own Tongue

Guerre de Course

The great end of a war fleet, however, is not to chase, nor to fly, but to control the seas.

Mahan: Lessons of the War with Spain, 1899

Guerrilla

The general of a large army may be defeated, but you cannot defeat the determined mind of a peasant.

Confucius, 551–478 B. C.

This country swarms with vile outragious men
That live by rapine and by lawless spoil.

Christopher Marlowe: Tamburlaine the Great, ii, 2, 1587

Having no fixed lines to guard or defined territory to hold, it was always my policy to elude the enemy when they came in search of me, and carry the war into their own camps.

John S. Mosby: War Reminiscences, vii, 1887

Boh da Thone was a warrior bold:
His sword and his Snider were bossed with gold . . .
He shot at the strong and he slashed at the weak
From the Salween scrub to the Chindwin teak:
He crucified noble, he sacrificed mean,
He filled old ladies with kerosene:
While over the water the papers cried,
"The patriot fights for his countryside!"

Rudyard Kipling: Ballad of Boh da Thone, 1888 (Burma War, 1883–1885)

In civil war, no matter to what extent guerrillas are developed, they do not produce the same results as when they are formed to resist an invasion of foreigners.

S.I. Gusev: Lessons of Civil War, 1918

The few active rebels must have the qualities of speed and endurance, ubiquity and independence of arteries of supply. They must have the technical equipment to destroy or paralyze the enemy's organized communications.

T.E. Lawrence, 1888–1935, The Science of Guerrilla Warfare

. . . innumerable gnats, which, by biting a giant both in front and in rear, ultimately exhaust him. They make themselves as unendurable as a group of cruel and hateful devils, and as they grow and attain gigantic proportions, they will find that their victim is not only exhausted but practically perishing.

Mao Tse-tung: On Guerrilla Warfare, 1937

Many people think it is impossible for guerrillas to exist for long in the enemy's rear. Such a belief reveals lack of comprehension of the relationship that should exist between the people and the troops. The former may be likened to water and the latter to the fish who inhabit it.

Mao Tse-tung: On Guerrilla Warfare, 1937

The ability to run away is the very characteristic of the guerrilla.

Mao Tse-tung: Strategic Problems in the Anti-Japanese Guerrilla War, 1939

A trained and disciplined guerrilla is much more than a patriotic peasant, workman, or student armed with an antiquated fowling-piece and home-made bomb. His endoctrination begins even before he is taught to shoot accurately, and it is unceasing. The end product is an intensely loyal and politically alert fighting man.

Brigadier General S.B. Griffith, USMC: Introduction to Mao Tse-Tung on Guerrilla Warfare, 1961

The enemy's rear is the guerrillas' front.

Brigadier General S.B. Griffith, USMC: Introduction to Mao Tse-tung on Guerrilla Warfare, 1961

Guerrillas never win wars but their adversaries often lose them.

Charles W. Thayer: Guerrilla, 1963

(*See also* Guerrilla Warfare, Insurgency Partisan Warfare.)

Guerrilla Warfare

Night and day we chased an enemy who never awaited our approach but to harm us, was never found sleeping. Each tree, each hole, each piece of rock hid from our unseeing eyes a cowardly assassin, who, if undiscovered, came to pierce our breasts;

but who fled or begged for mercy if we found him face to face.

Extract from My Odyssey, journal of an unknown French Creole during the Haitian War for Independence, May 1793

Militia and armed civilians cannot and should not be employed against the main force of the enemy, or even against sizeable units. They should not try to crack the core, but only nibble along the surface and on the edges. They should rise in provinces lying to one side of the main theater of war, which the invader does not enter in force, in order to draw these areas entirely from his grasp. These storm clouds, forming on his flanks, should also follow to the rear of his advance.

Clausewitz: On War, 1832

It is just as legitimate to fight an enemy in the rear as in the front. The only difference is in the danger.

John S. Mosby: War Reminiscences, iv, 1887

The word of a scout—a march by night—
A rush through the mist—a scattering fight—
A volley from cover—a corpse in the clearing—
A glimpse of a loin-cloth and heavy jade earring—
A flare of a village—the tally of slain—
And . . . the Boh was abroad on a raid again!

Rudyard Kipling: Ballad of Boh da Thone, 1888

[Cuba] is divided into two great military camps, one situated within the forts, the other scattered over the fields and mountains outside them. The Spaniards have absolute control over everything within the fortified places; that is, the cities, towns, seaports, and along the lines of the railroad; the insurgents are in possession of the rest. They are not in fixed possession, but they have control much as a mad bull may be said to have control of a ten-acre lot when he goes on the rampage. Some farmer may hold a legal right to the ten-acre lot . . . and the bull may occupy but one part of it at a time, but he has possession, which is better than the law.

Richard Harding Davis: Cuba in War Time, 1897

Air-power may be effective against elaborate armies: but against irregulars it has no more than moral value.

T.E. Lawrence: Letter to A.P. Wavell, 21 May 1923

Guerrilla war is far more intellectual than a bayonet charge.

T.E. Lawrence, 1888–1935, The Science of Guerrilla Warfare

[Guerrilla war] must have a friendly population, not actively friendly, but sympathetic to the point of not betraying rebel movements to the enemy. Rebellions can be made by two percent active in a striking force, and 98 percent passively sympathetic.

T.E. Lawrence, 1888–1935: The Science of Guerrilla Warfare

Guerrilla strategy is the only strategy possible for an oppressed people.

Kao Kang (source and date unknown; quoted by Mao Tse-tung in On Guerrilla War)

The advantages are nearly all on the side of the guerrilla in that he is bound by no rules, tied by no transport, hampered by no drill-books, while the soldier is bound by many things, not the least by his expectation of a full meal every so many hours. The soldier usually wins in the long run, but very expensively.

Sir A.P. Wavell: Critique on a counter-guerrilla exercise, Blackdown, 30 August 1932

When the enemy advances, we retreat.
When he escapes we harass.
When he retreats we pursue.
When he is tired we attack.
When he burns we put out the fire.
When he loots we attack.
When he pursues we hide.
When he retreats we return.

Mao Tse-tung, 1893–

Select the tactic of seeming to come from east and attacking from the west; avoid the solid, attack the hollow; attack, withdraw; deliver a lightning blow, seek a lightning decision . . . Withdraw when he advances; harass him when he stops; strike him when he is weary; pursue him when he withdraws.

Mao Tse-tung: On Guerrilla War, 1937

Guerrilla war is a kind of war waged by the few but dependent on the support of the many.

B.H. Liddell Hart: Foreword to Guerrilla Warfare, 1961

Guerrilla war is not dependent for success on the efficient operation of complex mechanical devices, highly organized logistical systems, or the accuracy of electronic computers. It can be conducted in any terrain, in any climate, in any weather; in swamps, in mountains, in farmed fields; its basic element is man, and man is more complex than any of his machines.

Brigadier General S.B. Griffith, USMC: Introduction to Mao Tse-tung on Guerrilla Warfare, 1961

If historical experience teaches us anything about revolutionary guerrilla war, it is that military measures alone will not suffice.

Brigadier General S.B. Griffith, USMC: Introduction to Mao Tse-tung on Guerrilla Warfare, 1961

There is another type of warfare—new in its intensity, ancient in its origin—war by guerrillas, subversives, insurgents, assassins; war by ambush instead of by combat, by infiltration instead of aggression, seeking victory by eroding and exhausting the enemy instead of engaging him. It is a form of warfare uniquely adapted to what have been strangely called "wars of liberation," to undermine the efforts of new and poor countries to maintain the freedom that they have finally achieved. It preys on unrest and ethnic conflicts.

John F. Kennedy: Address to the graduating class, U.S. Naval Academy, 6 June 1962

It is for all practical purposes impossible to win a guerrilla war if there is a privileged sanctuary behind the guerrilla fighters.

Walter Lippmann: In Washington Post, 5 September 1963

The front is everywhere.

Maxim of guerrilla war

(*See also* Guerrilla, Insurgency, Partisan Warfare.)

Gun

And, but for these vile guns,
He would himself have been a soldier.

Shakespeare: I King Henry IV, i, 3, 1597

Don't forget your great guns, which are the most respectable arguments of the rights of kings.

Frederick the Great: Letter to his brother, Prince Henry, 21 April 1759

And there we see a swampin' gun
As big as a log of maple,
Upon a deuced little cart,
A load for father's cattle.
And every time they shoot it off,
It takes a horn of powder,
And makes a noise like father's gun,
Only a nation louder.

Edward Bangs: Yankee Doodle, 1775

Where a goat can go, a man can go,
and where a man can go, he can drag a gun.

Colonel William Phillips: At Mount Defiance, adjacent to Ticonderoga, 1777

. . . the distant and random gun
That the foe was sullenly firing.

Charles Wolfe: The Burial of Sir John Moore at Corunna, 1817

As a general rule, the maxim of marching to the sound of the guns is a wise one.

Jomini: Précis Politique et Militaire de la Campagne de 1815, 1839

The gun is the rallying-point of the detachment, its point of honor, its flag, its banner. It is that to which the men look, by which they stand, with and for which they fight, by and for which they fall. As long as the gun is theirs, they are unconquered, victorious; when the gun is lost, all is lost.

Major Robert Stiles, CSA: Four Years Under Marse Robert, 1903

For you all love the screw-guns—the
screw-guns they all love you!
So when we call round for a few guns,
o' course you will know what to do—
Jest send in your Chief an' surrender—it's
worse if you fights or you runs:
You can go where you please, you can skid
up the trees, but you don't get away
from the guns!

Rudyard Kipling: Screw-Guns, 1890 ("Screw-Gun," a British Army term for pack artillery)

'Less you want your toes trod off, you'd
better get back at once,
For the bullocks are walking two by two,

The byles are walking two by two,
And the elephants bring the guns.
Ho! Yuss!
Great—big—long—black—forty-pounder guns
Jiggery-jolty to and fro,
Each as big as a launch in tow—
Blind—dumb—broad-breeched—beggars o' battering-guns.
Rudyard Kipling: My Lord the Elephant, 1893

. . . the Guns, thank Gawd, the Guns!
Rudyard Kipling: Ubique, 1903

The men behind the guns.
J. J. Rooney, title of poem, 1898

The columns should always support each other, and therefore, unless they have received a specific mission, or are in conflict themselves, should march to the sound of the guns.
Colonel G. A. Furse: The Art of Marching, 1901

Over the grasses of the ancient way
Rutted this morning by the passing guns.
John Masefield: August 1914

Only the monstrous anger of the guns,
Only the stuttering rifles' rapid rattle...
Wilfred Owen: Anthem for Doomed Youth, 1915

There can never be too many guns, there are never enough of them.
Ferdinand Foch: Precepts, 1919

Gun upon gun, ha! ha!
Gun upon gun, hurrah!
Don John of Austria
Has loosed the cannonade.
G. K. Chesterton, 1874–1936, Lepanto

What the colors are to the infantryman, the gun is, or ought to be, to the artilleryman. It is our emblem, our standard. When I joined, the last honor to a gunner was the burial service "over the metal." Time and habits may change, but the devotion to the gun must be inculcated into the heart of every gunner.
Field Marshal Sir George Francis Milne: 1866–1948

March to the sound of the guns.
Military maxim.

(*See also* Artillery, Cannon.)

Gun Salutes

No jocund health that Denmark drinks today,
But the great cannon to the clouds shall tell.
Shakespeare: Hamlet, i, 2, 1600

I shall ever be careful in keeping especially my Royal orders, which positively command me to salute neither garrison nor flagg of any forrainer except I am certaine to receave gunne for gunne.
Sir Cloudesley Shovell: Letter to Sir Martin Wescomb, 1688

I have just received your dispatch announcing the capture of Atlanta. In honor of your great victory I have ordered a salute to be fired with *shotted* guns from every battery bearing on the enemy. The salute will be fired within an hour, amid great rejoicing.
U. S. Grant: Telegram to Gen. Sherman, 4 September 1864

Gunboat Diplomacy

A man-of-war is the best ambassador.
Oliver Cromwell, 1599–1658

We want Perdicaris alive, or Raisuli dead.
John Hay: Telegram, while Secretary of State, to the Sultan of Morocco, after the kidnapping of Perdicaris, a U. S. citizen, by a Moroccan bandit, Raisuli, June 1904

War ministers frequently form a habit of mind which regards a frigate and a battalion of the Line as the normal messengers of . . . policy.
Philip Guedalla: Palmerston, 1927

(*See also* Diplomacy, National Power, Warship.)

Gunner

. . . the nimble gunner
With linstock now the devilish cannon touches, (*alarum and the chambers go off.*)
And down goes all before them.
Shakespeare: King Henry V, iii, 1598

A gunner ought to be a sober, wakeful, lusty, hardy, patient, prudent, and quick-spirited man; he ought also to have a good eyesight, a good judgment, a perfect

knowledge to select a convenient place in the day of service, to plant his ordnance where he may do most hurt to the enemies, and be least annoyed by them.
Niccolo Fontana Tartaglia, 1499–1559, Colloquies Concerning the Art of Shooting in Great and Small Pieces of Artillery.

Many a time it falleth out that most men employed for gunners are very negligent of the fear of God.
Unknown Puritan moralist quoted in Sound of the Guns, Fairfax Downey

No one accuses the Gunner of maudlin affection for anything except his beasts and his weapons . . . He serves at least three jealous gods—his horse and all its sadlery and harness; his gun, whose least detail of efficiency is more important than men's lives; and, when these have been attended to, the never-ending mystery of his art commands him.
Rudyard Kipling: The New Army. 1915

(*See also* Cannoneer, Gunnery.)

Gunnery

Gonners, to schew ther arte,
Into the towne in many a parte
Schote many a full grete stone.
Thankyd be God and Mary myld,
They hurt nothir man, woman, ne chyld,
To the howsis thow they did harm.
Ballad during the first siege of Calais, 1346

The first shot is for the Devil, the second for God, and only the third for the King.
Gunners' proverb, 16th century

Every gunner ought to know that it is a wholesome thing for him to eat and drink a little meat before he doth discharge any piece of artillery because the fumes of the saltpeter and brimstone will otherwise be harmful to his brain.
Niccolo Fontana Tartaglia, 1499-1599. Colloquies Concerning the Art of Shooting in Great and Small Pieces of Artillery.

The fact appears to be but too clearly established that the Americans have some superior mode of firing.
London Times, 1813 (after repeated U. S. frigate victories.)

The best protection against the enemy's fire is a well directed fire from our own guns.
David G. Farragut: General order for the attack on Port Hudson, 14 March 1863

I don't care if he drinks, gambles, and womanizes; *he hits the target.*
Sir John Fisher: Admiral Fisher's reply, c. 1908, to critics of Captain Sir Percy Scott, who revolutionized British naval gunnery.

Pick out the biggest one and fire.
Captain Edward J. Moran, USN: To his gunnery officer, USS Boise, *on sighting Japanese heavy cruisers off Cape Esperance, night of 11–12 October 1942*

Gunnery Sergeant

I know the disciplines of wars.
Shakespeare: King Henry V, iii, 2, 1958

(*See also* Noncommissioned Officer.)

Gunpowder

. . . hot as gunpowder.
Shakespeare: King Henry V, iv, 7, 1598

Such I hold to be the genuine use of gunpowder; that it makes all men alike tall.
Thomas Carlyle, 1795–1881

Sometimes gunpowder smells good.
R. W. Emerson: Expressing relief after the attack on Fort Sumter, 1861

(*See also* Ammunition.)

Gustavus Adolphus II (1594–1632)

Consider the great Gustavus Adolphus! In eighteen months he won one battle, lost a second, and was killed in the third. His fame was won at a bargain price.
Napoleon I: Letter to General Gaspard Gourgaud, St. Helena, 1818

Haitian War of Independence, 1804

Some doubt the courage of the Negro. Go to Haiti and stand on those fifty thousand graves of the best soldiers France ever had, and ask them what they think of the Negro's sword.

Wendell Phillips: Address on Toussaint l'Ouverture, 1861

Halleck, Henry Wager (1815–1872)

Originates nothing, anticipates nothing, takes no responsibility, plans nothing, suggests nothing, is good for nothing.

Gideon Welles: Diary entry, 1862

Hand-to-Hand Combat

As to the proposition advanced . . . that hereafter war will be made altogether with artillery, I consider that this observation is wholly erroneous . . . For whoever wishes to train a good army must . . . train his troops to attack the enemy sword in hand, and to seize hold of him bodily.

Niccolo Machiavelli: Discorsi, xvii, 1531

In battle most men expose themselves enough to satisfy the needs of honor; few wish to do more than this, or more than enough to carry to success the action in which engaged.

Francois de la Rochefoucauld: Maxims, 1665

If your bayonet breaks, strike with the stock; if the stock gives way, hit with your fists; if your fists are hurt, bite with your teeth.

General Mikhail Ivanovich Dragomirov: Notes for Soldiers, c. 1890

Hannibal (247–183 B. C.)

Hannibal knew how to gain a victory, but not how to use it.

Plutarch: Lives, c. 100 (Hannibal's cavalry commander, Maharbal, reportedly said this to Hannibal after Cannae when the latter did not immediately march on Rome.)

. . . the most audacious of all, probably the most stunning, so hardy, so sure, so great in all things.

Napoleon I, 1769–1821

...by many degrees the greatest soldier on record.

Wellington, 1769–1852

Harassment

When the enemy is at ease, be able to weary him; when well fed, to starve him; when at rest, to make him move.

Sun Tzu, 400–320 B. C., The Art of War

When the enemy is at ease, tire him.

Li Ch'uan, c. 750

Destroying their settlements, spreading alarms, showing and keeping up a spirit of enterprise that will oblige them to defend their extensive possessions at all points, is of infinitely more consequence to the United States of America than all the plunder that can be taken.

Robert Morris: Letter to John Paul Jones, 1 February 1777

I have often thought that their fierce hostility to me was more on account of the sleep I made them lose than the number we killed and captured.

John S. Mosby: War Reminiscences, iv, 1887

Harrison, William Henry (1773–1841)

The iron-armed soldier, the true-hearted soldier,
The gallant old soldier of Tippecanoe!

Campaign song, 1840

Headquarters

Everything was strikingly quiet and unostentatious...There was no throng of scented staff officers with plumed hats, orders and stars, no main guards, no crowd of contractors, actors, valets, cooks, mistresses, equipages, horses, dogs, forage and baggage wagons . . . Just a few aides-de-camp, who went about the streets alone and in their overcoats, a few guides, and a small staff guard; that was all.

A. L. F. Schauman, 1778–1840, On the Road with Wellington, the Diary of a Commissary

My headquarters will be in the saddle.

Major General John Pope, USA: To the press, on assuming command of the Army of the Potomac, June 1862 (on hearing of this statement, Lincoln commented, "A better place for his hindquarters.").

. . . a house with roomy offices, where telegraph and wireless, telephone and signalling instruments are at hand, while a fleet of automobiles and motorcycles, ready for the longest trips, wait for orders. Here, in a comfortable chair before a large table, the modern Alexander overlooks the whole battlefield on a map. From here he telephones inspiring words, and here he receives the reports from army and corps commanders and from balloons and dirigibles which observe the enemy's movements and detect his positions.

Alfred von Schlieffen: Cannae, 1913

Notice in [German Army] headquarters the presence of the King; the Commander-in-Chief of the Allied Forces with his general staff; of the German Princes; and also of the Minister of War, of the Minister for Foreign Affairs, and of the Federal Chancellor. There, indeed, you have an example of the command of nations going to war. The whole power of the Government accompanies the Commander-in-Chief, in order to put at his disposition all the resources of diplomacy, of finance, and of the national soil; in order that the military enterprise to which the nation has given all its energies . . . may proceed.

Ferdinand Foch: Precepts, 1919

Health

Daily practice of the military exercises is much more efficacious in preserving the health of an army than all the art of medicine.

Vegetius: De Re Militari, 378

There are no stores, no surgeons, no drugs, and the hammocks are infected and loathsome, and the men stink as they go, and the poor rags they have are rotten and ready to fall off.

Sir John Pennington: Of conditions in the British Fleet, c. 1626

No price is too great to preserve the health of the fleet.

Lord St. Vincent: Letter to the Admiralty, 1796

For the accomodation, comfort, & health of the Troops, the hair is to be cropped without exception, and the General will give the example.

General Order of the Army, Pittsburgh, 30 April 1801

The great thing in all military service is health.

Nelson: Letter to Dr. Moseley, October 1803

(*See also* Field Hospital, Sanitation, Surgeon.)

Helmet

Were it a casque composed by Vulcan's
skill,
My sword should bite it.

Shakespeare: Troilus and Cressida, v, 2, 1601

Henry of Navarre (1553–1610)

I make war, I make love, I build.

Statement by himself

Navarre shall be the wonder of the world.

Shakespeare: Love's Labor Lost, i, 1, 1594

Press where ye see my white plume shine,
Amidst the ranks of war,
And be your oriflamme today the
helmet of Navarre.

T. B. Macaulay, 1800–1859, Ivry

Heroism

Heroes as great have died, and yet shall fall.

Homer: Iliad, xv, c. 1000 B. C.

See the conquering hero comes!
Sound the trumpet, beat the drums!

Thomas Morell, 1703–1784, Judas Maccabaeus

Believe me, that every man you see in a military uniform is not a hero.

Wellington: Letter to a historian, 8 August 1815

Every hero becomes a bore at last.

R. W. Emerson: Representative Men, 1849

U. S. Navy

WELLINGTON

1769–1852

"Next to a battle lost, the greatest misery is a battle won."

Heroism is the brilliant triumph of the soul over the flesh—that is to say, over fear . . . Heroism is the dazzling and brilliant concentration of courage.
Henri-Fréderic Amiel: Journal, 1 October 1849

Hail, ye indomitable heroes, hail!
Despite of all your generals ye prevail.
Walter Savage Landor: The Crimean Heroes, 1856

Right in the van,
On the red rampart's slippery swell,
With heart that beat a charge, he fell . . .
James Russell Lowell: Memoriae Positum, August 1863 (in memory of Colonel Robert Gould Shaw, 54th Massachusetts Infantry, killed in the assault on Fort Wagner, 18 July 1863

After you, Pilot.
Captain T. A. M. Craven, USN: To his navigator at the foot of an escape hatch, USS Tecumseh, *sinking during the battle of Mobile Bay, 5 August 1864. Craven was lost; the navigator survived.*

When the will defies fear, when duty throws the gauntlet down to fate, when honor scorns to compromise with death—this is heroism.
Robert G. Ingersoll: Speech in New York, 29 May, 1882

Wars may cease, but the need for heroism shall not depart from the earth, while man remains man and evil exists to be redressed.
Mahan: Life of Nelson, 1897

At the grave of a hero we end, not with sorrow at the inevitable loss, but with the contagion of his courage; and with a kind of desperate joy we go back to the fight.
Oliver Wendell Holmes, Jr. 1841–1935

(*See also* Bravery, Courage, Daring, Gallantry.)

Hesitation

The god of war hates those who hesitate.
Euripides: Heraclidae, c. 425 B. C.

Hesitation and half measures lose all in war.
Napoleon I: Maxims of War, 1831

(*See also* Timidity.)

High Command

I do not desire to place myself in the most perilous of all positions:—*a fire upon my rear, from Washington, and the fire, in front, from the Mexicans.*
Winfield Scott: Letter to Secretary of War Marcy 21 May 1846

The first condition of supreme command is the full freedom of mind of the chief.
Lyautey: Address on admission to the Academie Française, 8 July 1920

There is required for the composition of a great commander not only massive common sense and reasoning power, not only imagination, but also an element of legerdemain, an original and sinister touch, which leaves the enemy puzzled as well as beaten.
Winston Churchill: The World Crisis, II, 1923

The high commands of the Army are not a club. It is my duty and that of His Majesty's Government to make sure that exceptionally able men, even though not popular with their military contemporaries, should not be prevented from giving their services to the Crown.
Winston Churchill: Note for the Secretary of State for War, 4 September 1942

I would consider only one who has handled large forces in an independent command in more than one campaign; and who had shown his qualities in adversity as well as in success. I then proposed to judge him by his worth as a strategist; his skill as a tactician; his power to deal tactfully with his Government and his allies; his ability to train troops or direct their training; and his energy and driving power in planning and in battle.
Sir A. P. Wavell: Draft lecture on Belisarius, 1946 (these were Wavell's qualifying standards for the highest command.)

He who reaches the top will often be misunderstood and the target for much criticism; this will produce at times a feeling of loneliness, which is accentuated by the fact that those with whom he would most like to talk will often avoid him because of his position. The only policy in high positions is an intense devotion to duty and the unswerving pursuit of the target, in spite of criticism.

Montgomery of Alamein: Memoirs, 1958

A national war machine is . . . the most exhilarating of all vehicles to drive.

Correlli Barnett: The Swordbearers, xi, 1963

(*See also* Command.)

Hill, Ambrose Powell (1825–1865)

Order A.P. Hill to prepare for action. Pass the infantry to the front.

Stonewall Jackson: Penultimate words while dying, 10 May 1863

Tell Hill he *must* come up.

R.E. Lee: Penultimate words while dying, 12 July 1870

Hill, D. H. (1821–1889)

This man has the heart of a lion and the tongue of an adder, but I would not trade him for a brigade.

Attributed to R. E. Lee, 1807–1870

History

Of the events of the war, I have not ventured to speak from any chance information, nor according to any notion of my own; I have described nothing but what I either saw myself, or learned from others of whom I made the most careful and particular inquiry. The task was a laborious one . . .

Thucydides: History of the Peloponnesian War, i,c. 404 B. C.

It is as impossible to write well on the operations of war, if a man has no experience on active service, as it is to write well on politics without having been engaged in political transactions and vicissitudes.

Polybius, 200–118 B. C. History

If we only act for ourselves, to neglect the study of history is not prudent; if we are entrusted with the care of others it is not just.

Samuel Johnson, 1709–1784

Don't you think, Madam, that it is pleasanter to read history than to live it? Battles are fought and towns taken in every page, but a campaign takes six or seven months to hear, and achieves no great matter at last. I dare say Alexander seemed to the coffee-houses of Pella a monstrous while about conquering the world.

Horace Walpole: Letter to the Countess of Ossory, 8 October, 1777

Only the study of military history is capable of giving those who have no experience of their own a clear picture of what I have just called the friction of the whole machine.

Clausewitz: Principles of War, 1812

The history of a battle is not unlike the history of a ball. Some individuals may recollect all the little events of which the great result is the battle won or lost; but no individual can recollect the order in which, or the exact moment at which, they occurred, which makes all the difference...But if a true history is written, what will become of the reputation of half of those who have acquired reputations, and who deserve it for their gallantry, but who, if their mistakes and casual misconduct were made public, would not be so well thought of?

Wellington: Letter to a historian, 8 August 1815

What experience and history teach is this—that peoples and governments never have learned anything from history or acted on principles derived from it.

Hegel: The Philosophy of History, 1827

Military history, accompanied by sound criticism, is indeed the true school of war.

Jomini: Précis sur l'Art de la Guerre, 1838

To men of a sedate and mature spirit, in whom is any knowledge or mental activity, the detail of battle becomes insupportably tedious and revolting.

R. W. Emerson: War, 1849

Only study of the past can give us a sense of reality, and show us how the soldier will fight in the future.

Ardant du Picq, 1821–1870, Battle Studies

The smallest detail, taken from the actual incident in war, is more instructive to me, a soldier, than all the Thiers and Jominis in the world. They speak, no doubt, for the heads of states and armies but they never show me what I wish to know—a battalion, a company, a squad, in action.

Ardant du Picq, 1821–1870, Battle Studies

. . . the most effective means of teaching war during peace.

Helmuth von Moltke ("The Elder"), 1800–1891

The study of history lies at the foundation of all sound military conclusions and practice.

Alfred Thayer Mahan, 1840–1914

The value of history in the art of war is not only to elucidate the resemblance of past and present, but also their essential differences.

Sir Julian Corbett, 1854–1922

It is in military history that we are to look for the source of all military science. In it we shall find those exemplifications of failure and success by which alone the truth and value of the rules of strategy can be tested.

Dennis Hart Mahan: Out-Post (1864 edition)

The real war will never get into the books.

Walt Whitman, 1819–1892, The Real War

Although the story of a campaign is made up of many details which cannot be omitted, since they are essential to the truth as well as the interest of the account, it is of paramount importance that the reader should preserve a general idea . . . To appreciate the tale it is less necessary to contemplate the wild scenes and stirring incidents, than thoroughly to understand the logical sequence of incidents which all tend to and ultimately culminate in a decisive trial of strength.

Winston Churchill: The River War, xi, 1899

A great war does not kill the past, it gives it new life. It may seem a catastrophe which renders all that went before insignificant and not worthy study for men of action . . . But it is not so. As time gives us distance we see the flood as only one more pool in the river as it flows down to eternity.

Sir Julian Corbett, 1854–1922, The Revival of Naval History

A boy who hears a history lesson ended by the beauty of peace, and how Napoleon brought ruin upon the world and that he should be forever cursed, will not long have much confidence in his teacher. He wants to hear more about the fighting and less about the peace negotiations.

William Lee Howard: Peace, Dolls, and Pugnacity, 1903

War makes rattling good reading; but Peace is poor reading.

Thomas Hardy: The Dynasts, 1906

Those who cannot remember the past are condemned to repeat it.

George Santayana: The Life of Reason, 1906

For the first time in my life I have seen "History" at close quarters, and I know that its actual process is very different from what is presented to posterity.

General Max Hoffman, 1869–1927, diary entry, World War I

History is written for schoolmasters and armchair strategists. Statesmen and warriors pick their way through the dark.

Sir John Fisher, c. 1915

I hope you have kept the enemy always in the picture. War-books so often leave them out.

T. E. Lawrence: Letter to Colonel A. P. Wavell, 9 February 1928

The real way to get value out of the study of military history is to take particular situations, and as far as possible get inside the skin of the man who made a decision, realize the conditions in which the decision was made, and then see in what way you could have improved on it.

Sir A. P. Wavell: Lecture to officers, Aldershot, c. 1930

Study the human side of military history, which is not a matter of cold-blooded formulas or diagrams, or nursery-book principles, such as: Be good and you will be happy. Be mobile and you will be victorious. Interior lines at night are the general's delight. Exterior lines at morning are the general's warning. And so on.

Sir A.P. Wavell: Lecture to officers, Aldershot, c. 1930

No other art or science [as war] has so definitely clung to the . . . past as guidance for the future, or so far ignored the alterations of circumstances brought about by time.

Alfred Vagts: A History of Militarism, 1937

It is true that full study of war will not seriously assist a subaltern on picket duty; but when it comes to understanding present war conditions and the probable origins of the next war, a deep and impartial knowledge of history is essential. Further still, as it is not subalterns or generals who make wars, but governments and nations, unless the people as a whole have some understanding of what war meant in past ages, their opinions on war . . . today will be purely alchemical.

J. F. C. Fuller: Decisive Battles, 1939

It is only possible to probe into the mind of a commander through historical examples.

B. H. Liddell Hart: Strategy, 1929

The practical value of history is to throw the film of the past through the material projector of the present onto the screen of the future.

B. H. Liddell Hart: Thoughts on War, i, 1944

Nothing remains static in war or in military weapons, and it is consequently often dangerous to rely on courses suggested by apparent similarities in the past.

E. J. King: A Naval Record, 1952

Most official accounts of past wars are deceptively well written, and seem to omit many important matters—in particular, anything which might indicate that any of our commanders ever made the slightest mistake. They are therefore useless as a source of instruction.

Montgomery of Alamein: Memoirs, xxxiii, 1958

Among professional soldiers, anti-intellectualism can also express itself in an uncritical veneration of the military treatises of the past which, with almost metaphysical reverence, are taken as permanent contributions to military doctrine.

Morris Janowitz: The Soldier and the State, xx, 1960

Dead battles, like dead generals, hold the military mind in their dead grip.

Barbara W. Tuchman: The Guns of August, ii, 1962

Military history is the account of how force served political ends and how man, individual hero or leader or aggregated professionals, conscripts, or irregulars, accomplished this service.

Colonel F. B. Nihart, USMC: In Military Affairs, February 1965

Hitler, Adolph (1899–1945)

Adolph Hitler is a bloodthirsty guttersnipe, a monster of wickedness, insatiable in his lust for blood and plunder.

Winston Churchill: On the German invasion of Russia, 22 June 1941

Home Front

We had been told, on leaving our native soil, that we were going to defend the sacred rights conferred on us by so many of our citizens settled overseas, so many years of our presence, so many benefits brought by us to populations in need of our assistance and civilization. We could verify that all this was true, and, because true, we did not hesitate to shed our share of blood, to sacrifice our youth and hopes. We regretted nothing, but whereas we over here are inspired by this frame of mind, I am told that in Rome factions and conspiracies are rife, that treachery flourishes, and that many in their uncertainty and confusion lend ready ear to the dire temptations of relinquishment and vilify our actions . . . Make haste to reassure me, I beg you, and tell me that our fellow citizens understand us, support us, and protect us, as we ourselves are protecting the glory of the Empire. If it should be otherwise, if we should have to leave our bleached bones on these desert sands in vain, then beware the anger of the Legions.

Marcus Flavinius: To his cousin, Tertullus, in Rome, 2d century A. D. (Flavinius was a centurion, 2d Cohort, Augusta Legion.)

How much better it is thus to deserve the thanks of the country . . . than to skulk at home as the cowardly exempts do. Some of these poor dogs have hired substitutes, as though money could pay the service every man owes his country. Others claim to own twenty negroes . . . Others are warlike militia officers, and their regiments cannot dispense with such models of military skill and valor. And such noble regiments they have. Three field officers, four staff officers, ten captains, thirty lieutenants, and one private with a misery in his bowels. Some are pill and syringe gentlemen, and have done their share of killing at home. Some are kindly making shoes for the army, and generously giving them to the poor soldiers, only asking two months' pay. Some are too sweet and delicate for anything but fancy duty; the sight of blood is unpleasant, and the roar of cannon shocks their sensibilities.
D.H. Hill: General Order, 24 April 1863

Any distinction between belligerents and nonbelligerents is no longer admissable today either in fact or theory . . . When nations are at war, everyone takes part in it: the soldier carrying his gun, the woman loading shells at a factory, the farmer growing wheat, the scientist experimenting in his laboratory . . . It begins to look now as if the safest place may be the trenches.
Giulio Douhet: The Command of the Air, 1921

I deprecate the policy of "misery first" . . . The value of all the various self-strafing proposals should be estimated in tonnage of imports. If there is a heavy economy to be achieved on any article, let us effect it; but it would be unwise to embark upon a lot of fussy restrictions.
Winston Churchill: Note for Lord Cherwell, 10 March 1942

The morale of the civilians in a city which has not yet been touched by war is seldom as high as it is among the soldiers in the frontline.
Alan Moorehead: Gallipoli, 1956

(*See also* Noncombatant.)

Honor

The love of honor alone is ever young.
Pericles: Funeral Oration over the Athenian dead, 431 BC

When dangers are great, there the greatest honors are to be won by men and states.
Thucydides: History of the Peloponnesian Wars, c. 404 BC

Leave not a stain in thine honor.
Ecclesiasticus, XXXIII, 22

What is left when honor is lost?
Publilius Syrus: Sententiae, 1st century B.C.

The best memorial for a mighty man is to gain honor ere death.
Beowulf, c. 1000 A.D.

Better a thousand times to die with glory than live without honor.
Louis VI of France, 1081–1137

There is nothing left to me but honor, and my life, which is saved.
Francis I of France: Letter to his mother after the battle of Pavia, 24 February 1525 (This is often contracted into the phrase, "All is lost save honor.")

My honor is dearer to me than my life.
Cervantes: Don Quixote, 1604

Mine honor is my life; both grow in one;
Take honor from me and my life is done.
Shakespeare: King Richard II, i, 1595

Can honor set a leg? no: or take away the grief of a wound? no. Honor hath no skill in surgery, then? no. What is honor? a word. What is that word honor? air. Who hath it? he that died o' Wednesday. Doth he feel it? no. Doth he hear it? no. It is insensible, then? Yes, to the dead. But will it not live with the living? no. Therefore, I'll none of it. Honor is a mere scutcheon; and so ends my catechism.
Shakespeare: I King Henry IV, v, 1597

If it be a sin to covet honor,
I am the most offending soul alive.
Shakespeare: King Henry V, iv, 1, 1598

Set honor in one eye and death i' the other
And I will look on both indifferently;
For let the gods so speed me as I love
The name of honor more than I fear death.
Shakespeare: Julius Caesar, i, 1599

Life every man holds dear; but the brave man
Holds honor far more precious-dear than life.
Shakespeare: Troilus and Cressida, v, 3, 1601

If I lose mine honor, I lose myself.
Shakespeare: Antony and Cleopatra, iii, 1606

If e'er my son
Follow the war, tell him it is a school
Where all the principles tending to honor
Are taught, if truly followed.
Philip Massinger, 1583–1640

I could not love thee, dear, so much,
Loved I not honor more.
Richard Lovelace, 1618–1658, To Lucasta, on Going to the Wars

He that is valiant and dares fight
Though drubbed, can lose no honor by't.
Samuel Butler: Hudibras, i, 3, 1663

Godlike erect, with native honor clad . . .
Milton: Paradise Lost, iv, 1667

When vice prevails and impious men bear sway,
The post of honor is a private station.
Joseph Addison: Cato, 1713

Better to die ten thousand deaths,
Than wound my honor.
Joseph Addison: Cato, 1713

If honor calls, where'er she points the way,
The sons of honor follow and obey.
Charles Churchill, 1731–1764, The Farewell

The sodger's wealth is honor.
Robert Burns, 1759–1796, The Sodger's Return

I would lay down my life for America, but I cannot trifle with my honor.
John Paul Jones: Letter to A. Livingson, 4 September 1777

That chastity of honor which felt a stain like a wound.
Edmund Burke: Reflections on the Revolution in France, 1790

Unfit as my ship was, I had nothing left for the honor of our country but to sail, which I did in two hours afterward.
Nelson: Letter from Naples, 16 September 1793

The better he behaves, the more honor we shall have in taking him.
Captain James Dacres, RN: To his first lieutenant, in HMS Guerrière, *on observing the bold approach of USS* Constitution, *19 August 1812. (Within 15 minutes after the first broadside,* Guerrière *was dismasted and sinking.)*

The first object which a general who gives battle should consider is the glory and honor of his arms; the safety and conservation of his men is but secondary.
Napoleon I: Maxims of War, 1831

Keep honor, like your sabre, bright,
Shame coward fear—and then
If we must perish in the fight,
Oh! let us die like men.
George Washington Patten, 1808–1882, Oh! Let Us Die Like Men

There may be danger in the deed,
But there is honor too.
W. E. Aytoun, 1813–1865, The Island of the Scots

A military, or a naval man, cannot go very far astray, who abides by the point of honor.
Raphael Semmes, 1809–1877

Posts of honor are evermore posts of danger and of care.
Josiah Gilbert Holland, 1819–1881, Gold-Foil

What is life without honor? Degradation is worse than death.
Stonewall Jackson: Letter, 1862

How many sacrifice honor, a necessity, to glory, a luxury.
Joseph Roux, 1834–1886, Meditations of a Parish Priest

The nation's honor is dearer than the nation's comfort; yes, than the nation's life itself.
Woodrow Wilson: Speech, 29 January 1916

Duty, honor, country.
Motto of the U. S. Military Academy, West Point

Horrors of War

The cruelty of war makes for peace.
Publius Statius: Thebaid, vii, c. 90

War is the greatest plague that can afflict humanity; it destroys religion, it destroys states, it destroys families. Any scourge is preferable to it.
Martin Luther: Table-Talk, dcccxxi, 1569

I have seen enough of one war never to wish to see another.
Thomas Jefferson: Letter to John Adams, 1794

War is cruel to the people, and terrible to the conquered.
Napoleon I: Political Aphorisms, 1848

The tumult of each sacked and burning village;
The shout that every prayer for mercy drowns;
The soldier's revels in the midst of pillage;
The wail of famine in beleaguered towns.
Henry Wadsworth Longfellow: The Arsenal at Springfield, 1841

War and its horrors, and yet I sing and whistle.
George E. Pickett: Letter to his wife, May, 1864

[War] should be "pure and simple" as applied to the belligerents. I would keep it so, till all traces of the war are effaced; till those who appealed to it are sick and tired of it, and come to the emblem of our nation, and sue for peace. I would not coax them, or even meet them half-way, but make them so sick of war that generations would pass away before they would again appeal to it.
W. T. Sherman: Letter to Major General H. W. Halleck, 17 September 1863

If the people raise a howl against my barbarity and cruelty, I will answer that war is war, and not popularity-seeking. If they want peace, they and their relatives must stop the war.
W. T. Sherman: Letter to Major General H. W. Halleck, 4 September 1864

War is cruelty and you cannot refine it.
W. T. Sherman: Letter to the Mayor of Atlanta, 12 September 1864

No civilized war, however civilized, can be carried on on a humanitarian basis.
Brigadier General Jacob H. Smith, USA: General Order to U. S. Forces on Samar, October 1901

Towns without people, ten times took,
An' ten times left an burned at last;
An' starvin' dogs that come to look
For owners when a column passed.
Rudyard Kipling: The Return, 1903

The battlefield is fearful. One is overcome by a peculiar sour, heavy and penetrating smell of corpses. Rising over a plank bridge, you find that its middle is supported only by the body of a long dead horse. Men that were killed last October lie half in swamp and half in the yellow-sprouting beet-fields. The legs of an Englishman, still encased in puttees, stick out of a trench, the corpse being built into the parapet; a soldier hangs his rifle on them. A little brook runs through the trench, and everyone uses the water for drinking and washing; it is the only water they have. Nobody minds the pale Englishman who is rotting away a few steps further up.
Rudolph Binding: A Fatalist at War, 1915

Better a dog in time of peace than a man in time of war.
Chinese Proverb

(*See also* Atrocities, War.)

Horse Artillery

The mass was rent asunder, and Norman Ramsay burst forth sword in hand at the head of his battery, his horses breathing fire, stretched like greyhounds along the plain, the guns bounded behind them like things of no weight, and the mounted gunners followed close, with heads bent low and pointed weapons, in desperate career.
Sir William Napier: History of the War in the Peninsula, III, 1850 (this celebrated passage describes the action of Captain Norman Ramsay's troop of horse artillery at Fuentes de Oñoro, 5 May 1811).

(*See also* Artillery, Cavalry, Warhorse.)

Hostages

Seize some burgomaster of a city where you have a garrison, or the mayor of a village where you camp, and force him to take a disguised man, who speaks the language of the country, and under some pretext to conduct him as his servant in the enemy army. Threaten him that if he does not bring your man back, you will cut the throat of his wife and children whom you hold under guard while waiting, and that you will have his house burned. I was obliged to employ this sad expedient in Bohemia and it succeeded.

Frederick the Great: Instructions for His Generals, ix, 1747

Hunger

The rebellions of the belly are the worst.

Sir Francis Bacon, 1561–1626

The greatest secret of war and the masterpiece of a skillful general is to starve his enemy. Hunger exhausts men more surely than courage, and you will succeed with less risk than by fighting.

Frederick the Great: Instructions for His Generals, iii, 1747

A starving army is actually worse than none. The soldiers lose their discipline and their spirit. They plunder even in the presence of their officers.

Wellington: Letter from Spain, August 1809

When one army is full and another starving, lead and steel are hardly needed to decide the victory.

Sir John Fortescue: History of the British Army, 1899–1930

(*See also* Famine)

Hussar

The young hussar,
The whisker'd votary of waltz and war . . .

Byron, 1788–1824, The Waltz

That drunken hussar. (Cet ivrogne de hussard.)

Napoleon I, 1769–1821 (of Blücher)

. . . A resolute hussar officer who is no deep thinker . . .

Clausewitz: On War, 1832

Impedimenta

It is necessary from time to time to inspect the baggage and force the men to throw away useless gear. I have frequently done this. One can hardly imagine all the trash they carry with them year after year . . . It is no exaggeration to say that I have filled twenty wagons with rubbish which I have found in the review of a single regiment.

Maurice de Saxe: Mes Rêveries, xi, 1732

We can get along without anything but food and ammunition. The road to glory cannot be followed with much baggage.

Richard S. Ewell: Orders, Valley Campaign, 1862

If you can't carry it, eat it, or shoot it, don't bring it.

U.S. Marines' saying

Imperialism

Like all other imperial powers, we have acquired our dominion by our readiness to assist anyone, whether Barbarian or Hellene, who may have invoked our aid.

Alcibiades: Speech to the Athenians, 415 B.C.

God of our fathers, known of old,
Lord of our far-flung battle line,
Beneath whose awful hand we hold
Dominion over palm and pine—
Lord God of hosts, be with us yet,
Lest we forget—lest we forget!

Far-called, our navies melt away;
On dune and headland sinks the fire:
Lo, all our pomp of yesterday
Is one with Nineveh and Tyre!
Judge of the nations, spare us yet,
Lest we forget—lest we forget!

Rudyard Kipling: Recessional, 1897

I have not become the King's First Minister to preside over the liquidation of the British Empire.

Winston Churchill: To the House of Commons, 9 November 1942

Impressment

I have misused the king's press damnably; I have got in exchange of a hundred and fifty soldiers, three hundred and odd pounds. I press me none but good householders, yeomen's sons; inquire me out contracted bachelors, such as had been asked twice on the banns; such a commodity of warm slaves, as had as lieve hear the devil as a drum; such as fear the report of the caliver worse than a struck fowl or a hurt wild-duck.

Shakespeare: I King Henry IV, iv, 2, 1597 (caliver = a light arquebus)

Thousands of American citizens, under the safeguard of public law and of their national flag, have been torn from their country and from everything dear to them; have been dragged on board ships of war of a foreign nation and exposed, under the severities of their discipline, to be exiled to the most distant and deadly climes, to risk their lives in the battles of their oppressors, and to be the melancholy instruments of taking away those of their own brethren.

James Madison: War message to Congress, 1 June 1812

Impressed seamen call on every man to do his duty.

Legend on the battle-flag of Commodore Thomas Macdonough USN, Lake Champlain, 11 September 1814

(*See also* Press Gangs.)

Inactivity

You will not be in a Worse Situation, nor your arms in less Credit if you should meet with Misfortune than if you were to Remain Inactive.

Anthony Wayne: Letter to George Washington, November 1777

Inchon (15 September 1950)

We shall land at Inchon, and I shall crush them.

Douglas MacArthur: To the Joint Chiefs of Staff, at Tokyo, 23 August 1950

. . . the most masterly and audacious strategic stroke in all history.

W.F. Halsey, 1882–1959

Indecision

What can be more detestable than to be continually changing our minds?

Cleon of Athens: Speech to the Athenians, 427 B.C.

Whilst a field should be despatched and fought,
You are disputing of your generals.
Shakespeare: I King Henry VI, i, 1, 1591

Our doubts are traitors,
And make us lose the good we oft might win
By fearing to attempt.
Shakespeare: Measure for Measure, i, 1604,

The good old Duke of York!
He had ten thousand men;
He marched them up the hill,
And marched them down again.
Nursery rhyme, 18th century

An irresolute general, acting without rule or plan, although at the head of an army superior in number to that of the enemy, finds himself almost always inferior on the field of battle.
Napoleon I: Maxims of War, 1831

. . . the fatal tendency of top-level formations to compromise between their own convictions and those of the commander on the spot.
Basil Collier: Brasshat, xxiv, 1961

Indian Wars

Indians spurr'd on by our inveterate Enemys the French, are the only Brutes & Cowards in the Creation who were ever known to exercise their Cruelties upon the Sex, and to scalp and mangle the poor sick Soldiers & defenceless Women.
Jeffrey Lord Amherst: Order before landing at Louisbourg, 1758

He [King George III] . . . has endeavoured to bring on the inhabitants of our frontiers the merciless Indian Savages, whose known rule of warfare, is an undistinguished destruction of all ages, sexes, and conditions.
Declaration of Independence, 4 July 1776

At night we march to Moqui o'er lofty hills of snow
To meet and crush the savage foe, bold Johnny Navajo.
Johnny Navajo. O, Johnny Navajo!
We'll first chastise, then civilize bold Johnny Navajo!
Song of the California Volunteers, Navajo War, 1864

The only good Indians I ever saw were dead.
P.H. Sheridan: To Comanche chief, Toch-a-Way, Indian Territory, January 1869

We must act with vindictive earnestness against the Sioux, even to their extermination, men, women, and children. Nothing less will reach the root of the case.
W.T. Sherman, 1831–1888: Despatch to President Grant

The American Indian commands respect for his rights only so long as he inspires terror for his rifle.
George Crook, 1828–1890

Strategy, when practiced by Indians, is called treachery.
Remark by an Army officer during the Indian Wars

Indiscipline

We started with an army in the highest order, and up to the day of battle nothing could get on better; but that event has, as usual, totally annihilated all order and discipline. The soldiers of the army have got among them about a million sterling in money . . . The night of the battle, instead of being passed in getting rest and food to prepare them for the pursuit of the following day, was passed by the soldiers in looking for plunder. The consequence was, that they were incapable of marching in pursuit of the enemy, and were totally knocked up . . . I am quite convinced that we now have out of the ranks double the amount of our loss in battle; and that we have lost more men in the pursuit than the enemy have; and have never in any one day made more than an ordinary march. This is the consequence of the state of discipline of the British army. We may gain the greatest victories; but we shall do no good until we so far alter our system, as to force all ranks to perform their duty. The new regiments are, as usual, the worst of all.
Wellington: After Vitoria, 21 June 1813

Laziness of mind leads to indiscipline, just as does insubordination.

Ferdinand Foch: Precepts, 1919

(*See also* Discipline, Obedience, Mutiny.)

Industrial Mobilization

Modern war is a death grapple between peoples and economic systems, rather than a conflict of armies alone.

Bernard M. Baruch, 1870–1965

It is a war of smokestacks as well as of men.

George C. Marshall: Address to the Chamber of Commerce of the United States, 29 April 1941

The new god of war was born in the workshop: he is Mars-Mechanized.

B. H. Liddell Hart: Thoughts on War, iv, 1944

Infantry

Ah, yes, mere infantry—poor beggars . . .

Plautus: The Braggart Captain, 3d century, B.C.

Footmen threatening shot,
Shaking their swords, their spears, and iron bills,
Environing their standard round, that stood
As bristle-pointed as a thorny wood.

Christopher Marlowe, 1564–1593

Infantry is the nerve of an army.

Francis Bacon: Essays, xxix, 1625

The stubborn spear-men still made good
Their dark impenetrable wood,
Each stepping where his comrade stood,
The instant that he fell.

Walter Scott: Lay of the Last Minstrel, 1805

It all depends upon that article whether we do the business or not. Give me enough of it, and I am sure.

Wellington: On being asked by Thomas Creevey in Brussels, whether he would beat Napoleon, May 1815 (At the moment of reply, a redcoated infantry private walked by.)

Infantry is the Queen of Battles.

Attributed to Sir William Napier, 1785–1860

First there is the All-Highest [the Kaiser] then the Cavalry Officer, and then the Cavalry Officer's horse. After that there is nothing, and after nothing the Infantry Officer.

Pre-1914 Imperial German Army apothegm

The contest is always man to man, to end up with; everything in national defense is designed for that purpose and it has got to be that.

William Mitchell: Testimony before House Appropriations Committee, 1921

The Infantry, the Infantry, with dirt behind their ears,
The Infantry, the Infantry, can drink their weight in beers;
The Cavalry, the Artillery, and the God-damned Engineers
Can never beat the Infantry in a hundred thousand years.

American infantry song, early 20th century

Infantry is the arm which in the end wins battles. The rifle and the bayonet are the infantryman's chief weapons. The battle can be won in the last resort only by these weapons.

British Army Field Service Regulations, 1924

I wouldn't give a bean to be a fancy-pants Marine,
I'd rather be a dogface soldier like I am:
I wouldn't trade my old ODs for all the Navy's dungarees
For I'm the walking pride of Uncle Sam.

American infantry song, c. 1925

One well-known Brigadier always phrases his requirements of the ideal infantryman as "athlete, stalker, marksman." I always feel a little inclined to put it on a lower plane and to say that the qualities of a successful poacher, cat burglar, and gunman would content me.

Sir Archibald Wavell, "The Training of the Army for War," February 1933

When the smoke cleared away, it was the man with the sword, or the crossbow, or the rifle, who settled the final issue on the field.

George C. Marshall: Talk before the National Rifle Association 3 February 1939

Franklin D. Roosevelt Library

WINSTON CHURCHILL

1874–1965

"We shall fight on for ever and ever and ever."

The poor devil in the Army marches tremendous distances; he is in the mud; he's filthy dirty; he hasn't had a full meal. . . and he fights in a place he's never seen before.

George C. Marshall, 1880–1959

Look at an infantryman's eyes and you can tell how much war he has seen.

William H. Mauldin: Up Front, 1944

(*See also* Rifleman, Soldier.)

Infiltration

If the brains and the arteries of an army are destroyed then that army must either surrender or be defeated in detail later. In other words, is it really necessary to have a battle at all in order to overcome one's enemy?

Lieutenant H.E. Fox-Davies, Durham Light Infantry: Letter to Sir A.P. Wavell, 1 October 1935

(*See also* Guerrilla Warfare, Partisan Warfare.)

Information and Education

It is not set speeches at the moment of battle that render soldiers brave. The veteran scarcely listens to them, and the recruit forgets them at the first shot. If discourses and harangues are useful at all, it is during the campaign; to do away with unfavorable impressions, to correct false reports, to keep alive a proper spirit in the camp, and to furnish materials and amusement for the bivouac. All printed orders of the day should keep these objects in view.

Napoleon I: Maxims of War, 1831

Our soldiers ought to be instructed about the course of each war. It is only when aggression is legitimate that one can expect prodigies of valor.

Marshal Ney: 1769–1815

I feel some anxiety regarding the scheme conducted by the new Army Bureau of Current Affairs. The test must be whether discussions of such matters conducted by regimental officers will weaken or strengthen that tempered discipline without which our armies can be no match for the highly trained forces of Germany. The qualities required for conducting discussions of the nature indicated are not necessarily those which fit for command in the field. Will not such discussions only provide opportunities for the professional grouser and agitator with a glib tongue?

Winston Churchill: Note for the Secretary of State for War, 10 October 1941

Initiative

The first blow is half the battle.

Oliver Goldsmith: She Stoops to Conquer, 1773

To do nothing was disgraceful; therefore I made use of my understanding.

Nelson: From Egypt after the battle of the Nile, August 1798

A general who stands motionless to receive his enemy, keeping strictly on the defensive, may fight ever so bravely, but he must give way when properly attacked.

Jomini: Précis de l'Art de la Guerre, 1838

Twice is he armed that hath his quarrel
just,
And three times he who gets his fist in fust.

Artemus Ward (Charles Farrar Browne), 1834-1867, Shakespeare Up-To-Date

Uncomfortable fellows who press for progress have a bad time of it in London.

Sir Ian Hamilton: The Soul and Body of an Army, viii, 1921

All sensible initiative of subordinates must be encouraged through all possible means, and must be exploited by the commander in the general interest of [success in] battle. Sensible initiative is based upon an understanding of the commander's intentions.

Field Service Regulations, Army of the USSR, 1936

No military leader is endowed by heaven with an ability to seize the initiative. It is the intelligent leader who does so after a careful study and estimate of the situation and arrangement of the military and political factors involved.

Mao Tse-tung: On Guerrilla War, 1937

Some officers require urging, others require suggestions, very few have to be restrained.

George S. Patton, Jr.: War As I Knew It, 1947

When so much was uncertain, the need to recover the initiative glared forth.

Winston Churchill: Their Finest Hour, 1949 (on the situation after Dunkirk, 1940)

Initiative means freedom to act, but it does not mean freedom to act in an offhand or casual manner. It does not mean freedom to . . . depart *unnecessarily* from standard procedures or practices or instructions.

Ernest J. King: A Naval Record, 1952

. . . the supreme military weapon of initiative.

David Holden: in Manchester Guardian, 14 May 1964

Inspection

When you are ordered to visit the barracks, I would recommend it to you to confine your inspection to the outside walls: for what can be more unreasonable to expect, that you should enter the soldier's dirty rooms, and contaminate yourself with tasting their messes? As you are not used to eat salt pork or ammunition bread, it is impossible for you to judge whether they are good or not. Act in the same manner, when you are ordered to visit the hospital.

Francis Grose: Advice to the Officers of the British Army, 1782

Insubordination

To resist him that is set in authority is evil.

Instruction of Ptahhotep, c. 2675 B.C.

You know, Foley, I have only one eye; I have a right to be blind sometimes—I really do not see the signal.

Nelson: To his Flag Captain at Copenhagen, after having had his attention invited to Sir Hyde Parker's signal, "Discontinue engagement," 2 April 1801

Insubordination may only be the evidence of a strong mind.

Napoleon I: Letter from St. Helena, 1817

. . . The splendid idiosyncrasy of insubordination.

Sir John Fisher: Memories, 1919

Insurgency

I bear orders from the Captain, "Get ye
ready, quick and soon,
With your pike upon your shoulder at the
Rising of the Moon."

The Rising of the Moon, Irish rebel song, 19th century

That rake up in the rafters, mother, why
hangs it there so long?
Its handle of the best of ash is smooth, and
straight and strong.
And, mother, will you tell me? Why did
my father frown,
When to make hay in the summertime, I
climbed to take it down?
He swung his first-born in the air, while
his eyes with light did fill,
"You'll shortly know the reason why,"
said Rory of the Hill.

Rory of the Hill, Fenian song, Irish Rebellion, 19th Century

. . . The primary war strategy of a people seeking to emancipate itself.

Chang Tso Hua, c. 1937 (quoted by Mao Tse-tung in On Guerrilla War)

The aim is not simply the expansion of frontiers and the acquisition of new territory, but at the same time the extension of the totalitarian revolutionary movement into other countries. All this is virtually the transfer (to the international sphere) of the modern technique of the coup d'etat.

Adolf Hitler: To Hermann Rauschning, 1939

Liberation wars will continue to exist as long as imperialism exists . . . Such wars are not only admissible, but inevitable . . . We recognize such wars, and we will help the peoples striving for their independence. The Communists fully support such wars and march in the front rank with peoples waging liberation struggles.

Nikita Khruschev: Speech, 1961

The People's army is the instrument of the Party and of the Revolutionary State for the accomplishment, in armed form, of the tasks of the revolution.

Vo-Nguyên-Giap: People's War—People's Army, 1962

(*See also* Counterinsurgency, Guerrilla, Guerrilla Warfare, Insurrection, Partisan Warfare.)

Insurrection

. . . the ugly form of base and bloody insurrection.
Shakespeare: II King Henry IV, i, 1, 1597

In a national insurrection the center of gravity to be destroyed lies in the person of the chief leader and in public opinion; against these points the blow must be directed.
Clausewitz: On War, 1832

Insurrection is an art as much as war . . . and subject to certain rules . . . First, never play with insurrection unless you are fully prepared to face the consequences of your play . . . Second, act with the greatest determination and on the offensive. The defensive is the death of every armed rising . . . Surprise your antagonist . . . Keep up the moral ascendancy which the first successful rising has give you.
V.I. Lenin: Letters from Finland To the Bolsheviks in Petrograd, 1917

Intelligence

If I am able to determine the enemy's dispositions while at the same time I conceal my own, then I can concentrate and he must divide.
Sun Tzu, 400–320 B.C., The Art of War

Therefore, determine the enemy's plans and you will know which strategy will be successful and which will not.
Sun Tzu, 400–320 B.C., The Art of War, vi

War, as the saying goes, is full of false alarms.
Aristotle: Nichomachean Ethic, iii, c. 340 B.C.

It is essential to know the character of the enemy and of their principal officers—whether they be rash or cautious, enterprising or timid, whether they fight on principle or from chance.
Vegetius: De Re Militari, iii, 378

Before the army is dispatched, calculations are made respecting the degree of difficulty of the enemy's land; the directness and deviousness of its roads; the number of his troops; the quantity of his war equipment and the state of his morale. Calculations are made to see if the enemy can be attacked and only after this is the populace mobilized and troops raised.
Ho Yen-Hsi, c. 1000

Nothing is more worthy of the attention of a good general than the endeavor to penetrate the designs of the enemy.
Niccolo Machiavelli: Discorsi, xviii, 1531

Intelligence is the Soul of all Publick business.
Daniel Defoe: To Robert Harley, 1704

One should know one's enemies, their alliances, their resources and nature of their country, in order to plan a campaign. One should know what to expect of one's friends, what resources one has, and foresee the future effects to determine what one has to fear or hope from political maneuvers.
Frederick The Great: Instructions for His Generals, 1747

Knowledge of the country is to a general what a musket is to an infantryman and what the rules of arithmetic are to a geometrician. If he does not know the country he will do nothing but make gross mistakes . . . Therefore study the country where you are going to act.
Frederick The Great: Instructions for His Generals, iv, 1747

The necessity of procuring good Intelligence is apparent & need not be further urged. All that remains for me to add, is, that you keep all the whole matter as secret as possible. For upon Secrecy, Success depends in most Enterprizes of the kind, and for want of it, they are generally defeated, however well planned & promising a favourable issue.
George Washington: Letter to Colonel Elias Dayton, 26 Jule 1777

Great part of the information obtained in war is contradictory, a still greater part is false, and by far the greatest part is of a doubtful character.
Clausewitz: On War, 1832

Nothing should be neglected in acquiring a knowledge of the geography and military statistics of other states, so as to know their material and moral capacity for attack and defense as well as the strategic advantages of the two parties. Distinquished officers should be employed in these scientific labors and should be rewarded when they demonstrate marked ability.

Jomini: Précis de l'Art de la Guerre, 1838

How can any man say what he should do himself if he is ignorant what his adversary is about?

Jomini: Précis de l'Art de la Guerre, 1838

I have been passing my life in guessing what I might meet with beyond the next hill, or round the next corner.

Wellington: To J.W. Croker, 1845 (There are many versions of this celebrated remark, all differing in detail; conceivably Wellington repeated himself on different occasions.)

The unknown is the governing condition of war.

Ferdinand Foch: Principles of War, 1920

When I took a decision, or adopted an alternative, it was after studying every relevant—and many an irrelevant—factor. Geography, tribal structure, religion, social customs, language, appetites, standards—all were at my finger-ends. The enemy I knew almost like my own side. I risked myself among them a hundred times, to *learn*.

T.E. Lawrence: Letter to Liddell Hart, 26 June 1933

The great thing is to get the true picture, whatever it is.

Winston Churchill: Note to Chief of the Imperial General Staff, 24 November 1940

In a battle nothing is ever as good or as bad as the first reports of excited men would have it.

Sir William Slim: Unofficial History, vi, 1959

To lack intelligence is to be in the ring blindfolded.

General D. M. Shoup, USMC: Remarks to the staff, Marine Corps Headquarters, 2 January 1960

Intelligence is the decisive factor in planning guerrilla operations . . . Because of superior information, guerrillas always engage under conditions of their own choosing; because of superior knowledge of terrain, they are able to use it to their advantage and the enemy's discomfiture.

Brigadier General S. B. Griffith, USMC: Introduction to Mao Tse-tung on Guerrilla Warfare, 1961

Nothing helps a fighting force more than correct information. Moreover it should be in perfect order, and done well by capable personnel.

Chê Guevara: Memorandum, 1963

(*See also* Enemy Capabilities, Espionage, Reconnaissance.)

Interior Lines

The unquestionable advantages of the interior line of operations are valid only as long as you retain enough space to advance against one enemy . . . gaining time to beat and pursue him, and then to turn against the other . . . If this space, however, is narrowed down to the extent that you cannot attack one enemy without running the risk of meeting the other who attacks you from the flank or rear, then the strategic advantage of interior lines turns into the tactical disadvantage of encirclement.

Helmuth von Moltke ("The Elder"), 1800–1891

Interior lines are lines shorter in time than those the enemy can use.

Mahan: Naval Strategy, 1911

Interior lines at night are the general's delight.

Sir A.P. Wavell: Lecture at Aldershot, 1930

Intervention

The ideal intervention is smartly executed.

John Paton Davies: In New York Times, 23 May 1965

Invasion

Thus far into the bowels of the land
Have we marched on without impediment.

Shakespeare: King Richard II, v, 2, 1592

It is always more difficult to defend a coast than to invade it.
Sir Walter Raleigh: Historie of the World, 1615

France is invaded; I go to put myself at the head of my troops.
Napoleon I: At Paris, 23 January 1814

Remoteness is not a certain safeguard against invasion.
Jomini: Précis de l'Art de la Guerre, 1838

Our first line of defense against invasion must be as ever the enemy's ports.
Winston Churchill: Minute to Chiefs of Staff Committee, 5 August 1940

We are waiting for the long promised invasion. So are the fishes.
Winston Churchill: Broadcast to the French people, 21 October 1940

The pages of history are strewn with the wreckage of empires which collapsed as a result of an unsuccessful attempt at invasion. The British Empire was almost wrecked on the shores of Gallipoli. Napoleon's empire never recovered from the failure of his attempts to invade England. The Spanish empire failed with the wrecking of the Great Armada.
Major General D.A.D. Ogden, USA, 1897–

(*See also* Amphibious Operations, Amphibious Warfare, Beach.)

Irish Rebellion (19th century)

Side by side for the cause when our forefathers battled,
When our hills never echoed the tread of the slave,
On many a green field where the leaden hail rattled,
Through the red gap of glory, they marched to the grave.
But we who inherit their name and their spirit,
March 'neath the banner of liberty then,
Give them back blow for blow,
Pay them back woe for woe,
Out and make way for the Bold Fenian Men!
The Bold Fenian Men, Irish rebel song, 19th century

Iron

Ay me! what perils do environ
That man that meddles with cold iròn!
Samuel Butler: Hudibras, 1663

Iron sleet of arrowy shower
Hurtles in the darken'd air.
Thomas Gray, 1716–1771, The Fatal Sisters

"Gold is for the mistress—silver for the maid—
Copper for the craftsman, cunning at his trade."
"Good!" said the Baron, sitting in his hall,
"But Iron—Cold Iron—is master of them all."
Rudyard Kipling: Cold Iron, 1909

Iron Curtain

From Stettin in the Baltic to Trieste in the Adriatic, an iron curtain has descended across the continent.
Winston Churchill: Speech at Westminster College, Fulton, Missouri, 1946

Irregulars

The virtue of irregulars lay in depth, not in force.
T. E. Lawrence, 1888–1935

Iwo Jima, (February-March 1945)

Among the men who fought on Iwo Jima, uncommon valor was a common virtue.
C. W. Nimitz: Pacific Fleet communique, 16 March 1945

Jackson, Thomas Jonathan (Stonewall Jackson) (1824–1863)

There is Jackson, standing like a stone wall. Let us determine to die here, and we will conquer.
Brigadier General Barnard E. Bee, CSA: To his brigade at First Bull Run, 21 July 1861 (This was the origin of Jackson's soubriquet, "Stonewall.")

I never saw one of Jackson's couriers approach, without expecting an order to assault the North Pole.
Richard S. Ewell: During the Valley Campaign, 1862

Such an executive officer the sun never shone on. I have but to show him my design, and I know that if it can be done it will be done. No need for me to send or watch him. Straight as a needle to the pole he advances to the execution of my purpose.
R. E. Lee: Of Jackson, 1863

You are better off than I am, for while you have lost your *left*, I have lost my *right* arm.
R. E. Lee: Note to Jackson, when mortally wounded after Chancellorsville, 4 May 1863

Let us cross over the river and rest under the shade of the trees.
Jackson's last words, 10 May 1863

He was a gallant soldier, and a Christian gentleman.
U.S. Grant: Remark in 1864, when, during the Wilderness campaign, he spent the night in the house where Jackson died.

Stonewall Jackson, wrapped in his beard and his silence.
Stephen Vincent Benet: John Brown's Body, iv, 1928

Jervis, Sir John (Lord St. Vincent) (1735–1823)

My old oak.
George IV, 1762–1830 (the King's habitual phrase in referring to Lord St. Vincent)

Where I would take a penknife Lord St. Vincent takes a hatchet.
Nelson, 1758–1805

Jingo

We don't want to fight, but, by jingo, if
we do,
We've got the ships, we've got the men,
we've got the money, too
G. W. Hunt, fl 1878 (British music-hall song incident to the Russo-Turkish War)

(*See also* Militarism.)

Joint Chiefs of Staff

Chiefs, who no more in bloody fights
engage,
But wise through time...
Homer: Iliad, iii, c. 1000 B. C.

It is a matter of record that the strategic direction of the war, as conducted by the Joint Chiefs of Staff, was fully as successful as were the operations which they directed . . . The proposals or the convictions of no one member were as sound, or as promising of success, as the united judgments and agreed decisions of all the members.
Ernest J. King: The U. S. Navy at War, 1945

We have devised in the Joint Chiefs of Staff really one of the great military mechanisms of all time...It is an arrangement whereby those who make the plans are clearly responsible for carrying them out, and I do not want to see an arrangement whereby an officer has the power to make plans but is not accountable for their execution.
Ferdinand Eberstadt: Testimony to Congress, 1953

The Joint Chiefs of Staff system is unique among all known systems for the strategic direction of a war. It has brought victory where other systems have failed. It has withstood the all important test of war.
Arleigh Burke: Speech in Minneapolis, 6 October 1956

(*See also* Defense Organization, High Command)

Joint Planning

I am increasingly impressed with the disadvantages of the present system of having Naval, Army, and Air Force officers equally represented at all points and on all combined subjects, whether in committees or in commands. This has resulted in a paralysis of the offensive spirit.

Winston Churchill: Note for the Chiefs of Staff Committee, 2 March 1942

Decisions which determine the success or failure of the strategic direction of global war have to be determined by the meeting of a number of minds, each of which contributes its own specialized knowledge, while also serving as a balance and a check on the others.

James Forrestal: Testimony, Senate Naval Affairs Committee, 1 May 1946

Jomini, Antoine Henri (1799–1869)

Jomini was not merely a military theorist, who saw war from the outside; he was a distinguished and thoughtful soldier.

Mahan: Naval Administration and Warfare, 1908

John Paul Jones (1747–1792)

I have not yet begun to fight.

Reply when asked if he had struck, battle off Flamborough Head, between USS Bonhomme Richard *and HMS* Serapis, *25 September 1779*

The English nation may hate me, but I will force them to esteem me, too.

Letter by Jones, 1780

Everybody works but John Paul Jones
He lies around all day,
Body pickled in alcohol
On a permanent jag, they say.
Middies stand around him
Doing honor to his bones;
Everybody works in Crabtown
But John Paul Jones!

Midshipmen's ditty, U. S. Naval Academy, to the tune, "Everybody Works but Father," after reinterment of Jones's remains in the Academy Chapel ("Crabtown"= Annapolis)

(*See also* Pearson.)

Judgment

If I am to be hanged for it, I cannot accuse a man who I believe has meant well, and whose error was one of judgment and not of intention.

Wellington: Letter to Wellesley Pole, 31 July 1809 (on the subject of General Craufurd's handling of the Light Division on the River Coa, 24 July 1809)

That's enough for any man. Keep a cool head.

Joseph Conrad: Typhoon, 1903

The greatest commander is he whose intuitions most nearly happen.

T. E. Lawrence, 1888–1935, The Science of Guerrilla Warfare

Jungle Warfare

The vulnerable artery is the line of communications winding through the jungle. Have no L. of C. on the jungle floor. Bring in the goods like Father Christmas, down the chimney.

Orde Wingate: Recommendations for wholly air-supplied operations, Myitkyina area, Burma, 1943

Nothing is easier in jungle or dispersed fighting than for a man to shirk. If he has no stomach for advancing, all he has to do is to flop into the undergrowth; in retreat he can slink out of the rearguard, join up later, and swear he was the last to leave. A patrol leader can take his men a mile into the jungle, hide there, and return with any report he fancies.

Sir William Slim: Defeat into Victory, 1956

Kearny, Philip (1815–1862)

Lay him low, lay him low,
In the clover or the snow!
What cares he? he cannot know.
George Henry Boker: Dirge for a Soldier (in memory of General Kearny, killed in action at Chantilly, Virginia, 1 September 1862)

Kempenfelt, Richard (1718–1782)

Toll for the brave—
Brave Kempenfelt is gone,
His last sea-fight is fought,
His work of glory done.
William Cowper: The Loss of the Royal George, 1782

King, Ernest Joseph (1878–1956)

I never had no trouble with him. He was always a good man to a good man.
Characterization of Admiral King by a chief petty officer who had often served with him.

Kitchener, Horatio Herbert (1850–1916)

Lord K. is playing hell with its lid off at the War Office—what the papers call "standing no nonsense," but which often means "listening to no sense."
Lady Jean Hamilton: Diary entry, 12 August 1914

Knight

Beau chevalier, qui partez pour la guerre,
Qu'allez-vous faire
Si loin d'ici?
Chanson de Barberine, date unknown

He was a verray parfit gentil knight.
Geoffrey Chaucer: Canterbury Tales, c. 1387

Men in armor clad,
Upon their prancing steeds disdainfully,
With wanton paces trampling on the ground.
Christopher Marlowe, 1564–1593

The lives of knight-errants are subject to a thousand hazards and misfortunes; but on the other side, they may at any time become kings and emperors.
Cervantes: Don Quixote, 1605

The knights neglected to live but were prepared to die, in the service of Christ.
Edward Gibbon: Decline and Fall of the Roman Empire, 1776

(*See also* Chivalry.)

Knox, Henry (1750–1806)

The resources of his genius supplied the deficit of means.
George Washington: Report to the Continental Congress after the fall of Yorktown, 1781

Korean War (1950–1953)

Here we fight Europe's war with arms while the diplomats there still fight it with words.
Douglas MacArthur: Letter to Representative Joseph Martin, 6 April 1951

... The wrong war, at the wrong place, at the wrong time, and with the wrong enemy.
Omar N. Bradley: Testimony to Congress regarding the Korean War, 15 May 1951

The Korean War began in a way in which wars often begin. A potential aggressor miscalculated.
John Foster Dulles: Speech at St. Louis, 2 September 1953

This was the toughest decision I had to make as President.
Harry S. Truman: Memoirs, 1956 (on his decision to intervene in Korea)

It was very easy to start a war in Korea. It was not so easy to stop it.
Nikita Khruschev, 1894–, speech before the Bulgarian Party leadership

(*See also* Inchon.)

Lafayette, Marquis de (1757–1834)

Lafayette, we are here.
Colonel C.E. Stanton, USA: At the grave of Lafayette, Picpus Cemetery, Paris, 4 July 1917

Lake Champlain, (11 September 1814)

The Almighty has been pleased to grant us a signal victory on Lake Champlain, in the capture of one frigate, one brig, and two sloops of war of the enemy.
Thomas MacDonough: Despatch to the Secretary of the Navy, 11 September 1814

This is a proud day for America—the proudest day she ever saw.
Remark by an unidentified British officer prisoner after the victory on Lake Champlain, 11 September 1814

Lake Erie, (10 September 1813)

We have met the enemy, and they are ours: Two ships, two brigs, one schooner and one sloop.
Oliver Hazard Perry: Despatch to General W.H. Harrison 10 September 1813

See! he quits the *Lawrence's* side,
And trusts him to the foaming tide,
While thundering navies round him ride,
And flash their red artillery.
Verse, author unknown, inspired by Commodore Perry's transfer of his flag from USS Lawrence *to USS* Niagara *at the height of the battle of Lake Erie*

Land Warfare

The art of land warfare is an art of genius, of inspiration. On the sea nothing is genius or inspiration; everything is positive or empiric. The admiral needs only one science, that of navigation. The general needs all the sciences.
Napoleon I: Maxims of War, 1831

War on land today is land-air warfare
Sir John Slessor: Strategy for the West, 1954

Landing Craft

Picture puzzles are child's play compared with this game of working an unheard-of number of craft to and fro, in and out, of little bits of beaches.
Sir Ian Hamilton: Gallipoli Diary, I, 1920

Prior to the present war I never heard of any landing craft except a rubber boat. Now I think about little else.
Attributed to George C. Marshall, late 1943

The whole of this difficult question, how to divide military resources between the Normandy invasion and the invasion of Southern France, only arises out of the absurd shortage of the LSTs. How it is that the plans of two great empires like Britain and the United States should be so much hamstrung and limited by a hundred or so of these particular vessels will never be understood by history.
Winston Churchill: Letter to General Marshall, 1944

(*See also* Amphibious Operations, Amphibious Warfare.)

Last Stand

When men find they must inevitably perish, they willingly resolve to die with their comrades and with their arms in their hands.
Vegetius: De Re Militari, iii, 378

There is one certain means by which I can be sure never to see my country's ruin: I will die in the last ditch.
William III (William of Orange), 1650–1702

(*See also* Resistance, Will to Fight.)

Law

The laws are silent in the midst of arms. (Silent leges inter arma.)
Cicero: Pro Milano, c. 50 B.C.

No Connections, Interests, or Intercessions . . . will avail to prevent strict execution of justice.
George Washington: General Order, 7 July 1775

Brown Brothers

JAMES WOLFE

1727–1759

"Do they run already? Then I die happy!"

An army is a collection of armed men obliged to obey one man. Every change in the rules which impairs the principle weakens the army.

W.T. Sherman, 1820–1891

[Soldiers] must for the sake of public freedom, in the midst of public freedom, be placed under a despotic rule . . . must be subject to a sharper penal code and to a more stringent code of procedure than are administered by the ordinary tribunals.

T.B. Macaulay, 1800–1859

One of the great defects in our military establishment is the giving of weak sentences for military offenses. The purpose of military law is administrative rather than legal . . . In justice to other men, soldiers who go to sleep on post, who go absent for an unreasonable time during combat, who shirk battle, should be executed; and Army or Corps commanders should have the authority to approve the death sentence. It is utterly stupid to say that General Officers, as a result of whose orders thousands of gallant and brave men have been killed, are not capable of knowing how to remove the life of one poltroon.

George S. Patton, Jr.: War As I Knew It, 1947

(*See also* Court Martial, Law Specialists, Laws of War.)

Law Specialists

They have no lawyers among them for they consider them as a sort of people whose profession it is to disguise matters.

Thomas More: Utopia (Of Law and Magistrates), 1516

They have been grand-jurymen since before Noah was a sailor.

Shakespeare: Twelfth Night, ii, 4, 1599

They say soldiers and lawyers could never thrive both together in one shire.

Barnabe Rich: The Anothomy of Ireland, 1615

The presence of one of our regular civilian judge-advocates in an army in the field would be a first-class nuisance.

W.T. Sherman: Memoirs, II, 1875

(*See also* Court Martial, Law, Uniform Code of Military Justice.)

Lawrence, James (1781–1813)

And the fellow died as well as he lived, but it is part of a sailor's life to die well. He had no talk, but he inspired all about him with ardor; he always saw the best thing to be done; he knew the best way to do it; and he had no more dodge in him than the mainmast.

Stephen Decatur: Of Lawrence after his death in the action between USS Chesapeake *and HMS* Shannon, 1813

Lawrence, T.E. (1888–1935)

Nelson slightly cracked after his whack on the head after the battle of the Nile, coming home and insisting on being placed at the tiller of a canal barge and on being treated as nobody in particular, would have embarrassed the Navy far less. A callow and terrified Marbot, placed in command of a sardonic Napoleon at Austerlitz and Jena, would have felt much as your superiors must in command of Lawrence the great.

G.B. Shaw: Letter to Lawrence, 17 December 1922

. . . I drew these tides of men into my hands
And wrote my will across the sky in stars.

Lawrence's dedicatory verses to Seven Pillars of Wisdom, 1926

Laws of War

War gives the right to the conquerors to impose any conditions they please upon the vanquished. (Jus belli, ut qui vicissent, iis quos vicissent, quemadmodum vellent, imperarent.)

Julius Caesar: De Bello Gallico, c. 51 B.C.

It is lawful for Christian men, at the commandment of the Magistrate, to wear weapons, and serve in the wars.

Book of Common Prayer (Articles of Religion, xxxvii),

Extortions which are intolerable in their nature become excusable from the necessities of war.

Richelieu: Letter during Thirty Years' War, 1633

We ought to spare those edifices which do honor to human society, and do not contribute to the enemy's strength, such as temples, tombs, public buildings, and all works of remarkable beauty.

Emerich De Vattel: Le Droit des Gens, 1758

No treaty of peace shall be esteemed valid in which is tacitly reserved matter for future war.

Immanuel Kant: Perpetual Peace, i, 1795

A state shall not, during war, admit of hostilities of a nature that would render reciprocal confidence in a succeeding peace impossible: such as employing assassins, poisoners, violation of capitulations, secret instigations to rebellion, etc.

Immanuel Kant: Perpetual Peace, i, 1795

We wage no war with women nor with priests.

Robert Southey: Madoc in Wales, xv, 1805

A war between the governments of two nations is a war between all the individuals of the one and all the individuals of . . . the other.

James Kent: Commentaries on American Law, i, 1826

Men who take up arms against one another in public do not cease on this account to be moral beings, responsible to one another and to God.

U.S. Army General Order No. 100, 1863

Scarcely a day passes in a modern campaign that does not give the lie to the rules laid down in the ponderous tomes of the international law-writers. It is said that Gustavus Adolphus always had with him in camp a copy of Grotius, as Alexander is said to have slept over Homer. The improbability of finding a copy of Grotius in a modern camp may be taken as an illustration of the neglect that has long since fallen on the restraints with which our publicists have sought to fetter our generals, and of the futility of all such endeavors.

James A. Farrer: Military Manners and Customs, 1885

. . . *self*-preservation—the first law of states even more than of men; for no government is empowered to assent to that last sacrifice, which the individual may make for the noblest motives.

Mahan: The Influence of Sea Power Upon History, II, 1892

Absolute good faith with the enemy must be preserved as a rule of conduct. Without it war will degenerate into excesses and violences, ending only in the total destruction of one or both of the belligerents.

Rules of Land Warfare, U.S. Army, 1914

The first law of war is to preserve ourselves and destroy the enemy.

Mao Tse-tung: On Guerrilla Warfare, 1937

Leadership

Respect yourself and others will respect you.

Confucius, 551–478 B.C., Analects

The superior man is firm in the right way, and not merely firm.

Confucius, 551–478 B.C., Analects

The leader must himself believe that willing obedience always beats forced obedience, and that he can get this only by really knowing what should be done. Thus he can secure obedience from his men because he can convince them that he knows best, precisely as a good doctor makes his patients obey him. Also he must be ready to suffer more hardships than he asks of his soldiers, more fatigue, greater extremes of heat and cold.

Xenophon, 430–350 B.C., Cyropaedia

An army of deer led by a lion is more to be feared than an army of lions led by a deer.

Attributed to Chabrias, 410–375 B.C. (also attributed to Philip of Macedon)

When troops flee, are insubordinate, distressed, collapse in disorder, or are routed, it is the fault of the general. None of these disorders can be attributed to natural causes.

Sun Tzu, 400–320 B.C., The Art of War, x

Because [a good] general regards his men as infants they will march with him into the deepest valleys. He treats them as his own beloved sons and they will die with him.

Sun Tzu, 400–320 B.C., The Art of War, x

He that ruleth over men must be just.
II Samuel, XXIII

Everyone is bound to bear patiently the results of his own example.
Phaedrus: Fabulae Aesopiae, i, 26, c. 8 A.D.

The value of a whole army—a mighty host of a million men—depends on the man alone: such is the influence of spirit.
Tsao Kwei, c. 10 A.D.

Be swift to hear, slow to speak, slow to wrath.
James, I, 19

. . . reason and calm judgment, the qualities specially belonging to a leader.
Tacitus, 55–117 A.D., History, iii

The wise man, before he speaks, will consider well what he speaks, to whom he speaks, and where and when.
St. Ambrose, 340–397, De Officiis Ministrorum, i, 10

You may pardon much to others, nothing to yourself.
Ausonius, fl. 4th century A.D., Epigrams

When the general lays on unnecessary projects, everyone is fatigued.
Ch'ên Hao, fl. 700

The responsibility for a martial host of a million lies in one man. He is the trigger of its spirit.
Ho Yen-Hsi, fl. 1000

When one treats people with benevolence, justice, and righteousness, and reposes confidence in them, the army will be united in mind and all will be happy to serve their leaders.
Chang Yu, fl. 1000

Be just and fear not. Let all the ends thou aim'st at be they country's, thy god's and truth's.
Shakespeare: King Henry VIII, iii, 2, 1612

I shall desire all and every officer to endeavor by love and affable carriage to command his souldiers, since what is done for fear is done unwillingly, and what is unwillingly attempted can never prosper.
Earl of Essex: At Worcester, 24 September 1642

Pay well, command well, hang well.
Sir Ralph Hopton: Maxims for the Management of an Army, 1643

If you choose godly, honest men to be captains of Horse, honest men will follow them.
Oliver Cromwell: Letter to Sir William Springe, September 1643

One can be exact and just, and be loved at the same time as feared. Severity must be accompanied by kindness, but this should not have the appearance of pretense, but of goodness.
Maurice de Saxe: Mes Rêveries, xviii, 1732

The commander should practice kindness and severity, should appear friendly to the soldiers, speak to them on the march, visit them while they are cooking, ask them if they are well cared for, and alleviate their needs if they have any. Officers without experience in war should be treated kindly. Their good actions should be praised. Small requests should be granted and they should not be treated in an overbearing manner, but severity is maintained about everything regarding duty. The negligent officer is punished; the man who answers back is made to feel your severity by being reprimanded . . . pillaging or argumentative soldiers, or those whose obedience is not immediate, should be punished.
Frederick The Great: Instructions for His Generals, viii, 1747

Miseries seem light to a soldier if the chief who imposes hardships on him also volunteers to share them.
Comte De Ségur, 1689–1751, Memoirs

Self-confidence is the first requisite to great undertakings.
Samuel Johnson: Lives of the Poets (Pope), 1779

French officers will always lead, if the soldiers will follow: and English soldiers will always follow, if their officers will lead.
Samuel Johnson: On the Bravery of the English Common Soldiers, c. 1760

Never to repent, and never to reproach others, these are the first steps to wisdom.
Diderot: Pensées Philosophiques, 1746

Noblesse oblige.
Gaston de Levis, 1720–1787, Maxims, 73

Hereafter, if you should observe an occasion to give your officers and friends a little more praise than is their due, and confess more fault than you can justly be charged with, you will only become the sooner for it, a great captain. Criticizing and censuring almost everyone you have to do with, will diminish friends, increase enemies, and thereby hurt your affairs.
Benjamin Franklin: Letter to John Paul Jones, 1780

Ignorance of your profession is best concealed by solemnity and silence, which pass for profound knowledge upon the generality of mankind. A proper attention to these, together with extreme severity, particularly in trifles, will soon procure you the character of a good officer.
Francis Grose: Advice to the Officers of the British Army, 1782

Impossible is a word I never utter.
Colin D'Harleville: Malice for Malice, 1793

Nothing gives one person so much advantage over another as to remain always cool and unruffled under all circumstances.
Thomas Jefferson, 1743–1826

That quality which I wish to see the officers possess, who are at the head of the troops, is a cool, discriminating judgment when in action, which will enable them to decide with promptitude how far they can go and ought to go, with propriety; and to convey their orders, and act with such vigor and decision, that the soldiers will look up to them with confidence in the moment of action, and obey them with alacrity.
Wellington: General Order, 15 May 1811

Correction does much, but encouragement does more. Encouragement after censure is as the sun after a shower.
Goethe, 1749–1832

To the timid and hesitating, everything is impossible because it seems so.
Walter Scott, 1771–1832, Rob Roy

The more a leader is in the habit of demanding from his men, the surer he will be that his demands will be answered.
Clausewitz: On War, 1832

As the forces in one individual after another become prostrated, and can no longer be excited and supported by an effort of his own will, the whole inertia of the mass gradually rests its weight on the will of the commander: by the spark in his breast, by the light of his spirit, the spark of purpose, the light of hope, must be kindled afresh in others.
Clausewitz: On War, 1832

Every means should be taken to attach the soldier to his colors. This is best accomplished by showing consideration and respect to the old soldier.
Napoleon I: Maxims of War, 1831

At the head of an army, nothing is more becoming than simplicity.
Napoleon I: Political Aphorisms, 1848

Victory and disaster establish indestructible bonds between armies and their commanders.
Napoleon I: Political Aphorisms, 1848

There are no bad regiments; there are only bad colonels.
Attributed to Napoleon I, 1769–1821

Private feelings must always be sacrified for the public service.
Frederick Marryat: Peter Simple, 1834

He never errs who sacrifices self.
Bulwer-Lytton, 1803–1873, New Timon, iv, 3

First find the man in yourself if you will inspire manliness in others.
Bronson Alcott, 1799–1888, Table Talk

Trust men, and they will be true to you.
R.W. Emerson, 1803–1882, Essays (Prudence)

This world belongs to the energetic.
R.W. Emerson, 1803–1882

The man who trusts men will make fewer mistakes than he who distrusts them.
Conde Di Cavour, 1810–1861

STAFF OFFICER: General, the day is going against us!
STONEWALL JACKSON: If you think so, Sir, you had better not say anything about it!
At First Bull Run, 21 July 1861

Our army would be invincible if it could be properly organized and officered. There never were such men in an army before. They will go anywhere and do anything if properly led. But there is the difficulty—proper commanders . . .
R.E. Lee: To Stonewall Jackson, 1862

No matter what may be the ability of the officer, if he loses the confidence of his troops, disaster must sooner or later ensure.
R.E. Lee: Letter to Jefferson Davis, 8 August 1863

The true way to be popular with troops is not to be free and familiar with them, but to make them believe you know more than they do.
W.T. Sherman: To the Right Reverend Henry C. Lay, 11 November, 1864

. . . Lions led by asses.
Remark made on the French Army after the Franco-Prussian War

Ten good soldiers, wisely led,
Will beat a hundred without a head.
d'Arcy Thompson, 1829–1902, Paraphrase of Euripides

Those who appreciate true valor should in their daily intercourse set gentleness first and aim to win the love and esteem of others. If you affect valor and act with violence, the world will in the end detest you and look upon you as wild beasts. Of this you should take heed.
Emperor Meiji: Rescript to Soldiers and Sailors, 4 January 1883

In enterprise of martial kind,
When there was any fighting,
He led his regiment from behind,
He found it less exciting.
W.S. Gilbert: The Gondoliers, i, 1889

A body of soldiers, as I cannot too often repeat, is worth only what its leaders are worth. The latter must constantly consider the soldier's needs, and keep his mind absorbed by drills and exercises interrupted by amusements so that he is not unduly isolated and confronted by depressing thoughts.
Gustave le Bon, 1841–1931, World in Revolt

Learn to consume your own smoke.
Sir William Osler, 1849–1919

Executive ability is deciding quickly and getting somebody else to do the work.
John Garland Pollard, 1871–1937

A competent leader can get efficient service from poor troops, while on the contrary an incapable leader can demoralize the best of troops.
John J. Pershing: Experiences in the World War, II, 1931

The relation between officers and men should in no sense be that of superior and inferior nor that of master and servant, but rather that of teacher and scholar. In fact, it should partake of the nature of the relation between father and son, to the extent that officers, especially commanding officers, are responsible for the physical, mental, and moral welfare, as well as the discipline and military training of the young men under their command.
John A. Lejeune: Marine Corps Manual, 1920

The young American responds quickly and readily to the exhibition of qualities of leadership on the part of his officers. Some of these qualities are industry, energy, initiative, determination, enthusiasm, firmness, kindness, justness, self-control, unselfishness, honor, and courage.
John A. Lejeune: Reminiscences of a Marine, 1930

Discipline apart, the soldiers' chief cares are, first, his personal comfort, i.e., regular rations, proper clothing, good billets, and proper hospital arrangements (square meals and a square deal in fact); and secondly, his personal safety, i.e., that he shall be put into a fight with as good a chance as possible for victory and survival.
Sir A.P. Wavell, 1883–1950

The task of leadership is not to put greatness into humanity, but to elicit it, for the greatness is already there.
John Buchan, 1875–1940

The real leader displays his quality in his triumphs over adversity, however great it may be.

George C. Marshall: Address to 1st Officer Candidate Class, Fort Benning, 18 September 1941

The commander must try, above all, to establish personal and comradely contact with his men, but without giving away an inch of his authority.

Erwin Rommel, 1891–1944

A piece of spaghetti or a military unit can only be led from the front end.

George S. Patton, Jr.: In North Africa, 1942

Never tell people how to do things. Tell them *what* to do and they will surprise you with their ingenuity.

George S. Patton, Jr.: War As I Knew It, 1947

Men are neither lions nor sheep. It is the man who leads them who turns them into either lions or sheep.

Jean Dutourd: Taxis of the Marne, 1957

An army cannot be administered. It must be led.

Franz-Joseph Strauss: To the German Bundestag, 1957

A leader is a man who has the ability to get other people to do what they don't want to do, and like it.

Harry S. Truman: Memoirs, 1955

The first thing a young officer must do when he joins the Army is to fight a battle, and that battle is for the hearts of his men. If he wins that battle and subsequent similar ones, his men will follow him anywhere; if he loses it, he will never do any real good.

Montgomery of Alamein: Memoirs, xxxiii, 1958

All the regulations and gold braid in the Pacific Fleet cannot enforce a sailor's devotion. This, each officer in command must earn on his own.

Lieutenant Commander Arnold S. Lott: Brave Ship, Brave Men, 1965

Nothing has been accomplished if there is still anything to be done.

Latin Proverb

(*See also* Command, Generalship.)

Leave

Always ask for leave at all times and in all places, and in time you will acquire a right to it.

Francis Grose: Advice to the Officers of the British Army 1782

A certain amount of leave, although not in any contract of service, is a recognized part of a soldier's life.

Winston Churchill: Note to Minister of Pensions, 23 April 1943

Lee, Robert E. (1807–1870)

The very best soldier that I ever saw in the field.

Winfield Scott: Letter regarding Lee to Secretary of War Floyd, 8 May 1857

Lee is the only man I know whom I would follow blindfold.

Stonewall Jackson: Letter, May 1862

An angel's heart, an angel's mouth,
Not Homer's, could alone for me
Hymn well the great Confederate South,
Virginia first, and *Lee*!

Philip Stanhope Worsley: Verses inscribed in presentation copy of his translation of the Iliad to General Lee, January 1866

Strike the tent.

R.E. Lee: Last words, 12 October 1870

He was a foe without hate, a friend without treachery, a soldier without cruelty, and a victim without murmuring. He was a public officer without vices, a private citizen without wrong, a neighbor without reproach, a Christian without hypocrisy, and a man without guilt. He was Caesar without his ambition, Frederick without his tyranny, Napoleon without his selfishness, and Washington without his reward.

Benjamin H. Hill, 1832–1882, tribute to R. E. Lee

He never rose to the grand problem which involved a continent . . . His Virginia was to him the world . . . He stood at the front porch battling with the flames whilst the kitchen and the house were burning, sure in the end to consume the whole.

W.T. Sherman: In North American Magazine, March 1887

Legion

The Roman soldiers, bred in war's alarms,
Bending with unjust loads and heavy arms,
Cheerful their toilsome marches undergo,
And pitch their sudden camp before the foe.

Virgil 70–19 B.C.

Varus, give me back my legions.

Augustus Caesar: After the defeat and annihilation of Varus's column by Arminius in the Teutoberg Forest, 9 A.D.

The Roman, a politician above all, with whom war was only a means, wanted perfect means. He took into account human weakness and he discovered the legion.

Ardant du Picq, 1821–1870, Battle Studies

To the legion of the lost ones, to the cohort of the damned . . .

Rudyard Kipling: Gentlemen-Rankers, 1892

Lexington, (19 April 1775)

Stand your ground, men. Don't fire unless fired upon; but if they mean to have a war, let it begin here!

Captain Jonas Parker: To the Minute Men, Lexington Green, 19 April 1775, before the redcoats opened fire

Leyte Gulf, (25–26 October 1944)

Our ships have been salvaged and are retiring at high speed toward the Japanese fleet.

W.F. Halsey: Commenting on enemy report that the U.S. Third Fleet had been sunk or was fleeing, 26 October 1944

The whole world wants to know where is Task Force 34.

Message received by Admiral Halsey during Battle of Leyte Gulf, 24 October 1944 (this language, which infuriated Halsey, was in fact a cryptographer's padding in an unrelated message; Task Force 34 was the battleship force of the Third Fleet.)

Liberty

A sailor's liberty is but for a day; yet while it lasts it is perfect.

R.H. Dana, Jr.: Two Years Before the Mast, xvi, 1840

Limited War

A great country cannot wage a little war.

Wellington: to the House of Lords, 16 January, 1838

The pre-requisite for a policy of limited war is to reintroduce the political element into our concept of warfare and to discard the notion that policy ends when war begins or that war can have goals distinct from those of national policy.

Henry A. Kissinger; Nuclear Weapons and Foreign Policy, 1957

If we are to retain . . . a choice other than nuclear holocaust or retreat, we must be ready to fight a limited war for a protracted period of time anywhere in the world.

John F. Kennedy: Message on the Budget, fiscal year 1963

Lines of Communication

An army must have but one line of operations. This must be maintained with care and abandoned only for major reasons.

Napoleon I: Maxims of War, 1831

The line that connects an army with its base of supplies is the heel of Achilles—its most vital and vulnerable point.

John S. Mosby: War Reminiscences, ix, 1887

Free supplies and open retreat are two essentials to the *safety* of an army or a fleet.

Mahan: Naval Strategy, 1911

All military organizations, land or sea, are ultimately dependent upon open communications with the basis of national power.

Mahan: Naval Strategy, 1911

Nine times out of ten an army has been destroyed because its supply lines have been severed.

Douglas Mac Arthur: Remarks to

members of the Joint Chiefs of Staff, 23 August 1950, in Tokyo

(*See also* Communications.)

Logistics

There must be great care taken to send us munition and victual whithersoever the enemy goeth.
Francis Drake: To Walsyngham, during operations against the Armada, 29 July 1588

War is quite changed from what it was in the days of our forefathers; when in a hasty expedition and a pitch'd field, the matter was decided by courage; but now the whole art of war is in a manner reduced to money; and nowadays that prince who can best find money to feed, clothe, and pay his army, not he that hath the most valiant troops, is surest to success and conquest.
Charles Davenant: Essay on Ways and Means of Supplying the War, 1695

In order to make assured conquests it is necessary always to proceed within the rules: to advance, to establish yourself solidly, to advance and establish yourself again, and always prepare to have within reach of your army your resources and your requirements.
Frederick the Great: Instructions for His Generals, ii, 1747

It is very necessary to attend to all this detail and to trace a biscuit from Lisbon into a man's mouth on the frontier and to provide for its removal from place to place by land or by water, or no military operations can be carried out.
Attributed to the Duke of Wellington, Peninsular Campaign, 1811

What makes the general's task so difficult is the necessity of feeding so many men and animals. If he allows himself to be guided by the supply officers he will never move and his expedition will fail.
Napoleon I: Maxims of War, 1831

Logistics comprises the means and arrangements which work out the plans of strategy and tactics. Strategy decides where to act; logistics brings the troops to this point.
Jomini: Précis de l'Art de la Guerre, 1838

A general should be capable of making all the resources of the invaded country contribute to the success of his enterprise.
Jomini: Précis de l'Art de la Guerre, 1838

We must live upon the enemy's country as much as possible and destroy his supplies. This is cruel warfare, but the enemy has brought it upon himself.
Major General H.W. Halleck, USA: Letter to Major General S.A. Hurlbut, USA, April 1863

The army will forage liberally on the country.
W.T. Sherman: Special Field Orders, No. 120, 9 November 1864 (Sherman's administrative order for the march from Atlanta to the sea)

We are so ready that if the war should last two years, not a gaiter-button would be wanting.
Edmond Leboeuf: As Minister of War, to the French Corps Legislatif, just prior to outbreak of the Franco-Prussian War, 1870

Supply and transport stand or fall together; history depends on both.
Winston Churchill: The River War, viii, 1899

So long as the fleet is able to face the enemy at sea, communicatiosn mean, essentially, not geographical lines . . . but supplies which the ships cannot carry in their own hulls beyond a limited amount.
Mahan: Naval Strategy, 1911

We have a claim on the output of the arsenals of London as well as of Hanyang, and what is more, it is to be delivered to us by the enemy's own transport corps. This is the sober truth, not a joke.
Mao Tse-Tung: On Guerrilla Warfare, 1937

I don't know what the hell this "logistics" is that Marshall is always talking about, but I want some of it.
E. J. King: To a staff officer, 1942

The soldier cannot be a fighter and a pack animal at one and the same time, any more than a field piece can be a gun and a supply vehicle combined.

J.F.C. Fuller; Letter to S.L.A. Marshall, c. 1948

. . . billets and lines of communications, an inevitable source of friction between neighboring armies.
Barbara W. Tuchman: The Guns of August, xiii, 1962

(*See also* Lines of Communication, Supply, Transportation.)

Longstreet, James (1821–1904)

Here is my old war horse at last.
R.E. Lee: At Antietam, 17 September 1962

Lookout Mountain, (23 November 1863)

"Give me but two brigades," said Hooker, frowning at fortified Lookout . . .
George Henry Boker, 1823–1890, Battle of Lookout Mountain

Love and War

As it is base for a soldier to love; so am I in love with a base wench.
Shakespeare: Love's Labor Lost, i, 2, 1594

Old soldiers, sweetheart, are surest, and old lovers are soundest.
John Webster: Westward Hoe, ii, 2, 1638

None but the brave deserve the fair.
John Dryden: Alexander's Feast, 1697

The same heat that stirs them up to Love, spurs them on to Battle.
George Farquhar: The Recruiting Officer, 1706

If upon service you have any ladies in your camp, be valiant in your conversation before them. There is nothing pleases the ladies more than to hear of storming breaches, attacking the covert-way sword in hand, and such like martial exploits.
Francis Grose: Advice to the Officers of the British Army, 1782

* * * wants to go to Lisbon, and I have told him he may stay there 48 hours which is as long as any reasonable man can wish to stay in bed with the same woman.
Wellington: Letter from Portugal, 1811

He who is full of courage and sang-froid before an enemy battery, amid the bullets, sometimes trembles and loses his head before a skirt or a peruke.
Napoleon I, 1769–1821, Conversation with Captain Poppleton, St. Helena

In war, as in love, we must achieve contact ere we triumph.
Napoleon I: Political Aphorisms, 1848

. . . the scandal of old men at war and old men in love—but at what age a general ceases to be a danger to the enemy and a Don Juan, is not easy to determine.
Sir Archibald Wavell, 1883–1950

To force a woman's resistance by seducing her is peace; to do so by violating her is war.
Denis de Rougemont: Love in the Western World, 1956

He had discovered, like so many soldiers returning from the wars, that his girl had abandoned him for another man.
Alan Moorehead: The Blue Nile, iii, 1962

Loyalty

Blood is thicker than water.
Josiah Tattnall: Despatch to the Secretary of the Navy, after bringing his flagship to the aid of hard-pressed British ships in the Peiho River, China, 25 June 1859

Grant stood by me when I was crazy, and I stood by him when he was drunk, and now we stand by each other.
Attributed to W.T. Sherman, c. 1870

Loyalty is the marrow of honor.
Paul von Hindenburg: Out of My Life, 1920

An officer . . . should make it a cardinal principle of life that by no act of commission or omission on his part will he permit his immediate superior to make a mistake.
General Malin Craig, USA: Address to the graduating class, West Point, 12 June 1937

"Loyalty," analyzed, is too often a polite word for what would be more accurately described as "a conspiracy for mutual inefficiency."

B.H. Liddell Hart, 1895: Why We Don't Learn from History

There is a great deal of talk about loyalty from the bottom to the top. Loyalty from the top down is even more necessary and much less prevalent.

George S. Patton, Jr.: War As I Knew It, 1947

Luck

To a good general luck is important.

Livy: History of Rome, xxii, c. 110

Luck in the long run is given only to the efficient.

Helmuth Von Moltke, 1800–1891

Luck is like a sum of gold, to be spent.

Allenby, 1861–1936

A stout heart breaks bad luck.

Spanish Proverb.

(*See also* Chance, Fortunes of War.)

Lundy's Lane, (25 July 1814)

I'll try, Sir.

Colonel James Miller, USA: Reply when ordered to lead the 21st Infantry against the British works at Lundy's Lane. (This phrase was later adopted as the regimental motto.)

Magdala, Campaign of (1867–1868)

We have hoisted the standard of St. George upon the mountains of Rasselas.

Benjamin Disraeli: To the House of Commons after the capture of Magdala, 13 April 1868

Mahan, Alfred Thayer (1840–1914)

It is not the business of a Naval officer to write books.

Rear Admiral F. M. Ramsay, USN: endorsement on an unfavorable fitness report rendered on Mahan, 1893

Mahan was the only great naval writer who also possessed the mind of a statesman of the first class.

Theodore Roosevelt: in The Outlook, 13 January 1915

. . . Mahan, the maritime Clausewitz, the Schlieffen of the sea.

Barbara W. Tuchman: The Guns of August, xviii, 1962

Maine, USS

Remember the Maine!

Spanish-American War slogan (USS Maine *was sunk by a suspicious explosion in Havana Harbor, 15 February 1898.)*

(*See also* Spanish American War.)

Maintenance

Keep up your bright swords, for the dew will rust them.

Shakespeare: Othello, i, 2, 1604

For want of a nail the shoe is lost; for want of a shoe the horse is lost; for want of a horse the rider is lost; for want of a rider the battle is lost; for want of a battle the kingdom is lost.

George Herbert: Outlandish Proverbs, 1640

It has been well said that Nelson took more care of his topgallant masts, in ordinary cruising, than he did of his whole fleet when the enemy was to be checked or beaten.

Mahan: Lessons of the War with Spain, 1899

Man

Man, not men, is the most important consideration.

Napoleon I: Maxims of War, 1831

Remember also that one of the requisite studies for an officer is *man.* Where your analytical geometry will serve you once, a knowlege of men will serve you daily. As a commander, to get the right man in the right place is one of the questions of success or defeat.

Farragut: Letter to his son, 13 October 1864

On foot, on horseback, on the bridge of a vessel, at the moment of danger, the same man is found. Anyone who knows him well, deduces from his action in the past what his future action will be.

Ardant du Picq, 1821–1870, Battle Studies

Man is the foremost instrument of combat.

Ardant du Picq, 1821–1870, Battle Studies

Men, they are the first and best instruments of war.

General Mikhail Ivanovich Dragomirov, 1830–1905

Wars may be fought with weapons, but they are won by men. It is the spirit of the men who follow and the man who leads that gains the victory.

George S. Patton, Jr.: In Cavalry Journal, September 1933

We are apt now in this mechanistic age to forget the simple truths of military history, that Man and *not* machines dominates the battlefields of the world. One cannot chart the frenetic fever of human emotions on a graph; one cannot plumb the depths of the human soul with a calculating machine. Nor can one estimate with certainty how men react in mass and under stress. It is Man, in his infinite variety—stubborn, brave, cowardly, ignorant, brilliant Man—who provides the *forever new,* as well as the *old,* frontiers of our world.

Hanson W. Baldwin: Critical Tomorrows, December 1962

(*See also* Manpower.)

Man on Horseback

The presence of a fortunate soldier is dangerous to newly constituted states.

Jose de San Martin: Farewell Proclamation to the people of Peru, September 1822

I have heard, in such a way as to believe it, of your recently saying that both the army and the government needed a dictator . . . Only those generals who gain successes can set up dictators. What I ask of you now is military success, and I will risk the dictatorship.

Abraham Lincoln: Letter to Major General Joseph Hooker, appointing him to command the Army of the Potomac, 26 January 1863

Maneuver

Both advantage and danger are inherent in maneuver.

Sun Tzu, 400–320 B. C., The Art of War, vi

Maneuvers are threats; he who appears most threatening, wins.

Ardant du Picq, 1821–1870, Battle Studies

Nearly all the battles which are regarded as masterpieces of the military art, from which have been derived the foundation of states and the fame of commanders, have been battles of maneuver.

Winston Churchill: The World Crisis, ii, 1923

(*See also* Tactics.)

Manila Bay (1 May 1898)

Those fellows do not have any naval manners and I propose to teach them some.

Attributed to Captain Sir Edward Chichester, RN: On stationing his flagship, HMS Immortalité, *on the flank of the American line so that she could cover the unfriendly German squadron in Manila Bay, 13 August 1898*

The battle of Manila Bay was one of the most important ever fought. It decided that the United States should start in a direction in which it had never traveled before. It placed the United States in the family of great nations.

Bradley A. Fiske: Letter to Mark Sullivan, 1925

Manpower

Ye gods, what dastards would our host command?
Swept to the war, the lumber of a land.

Homer: Iliad, ii, c. 1000 B.C.

Men make the city and not walls or ships without men in them.

Nicias of Athens: To the Athenian expeditionary force at Syracuse, 414 B. C.

Peasants are the most fit to bear arms for they from their infancy have been exposed to all kinds of weather and have been brought up to the hardest labor. They are able to endure the most intense heat of the sun, are unacquainted with use of baths, and they are strangers to the other luxuries of life. They are simple, content with little, inured to fatigue, and prepared in some measure for a military life.

Vegetius: de Re Militari, i, 378

When service happeneth we disburthen the prisons of thieves, we rob the taverns and alehouses of tosspots and ruffians, we scour both town and country of rogues and vagabonds.

Barnabe Rich: A Pathway to Military Practice, 1587

Food for powder, food for powder; they'll fill a pit as well as better.

Shakespeare: I King Henry IV, iv, 2, 1591

At my first going out into this engagement [Edgehill] I saw our men beaten at every hand . . . "Your troops," said I, "are most of them old decayed serving-men, and tapsters, and such like of fellows; and," said I, "their troops are gentlemen's sons, younger sons and persons of quality: do you think that the spirits of such base mean fellows will ever be able to encounter gentlemen, that have honor and courage and resolution in them? . . . You must get men of spirit: and take it not ill what I say,—I know you will not,—of a spirit that is likely to go on as far as gentlemen will go: or else you will be beaten still."

Oliver Cromwell: To John Hampden, after Edgehill, October 1642

A few honest men are better than numbers.

Oliver Cromwell: Letter to Sir William Springe, September 1643

As the army now stands it is only a receptacle for ragamuffins.
Major General Henry Knox, USA: Recommending establishment of a military academy, 1776

The worst men are the best soldiers.
English saying, quoted in parliamentary debate on the abolition of flogging in the Army, 1808

You cannot stop me; I spend thirty thousand men a month.
Napoleon I: To Metternich, 1810

When defending itself against another country, a nation never lacks men, but too often, *soldiers.*
Napoleon I: Political Aphorisms, 1848

The lower people everywhere desire war. Not so unwisely; there is then a demand for lower people—to be shot!
Thomas Carlyle: Sartor Resartus, iii, 1836

Historically, good men with poor ships are better than poor men with good ships.
Commander J.K. Taussig, USN: 1916

A soldier who habitually breaks regulations must be dismissed from the army. Vagabonds and vicious people must not be accepted for service. The opium habit must be forbidden, and soldier who cannot break himself of it should be dismissed.
Mao Tse-tung: On Guerrilla Warfare, 1937

Two sets of considerations may induce manpower economy: (1) the good civil politician wants to save the men among the people who elect and re-elect him to a place in the government; (2) the good general wants to save soldiers who have learned the business of fighting.
Alfred Vagts: A History of Militarism, viii, 1959

People and not things are the fundamental factor determining the outcome of war.
General Lo Jui-Ching: In Red Flag, May 1965

Any blockhead is good enough to be shot at.
English Proverb

Good iron is not used for nails; good men do not become soldiers.
Chinese Proverb

(*See also* Casualties, Conscript, Conscription, Impressment, Man, Recruiting, Universal Military Training.)

Map

The commander must acquaint himself beforehand with the maps so that he knows dangerous places for chariots and carts, where the water is too deep for wagons; passes in famous mountains, the principal rivers, the locations of highlands and hills; where rushes, forests, and reeds are luxuriant; the road distances; the size of cities and towns; well-known cities and abandoned ones, and where there are flourishing orchards. All this must be known, as well as the way boundaries run in and out. All these facts the general must store in his mind.
Tu Mu, 803–852, Wei Liao Tzu

But thou at home, without tide or gale,
Canst in thy map securely sail.
Robert Herrick, 1591–1674

In Afric maps
With savage pictures fill their gaps,
And o'er uninhabitable downs
Place elephants for want of towns.
Jonathan Swift, 1667–1745

I am told that there are poeple who do not care for maps, and I find it hard to believe.
R.L. Stevenson, 1850–1894

A study of the map will indicate where critical situations exist or are apt to develop, and so indicate where the commander should be.
George S. Patton, Jr.: Was As I Knew It, 1947

Battle is a process which always takes place at the junction of two maps.
Military saying

Marathon, (490 B. C.)

That man is little to be envied whose patriotism would not gain force upon the plain of Marathon.
Samuel Johnson: Journey to the Western Isles of Scotland, 1775

The Bettman Archive

LORD ST. VINCENT

1735–1823

"Discipline is summed up in the one word obedience."

March

Maintain discipline and caution above all things, and be on the alert to obey the word of command.

Archidamus of Sparta: Orders to the Spartan expedition invading Athenian territory, 431 B.C.

Those who do not know the condition of mountains and forests, hazardous defiles, marshes and swamps, cannot conduct the march of an army. Those who do not use native guides are unable to obtain the advantage of the ground.

Sun Tzu, 400–320 B.C., The Art of War, vii

When the regiment is on the march, gallop from front to rear as often as possible, especially if the road is dusty . . . It is diverting enough to dust a parcel of fellows already half choked, and to see a poor devil of a soldier, loaded like a jack-ass, endeavouring to get out of the way.

Francis Grose: Advice to the Officers of the British Army, 1782

I have destroyed the enemy merely by marches.

Napoleon I: After the Austrian Campaign, 1805

The strength of an army, like power in mechanics, is reckoned by multiplying the mass by the rapidity; a rapid march increases the morale of an army, and increases its means of victory. Press on!

Napoleon I: Maxims of War, 1831

I had rather lose one man in marching than five in fighting.

Stonewall Jackson, c. 1862

One who never turned his back but marched breast forward.

Robert Browning, 1812–1899, Epilogue to Asolando

Boots—boots—boots—boots—movin' up
and down again!
There's no discharge in the war!

Rudyard Kipling: Boots, 1903

Mariner

Ye Mariners of England!
That guard our native seas:
Whose flag has braved a thousand years,
The battle and the breeze!

Thomas Campbell, 1777–1844, Ye Mariners of England

(*See also* Sailor, Seaman.)

Marines

We have compelled every land and every sea to open a path for our valor, and have everywhere planted eternal memorials of our friendship and of our enmity.

Pericles: Funeral Oration over the Athenian dead, 431 B.C.

The Marines too were full of anxiety that, when ship struck ship, the deeds on deck should not fall short of the rest.

Thucydides: Of the battle in the Great Harbor of Syracuse, 414 B.C.

That twelve hundred land Souldjers be forthwith raysed, to be in readinesse, to be distributed into his Ma[ts] Fleets prepared for Sea Service . . .

Charles II: Order in Council, 28 October 1664 (the order establishing the world's first permanent corps of Marines, the Royal Marines)

A health to brave Sea-Soldiers all,
Let cans a-piece go round-a;
Pell-mell let's to the battle fall
And lofty music sound-a.

Wit and Drollery, verse by unknown author, 1682

That two Battalions of Marines be raised consisting of one Colonel two lieutenant Colonels, two Majors & Officers as usual in other regiments, that they consist of an equal number of privates with other battalions; that particular care be taken that no person be appointed to office or inlisted into said Battalions, but such as are good seamen, or so acquainted with maritime affairs as to be able to serve to advantage by sea . . . .

Resolution of the Continental Congress creating the U.S. Marine Corps, 10 November 1775

Land forces are nothing. Marines are the only species of troops proper for this nation. A powerful fleet and 30,000 Marines will save us from destruction, and nothing else.

Major General Henry Lloyd: The History of the Late War in Germany 1779

How much might be done with a hundred thousand soldiers such as these.

Napoleon I: While inspecting the Marine guard, HMS Bellerophon, *after surrendering, 15 July 1815*

A very considerable corps [of Marines] should be kept up, and I hope to see the day when there is not another foot-soldier in the Kingdom, in Ireland, or in the Colonies . . .

Lord St. Vincent: Letter to Lord Spencer, 30 June 1797

I never knew an appeal made to them for honor, courage, or loyalty that they did not more than realize my highest expectations. If ever the hour of real danger should come to England they will be found the Country's Sheet Anchor.

Lord St. Vincent: Of the Royal Marines, 1802

The public should be kept alive upon the subject of our monstrous Army . . . showing that the Marine Corps is best adapted to the security of our dockyards; and that no soldier, of what is termed the line, shall approach them; our Colonies ought to have no other infantry to protect them, and the Corps of Marine Artillery should be substituted for the old Artillery.

Lord St. Vincent: Letter to Benjamin Tucker, 1818

Yet 'ere I venture in an arduous strain,
To sketch the hardy native of the main;
Permit my fair impartial verse to raise
Another tribe to due and well-earned praise;
A tribe full oft in Honour's Causes seen,
Nor idle there—Stand forth thou bold MARINE!
When 'gainst th' hostile shore th' attack is plann'd;
To storm the batteries that guard the land;
Destroy the magazine, the tower, the fort,
And open and defenceless leave the port;
Then this amphibious hero gives to fame
At once the sailor's and the soldier's name.

George Woodley: Britain's Bulwarks, 1811

That will do for the Marines.

Byron: The Island, ii, 1823

Tell that to the Marines—the sailors won't believe it.

Walter Scott: Redgauntlet, II, vii, 1824

The words Marine and Mariner differ by one small letter only; but no two races of men, I had well nigh said no two animals, differ from one another more completely.

Captain Basil Hall, RN: Journal, 1832

The Marines are properly the garrisons of His Majesty's ships, and upon no pretence ought they to be moved from a fair and safe communication with the ships to which they belong.

Wellington: To the House of Lords, 21 April 1837

It is a Corps which never appeared on any occasion or under any circumstances without doing honor to itself and its country.

Marquis of Anglesey: Speech, 5 August 1841, at Portsmouth, England.

A life on the ocean wave,
A home on the rolling deep;
Where the scattered waters rave,
And the winds their revels keep!

Epes Sargent: A Life on the Ocean Wave, 1847 (regimental march, Royal Marines)

I should not deem a man-of-war complete without a body of Marines . . . imbued with that *esprit* that has so long characterized the "old Corps."

Commodore Joshua R. Sands, USN: Letter to Brigadier General Archibald Henderson, USMC, 1852

The Marines . . . will never disappoint the most sanguine expectations of their country—never! I have never known one who would not *readily advance in battle.*

Captain C.W. Morgan, USN: Letter to Brigadier General Archibald Henderson, USMC, 1852

A ship without Marines is like a garment without buttons.

Admiral David D. Porter, USN: Letter to Colonel John Harris, USMC, 1863

Connected with the Navy, there is the finest body of troops in the World, and that is those gallant Marines who are ever ready to devote themselves to the interests of their country.

Benjamin Disraeli: Speech 18 September 1879

The Marines have landed and the situation is well in hand.

Attributed to Richard Harding Davis, 1864–1916

From the halls of Montezuma to the shores of Tripoli,
We fight our country's battles in the air, on land and sea,
First to fight for right and freedom,
And to keep our honor clean,
We are proud to claim the title
Of United States Marine . . .
If the Army and the Navy
Ever gaze on Heaven's scenes,
They will find the streets are guarded
By United States Marines.

The Marines' Hymn, author unknown, late 19th century

Sez 'e, "I'm a Jolly—'Er Majesty's Jolly—soldier an' sailor too!"

Rudyard Kipling: "Soldier 'an Sailor Too," 1896

Their record is Second to None. I have been with them on Active Service, on Police Service, in Daily Routine and in Gales of Wind—I have had them with me everywhere, and I tell you there is nothing like the Royal Marines.

Lord Charles Beresford: Speech, 30 June 1909

No-one can ever say that the Marines have ever failed to do their work in handsome fashion.

Major General Johnson Hagood, USA: We Can Defend America, 1937

The raising of that flag on Suribachi means a Marine Corps for the next 500 years.

James Forrestal: To Lieutenant General H.M. Smith, USMC, as the Marines raised the Colors on Mt. Suribachi, 23 February 1945

The bended knee is not a tradition of our Corps.

A.A. Vandergrift: To the Senate Naval Affairs Committee, 5 May 1946 (regarding U.S. Army proposals for abolition of the Marine Corps).

The British Marine represents every admiral's embodied idea of the perfect Marine: heel-clicking, loyal, immaculately turned out, wise in his way like a greying family retainer—and, like a family retainer, carefully restricted in latitude of opinion and activity.

Holland M. Smith: Coral and Brass, iii, 1949

The Marine Corps is the Navy's police force and as long as I am President that is what it will remain. They have a propaganda machine that is almost equal to Stalin's.

Harry S. Truman: Letter to Representative Gordon L. McDonough 29 August 1950

I sincerely regret the unfortunate choice of language which I used in my letter of August 29 to Congressman McDonough concerning the Marine Corps . . .

Harry S. Truman: Letter to General C.B.Cates, USMC, 6 September 1950

I have just returned from visiting the Marines at the front, and there is not a finer fighting organization in the world.

Douglas MacArthur: In the outskirts of Seoul, 21 September 1950

In the vast complex of the Department of Defense, the Marine Corps plays a lonely role.

John Nicholas Brown: To the Senate Armed Forces Committee, 17 April 1951

. . . the Marines, a curiously amphibian force, which owes its origin, if not its continued existence, more to political considerations than to military designs.

Alfred Vagts: A History of Militarism, 1959

Glamor, caste, and unpopularity are among the gifts of the sea, and their influence on Marines is not dead yet.

Marc Parrott: Hazard, 1962

The Marines' best propaganda has usually been the naked event.

Marc Parrott: Hazard, 1962

The appearance of Marines on foreign soil has always in the past indicated the beginning of extremely dangerous military adventures.

Article in Krasnaya Zvezda (Red Star), 16 March 1965

Once a Marine, always a Marine.

Marine saying

Marksmanship

He that rides at high speed and with his pistol kills a sparrow.
Shakespeare: I King Henry IV, ii, 4, 1597

It is not sufficient that the soldier must shoot, he must shoot well.
Napoleon I, 1769–1821

All a soldier needs to know is how to shoot and salute.
John J. Pershing, 1860–1948

You don't hurt 'em if you don't hit 'em.
Lewis B. Puller: Marine, 1962

The shots that hit are the shots that count.
Military maxim

Marlborough, Duke of (1650–1722)

He never rode off any field except as a victor. He quitted war invincible; and no sooner was his guiding hand withdrawn than disaster overtook the armies he had led.
Winston Churchill: Marlborough, I, 1933

Marriage

Let soldiers marry; they will no longer desert. Bound to their families, they are bound to their countries.
Voltaire: Satirical Dictionary, 1751

Sir,
You, having thought fit to take to yourself a wife, are to look for no further attentions from your humble servant,
Lord St. Vincent: Letter to Lieutenant Bayntun, 1795

Marriage is good for nothing in the military profession.
Napoleon I: Political Aphorisms, 1848

(*See also* Dependents, Love and War.)

Marshall, George Catlett (1880–1959)

. . . the organizer of Victory.
Winston Churchill, 1874–1965 (this remark was also made of Carnot, q.v., in 1795).

Martial Music

Noble and manly music invigorates the spirit, strengthens the wavering man, and incites him to great and worthy deeds.
Homer: The Iliad, c. 1000 B.C.

Thou hast heard, O my soul, the sound of the trumpet, the alarm of war.
Jeremiah, IV, 19

Yemen on foot, and knaves many oon,
With shorte staves, as thikke as they may goon;
Pypes, trompes, nakers, and clariouness,
That in the batail blewe bloody sownes.
Geoffrey Chaucer: Canterbury Tales, c. 1387 ("Naker" = drum)

The trumpet's loud clangor
Excites us to arms,
With shrill notes of anger
And mortal alarms.
The double double double beat
Of the thundering drum
Cries Hark! the foes come!
John Dryden: A Song for St. Cecilia's Day, 1687

A song is as necessary to sailors as the drum and fife to a soldier.
R.H. Dana, Jr.: Two Years Before the Mast, xxviii, 1840

How good bad music and bad reasons sound when we march against an enemy.
Friedrich Nietzsche, 1844–1900

(*See also* Bagpipe, Bands, Bugle, Drum, Trumpet.)

Martial Spirit

The security of every society must always depend, more or less, upon the martial spirit of the great body of the people . . . Martial spirit alone, and unsupported by a well-disciplined standing army, would not perhaps, be sufficient for the defense and security of any society. But where every citizen had the spirit of a soldier, a smaller standing army would surely be necessary.
Adam Smith: An Inquiry into the Nature and Causes of the Wealth of Nations, V,1, 1776

(*See also* Militarism, Will to Fight.)

Martinet

Be sure also to stigmatize every officer,

who is attentive to his duty, with the appellation of *Martinet*; and say he has been bitten by a mad adjutant. This will discourage others from knowing more than yourself, and thereby keep you on an equality with them.

Francis Grose: Advice to the Officers of the British Army, 1782

Mass

Use the most solid to attack the most empty.

Ts'ao Ts'ao, 155-220

Not by rambling operations, or naval duels, are wars decided, but by force massed and handled in skillful combinations.

Mahan, 1840–1914

(*See also* Concentration.)

Massed Fires

In battle, as in a siege, skill consists in converging a mass of fire upon a single point. After the combat has started, he that has the power to bring a sudden, unexpected concentration of artillery to bear upon a selected point is sure to capture it.

Napoleon I: Maxims of War, 1831

Materiel

In our day wars are not won by mere enthusiasm, but by technical superiority.

V.I. Lenin: Speech, 1918

The gun, the missile, the ship, the plane, the spaceship is no better than the Man who operates it.

Hanson W. Baldwin: Critical Tomorrows, December 1962

Maxims, Military

Nothing so comforts the military mind as the maxim of a great but dead general.

Barbara W. Tuchman: The Guns of August, 1962

McClellan, George B. (1826–1885)

[General McClellan] is an admirable Engineer, but he seems to have a special talent for the stationary engine.

Abraham Lincoln, 1809–1865 (McClellan was originally an officer of the Corps of Engineers)

Mechanized Warfare

The officers of a panzer division must learn to think and act independently within the framework of the general plan and not wait until they receive orders.

Erwin Rommel: The Rommel Papers, i, 1953

(*See also:* Blitzkrieg, Tanks.)

Medals

The number of medals on an officer's breast varies in inverse proportion to the square of the distance of his duties from the front lines.

C.E. Montague: Fiery Particles, 1915

If medals were ordained for drinks,
Or soft communings with a minx,
Or being at your ease belated,
By heavens, you'd be decorated!

Oliver St. John Gogarty, 1878–

Who ever saw a dirty soldier with a medal?

Old Army saying, c. 1925

(*See also* Awards, Decorations.)

Medicare

If any of the soldiers' wives or children happen to be taken ill, never give them any assistance. You receive no pence from them, and you know *ex nihilo nihil sit.*

Francis Grose: Advice to the Officers of the British Army, 1782

(*See also* Dependents, Health.)

Meeting Engagement

When armies approach each other, it makes all the difference which owns only the ground on which it stands or sleeps and which one owns all the rest.

Winston Churchill: Their Finest Hour, 1949

Mercenary

I confess when I went into arms at the

beginning of this war, I never troubled myself to examine sides; I was glad to hear the drums beat for soldiers, as if I had been a mere Swiss, that had not cared which side went up or down, so I had my pay.

Memoirs of a Cavalier, author unknown, 17th century

No pay, no Swiss. (Point d'argent, point de Suisse.)

Racine: Les Plaideurs, i, 1668

They know no country, own no lord,
Their home the camp, their law the sword.

Silvio Pellico, 1788–1854, Enfernio de Messina, v, 2

A Freeman contending for *Liberty* on his own ground is superior to any slavish mercenary on earth.

George Washington: General Order to the Continental Army, 2 July 1776

Indifferent what their banner, whether 'twas
The Double Eagle, Lily or the Lion . . .

Lines regarding mercenaries, author unknown

These, in the day when heaven was falling,
The hour when earth's foundations fled,
Followed their mercenary calling
And took their wages and are dead.

A.E. Housman: Epitaph on an Army of Mercenaries (i.e., the Regulars of the British Expeditionary Force in 1914), 1922

The fourteen years of war and, before them the long period of rearmament had created all over Europe a class of military adventurers, landless, homeless, without family, without any of the natural pieties, without religion or scruple, without knowledge of any trade but war, and incapable of anything but destruction . . . To any hint of peace they reacted with all the dismay and fury of bishops threatened with disestablishment, or of mill owners at the prospect of a law to regulate child labor.

Aldous Huxley: Grey Eminence, ix, 1941 (on the Thirty Years' War)

Merchant Marine

If you had seen that which I have seen, of the simple service that hath been done by the merchant and coast ships, you would have said that we had been little holpen by them, otherwise than that they did make a show.

Sir William Wynter: Letter to Walsingham, 1 August 1588, after the defeat of the Spanish Armada.

Sea power in the broad sense . . . includes not only the military strength afloat that rules the sea or any part of it by force of arms, but also the peaceful commerce and shipping from which alone a military fleet naturally and healthfully springs, and on which it securely rests.

Mahan: The Influence of Sea Power Upon History, 1890

To the spread of our trade in peace and the defense of our flag a great and prosperous merchant marine is indispensable.

Theodore Roosevelt: To Congress, 7 December 1903

Dirty British coaster with a salt-caked smoke stack
Butting through the Channel in the mad March days,
With a cargo of Tyne coal,
Road-rail, pig-lead,
Firewood, iron-ware, and cheap tin trays.

John Masefield, 1878– , Cargoes

Trade follows the flag.

British proverb, 19th century

Mess Call

Soup-y, soup-y, soup-y
Not a single bean;
Pork-y, pork-y, pork-y,
Not a streak of lean;
Coffee, coffee, coffee,
Not a drop of cream.

Traditional words of Mess call

Message

Whenever the general sends you with a message in the field, though ever so trifling, gallop as fast as you can up to and against the person, to whom it is addressed. Should you ride over him, it would show your alertness in the performance of your duty. In delivering the message, be as concise as possible whether you are understood or not.

Francis Grose: Advice to the Officers of the British Army, 1782

It is not book learning young men need, nor instruction about this and that, but a stiffening of the vertebrae which will cause them to be loyal to a trust, to act promptly, concentrate their energies, do a thing—"Carry a message to Garcia."

Elbert Hubbard: A Message to Garcia, March 1900

There is no rest for a messenger till the message is delivered.

Joseph Conrad: The Rescue, 1920

Get the message through.

Motto of the U.S. Army Signal Corps

Mexican War (1846–1848)

If I were a Mexican, I would tell you, "Have you not enough room in your own country to bury your dead men? If you come into mine, we will greet you with bloody hands and welcome you to hospitable graves."

Thomas Corwin: During Senate debate before declaration of war, 11 February 1846

Militarism

The prolongation of military commands caused Rome the loss of her liberty.

Niccolo Machiavelli: Discorsi, xxiv, 1531

I conclude saying, I wish there was a war.

Alexander Hamilton: Letter (at age 12) to a schoolfellow, 1767

The great increase of commerce and manufactures hurts the military spirit of a people; because it gives them a competition for something else than martial honors, a competition for riches.

Samuel Johnson: To James Boswell, 13 April 1773

Prussia was hatched from a cannon ball.

Napoleon I, 1769–1821

Man is a military animal,
Glories in gunpowder, and loves parades.

Philip James Bailey: Festus, 1839

There is no military spirit in a democratic society, where there is no aristocracy, no military nobility. A democratic society is antagonistic to the military spirit.

Ardant du Picq, 1821–1870, Battle Studies

No great art yet rose on earth but among a nation of soldiers.

John Ruskin: The Crown of Wild Olive, iii, 1866

All great nations learned their truth of word, and strength of thought, in war; they were nourished in war, and wasted by peace; taught by war, and deceived by peace; trained by war, and betrayed by peace;—in a word, they were born in war and expired in peace.

John Ruskin: The Crown of Wild Olive, iii, 1866

Man shall be framed for War, and Woman for the entertainment of the warrior: all else is folly.

F.W. Nietzche: Thus Spake Zarathustra, i, 18, 1885

Ye shall love peace only as a means to new wars—and the short peace more than the long.

F.W. Nietzche: Thus Spake Zarathustra, x, 1885

Ye say, a good cause will hallow even war? I say unto you: a good war halloweth every cause.

F.W. Nietzche: Thus Spake Zarathustra, iii, 1885

So we are bound together—I and the Army—so we are born for one another, and so we shall hold together indissolubly, whether, as God wills, we shall have peace or storm.

Kaiser Wilhelm II: Proclamation to the Imperial German Army, 1888

But, for my part, I'll go on trusting and appealing to God and my sharp sword! And damn the whole concern.

Kaiser Wilhelm II: Of the Hague Peace Conference, 1899

No triumph of peace is quite so great as the supreme triumphs of war. The courage of the soldier, the courage of the states-man who has to meet storms which can be quelled only by soldierly virtues—this stands higher than any quality called out merely in time of peace.

Theodore Roosevelt: Speech at the Naval War College, Newport, R.I., 2 June 1897

In military affairs only military men should be listened to.

Theodore Roosevelt: Letter to

Lieutenant W.F. Fullam, USN, 28 June 1897

There is one form of centralized government which is almost entirely unprogressive and beyond all other forms costly and tyrannical—the rule of an army.

Winston Churchill: The River War, iii, 1899

An occasional fight is a good thing for a nation. It strengthens the race . . . Let war cease altogether and a nation will become effeminate.

Lieutenant General Adna R. Chaffee, USA: On War, 1902

We oppose militarism. It means conquest abroad and intimidation and oppression at home. It means the strong arm which has ever been fatal to free institutions. It is what millions of our citizens have fled from in Europe.

Democratic National Platform, 1900

Militarism does not consist in the existence of any army, nor even in the existence of a very great army. Militarism is a spirit. It is a point of view. It is system. It is a purpose. The purpose of militarism is to use armies for aggression.

Woodrow Wilson: Address to the graduating class, West Point, 13 June 1916

You cannot organize civilization around the core of militarism and at the same time expect reason to control human destiny.

Franklin D. Roosevelt: Speech, 25 October 1938

. . . the bellicose frivolity of senile empires.

Barbara W. Tuchman: The Guns of August, 1962

(*See also* Martial Spirit, War, Will to Fight.)

Military Attache

The purpose of the assignment of these officers is accurate knowledge of states from the purely military viewpoint. Their purpose is absolutely apolitical, and they must avoid any meddling in politics and must, above all, observe the utmost caution and circumspection in their behavior.

C.W.G. von Grolman: General order establishing Prussia's system of military attaches, 14 April 1816 (this order marks the establishment of the first modern military attache system in the world).

Military Government

Whoever conquers a free town and does not demolish it commits a great error, and may expect to be ruined himself; because whenever the citizens are disposed to revolt, they betake themselves of course to that blessed name of liberty, and the laws of their ancestors, which no length of time nor kind usage whatever will be able to eradicate.

Niccolo Machiavelli: The Prince, v, 1513

If you can win over the whole country so much the better. At least organize your partisans. The friendship of the neutral country is gained by requiring the soldiers to observe good discipline and by picturing your enemies as barbarous and badly intentioned; if they are Catholic, do not speak about religion; if they are Protestant, make the people believe that a false ardor for religion attaches you to them . . . However, move carefully with your partisans and always play a sure game.

Frederick The Great: Instructions for His Generals, x, 1747

Ths use of force alone is but *temporary*. It may subdue for a moment; but it does not remove the necessity of subduing again: and a nation is not governed, which is perpetually to be conquered.

Edmund Burke: Second speech on conciliation with America, 22 March 1775

We come to give you liberty and equality. But don't lose your heads about it—the first person who stirs without my permission will be shot.

Marshal Lefebvre: On occupying a Franconian town, 1807

Although I have served in my profession in several countries, and among foreigners, some of whom professed various forms of the Christian religion, while others did not profess it at all; I never was in one in which it was not the bounden duty of the soldier to pay proper deference and respect to whatever happened to be the religious institutions or ceremonies of the place.

Wellington: To the House of Lords, 8 April 1829

The conduct of a general in a conquered country is full of difficulties. If severe, he irritates and increases the number of his enemies. If lenient, he gives birth to expectations which only render the abuses and vexations inseparable from war the more intolerable. A victorious general must know how to employ severity, justness, and mildness by turns, if he would allay sedition or prevent it.

Napoleon I: Maxims of War, 1831

As the officers and soldiers of the United States have been subject to repeated insults from the women (calling themselves ladies) of New Orleans . . . it is ordered that hereafter when any female shall, by word, gesture, or movement, insult or show contempt for any officer or soldier of the United States, she shall be regarded and held liable to be treated as a woman of the town plying her avocation.

Major General B. F. Butler, USA: General Order No. 28, New Orleans, 15 May 1862

I have your letter of the 11th, in the nature of a petition to revoke my orders removing all inhabitants from Atlanta. I have read it carefully, and give full credit to your statements of the distress that will be occasioned, and yet shall not revoke my orders, because they were not designed to meet the humanities of the case, but to prepare for the future struggles . . .

W.T. Sherman: Letter to the Mayor of Atlanta, 12 September 1864

A nation cannot be kept permanently interned.

B.H. Liddell Hart: Defense of the West, 1950

. . . bureaucracy enforced by military law.

Alan Moorehead: The Blue Nile, vii, 1962

Military Mind

For even soldiers sometimes think—
Nay, Colonels have been known to
reason—
And reasoners, whether clad in pink,
Or red, or blue, are on the brink
(Nine cases out of ten) of treason.

Thomas Moore, 1779–1852

Nine soldiers out of ten are born fools.

George Bernard Shaw: Arms and the Man, i, 1894

I never expect a soldier to think.

George Bernard Shaw: The Devil's Disciple, iii, 1897

The professional military mind is by necessity an inferior and unimaginative mind; no man of high intellectual quality would willingly imprison his gifts in such a calling.

H.G. Wells: Outline of History, xl, 1920

The mind of the soldier, who commands and obeys without question, is apt to be fixed, drilled, and attached to definite rules.

Sir Archibald Wavell: Generals and Generalship, 1939

The soldier serves in small garrisons and exercises in cramped areas, while the sailor traverses the wide oceans and learns navigation as the staple of his craft. For him geography precedes gunnery.

B.H. Liddell Hart: Thoughts on War, v, 1944

The only thing harder than getting a new idea into the military mind is to get an old one out.

B.H. Liddell Hart: Thoughts on War, v, 1944

Prejudice against innovation is a typical characteristic of an Officer Corps which has grown up in a well-tried and proven system.

Erwin Rommel: Rommel Papers, ix, 1953

Civilians find it hard to credit soldiers with ordinary mental processes.

Barbara W. Tuchman: The Guns of August, xii, 1962

We soldiers, sailors, and airmen regard a military mind as something to be sought and developed—an indispensable professional asset which can only be acquired after years of training in, reflecting and acting on military and related problems.

General Maxwell D. Taylor, USA: Speech to the American Bar Association, Chicago, February 1964

(*See also* Militarism, Profession of Arms.)

Military Necessity

Extortions which are intolerable in their nature become excusable from the necessities of war.

The Bettman Archive

T. E. LAWRENCE

1888–1935

"Curse the Brass hats: poor reptiles."

Richelieu: Letter during Thirty Years' War, 1633

War, like the thunderbolt, follows its laws and turns not aside even if the beautiful, the virtuous and charitable stand in its path.

W.T. Sherman: Letter to Charles A. Dana, April 1864

Make war support war.

Military maxim

(*See also* Necessity.)

Militia

Because such as are apt to become men of war are to be of a perfect age most apt for all manner of services and best able to support and endure the infinite toils and continual hazards of wars, I have chosen all between the ages of eighteen and fifty to become trained soldiers.

Henry Knyvett: The Defence of the Realme, 1596

Raw in the fields the rude militia swarms,
Mouths without hands; maintain'd at vast expense,
In peace a charge, in war a weak defense.

John Dryden: Cyman and Iphigenia, 1699

Every citizen shall be a soldier from duty; none by profession. Every citizen shall be ready, but only when need calls for it.

Jean-Jacques Rousseau, 1712–1778

No militia will ever acquire the habits necessary to resist a regular force . . . The firmness requisite for the real business of fighting is only to be attained by a constant course of discipline and service. I have never yet been witness to a single instance that can justify a different opinion, and it is most earnestly to be wished that the liberties of America may no longer be trusted, in any material degree, to so precarious a dependence.

George Washington, 1732–1799

To place any dependence upon militia is assuredly resting upon a broken staff.

George Washington: Letter to the President of Congress, 24 September 1776

I am fully persuaded, that the Fensibles, Fusileers, or Train Bands formed of the Inhabitants of Cities and Incorporated towns will not afford that prompt and efficacious resistance to an Enemy, which might be expected from regularly established Light Infantry Companies.

George Washington: Sentiments on a Peace Establishment, 1783

The Congress shall have power . . .
To provide for calling forth the Militia to execute the Laws of the Union, suppress Insurrections and repel Invasions; To provide for organizing, arming, and disciplining, the Militia, and for governing such Part of them as may be employed in the Service of the United States, reserving to the States respectively, the Appointment of the Officers, and the Authority of training the Militia according to the discipline prescribed by Congress.

Constitution of the United States, I, 8, 1789

A well regulated Militia, being necessary to the security of a free State, the right of the people to keep and bear Arms, shall not be infringed.

Constitution of the United States, Amendment II, 15 December 1791

For a people who are free, and who mean to remain so, a well organized and armed militia is their best security.

Thomas Jefferson: Message to Congress, November 1808

The Greeks by their laws, and the Romans by the spirit of their people, took care to put into the hands of their rulers no such engine of oppression as a standing army. Their system was to make every man a soldier, and oblige him to repair to the standard of his country whenever that was reared. This made them invincible; and the same remedy will make us so.

Thomas Jefferson: Letter to Thomas Cooper, 1814

Such noble regiments they have. Three field officers, four staff officers, ten captains, thirty lieutenants, and one private with a misery in his bowels.

D. H. Hill; of the Confederate Militia, 24 April 1863

Every member of society who is fit for war can be taught, along with his other activities, to master the use of weapons, as much as is needed, not for taking part in parades, but for defending the country.

Frederich Engels, 1820–1895, Letter to Karl Marx

Every Man a Soldier.

Chinese Communist People's Militia slogan

(*See also* Reservists.)

Mines (Land)

Gentleman, I don't know whether we will make history tomorrow, but we will certainly change geography.

Sir Herbert Plumer: To press conference the day before the blowing up of Messines Ridge, 6 June 1917

Everything that is shot or thrown at you or dropped on you in war is most unpleasant but, of all horrible devices, the most terrifying . . . is the land mine.

Sir William Slim: Unofficial History, vi, 1959

Mines (Submarine)

Dawn off the Foreland—the young flood making
Jumbled and short and steep—
Black in the hollows and bright where it's breaking—
Awkward water to sweep.
"Mines reported in the fairway,
"Warn all traffic and detain.
" 'Sent up *Unity, Claribel, Assyrian, Stormcock*, and *Golden Gain*."

Rudyard Kipling: Mine Sweepers, 1915

The mine issues no official communiques.

Admiral William V. Pratt, USN: In Newsweek magazine, 5 October 1942

Mines never surrender.

Lieutenant Commander Arnold S. Lott, USN: Most Dangerous Sea, 1959

Minute Men

Stand your ground, men. Don't fire unless fired upon. But if they mean to have a war, let it begin here!

Captain Jonas Parker: To his company of Minute Men, Lexington Green, 19 April 1775

Miss

A miss is as good as a mile.

Walter Scott: Journal, 3 December 1825

Missiles

Could not explosives even of the existing type be guided automatically in flying machines by wireless or other rays, without a human pilot, in ceaseless procession upon a hostile city, arsenal, camp, or dockyard?

Winston Churchill: Thoughts and Adventures, 1925

It is clear that nowadays the Ground Forces cannot play their former decisive role, and the Queen of the battlefield has surrendered her crown to the Strategic Rocket Forces.

Colonel General S. Shtemenko: In Nedelya, 31 January 1965

Mistake

In war there is never any chance for a second mistake.

Lamachus, 465–414 B.C.

I am more afraid of our own mistakes than of our enemies' designs.

Pericles: Speech to the Athenians, 432 B.C.

When a man has committed no faults in war, he can only have been engaged in it but a short time.

Turenne: After the battle of Marienthal, 1645

I am not sorry that I went, notwithstanding what has happened. One may pick up something useful from among the most fatal errors.

James Wolfe: Of the Rochefort expedition, 1757

It must be a rare occurrence if a battle is fought without many errors and failures, but for which more important results would have been obtained, and the exposure of these diminishes the credit due, impairs the public confidence, undermines the morale of the army, and works evil to the cause for which men have died.

Jefferson Davis: Letter to Major General D.H. Hill, 3 October 1863, after Chickamauga

Errors toward the enemy must be lightly judged.

Winston Churchill: Their Finest Hour, 1949 (of the Dakar operation, 1940)

It is always a bad sign in an army when scapegoats are habitually sought out and brought to sacrifice for every conceivable mistake. It usually shows something wrong in the very highest command. It completely inhibits the willingness of junior commanders to make decisions, for they will always try to get chapter and verse for everything they do, finishing up more often than not with a miserable piece of casuistry instead of the decision which would spell release.

Erwin Rommel: The Rommel Papers, xviii, 1953

Happily for the result of the battle—and for me—I was, like other generals before me, to be saved from the consequences of my mistakes by the resourcefulness of my subordinate commanders and the stubborn valor of my troops.

Sir William Slim: Defeat into Victory, 1956

To inquire if and where we made mistakes is not to apologize. War is replete with mistakes because it is full of improvisations. In war we are always doing something for the first time. It would be a miracle if what we improvised under the stress of war should be perfect.

Vice Admiral H.G. Rickover, USN: Testimony before House Military Appropriations Subcommittee, April 1964

(*See also* Blunder.)

Mob

I am not fond of mobs, Madam.

Horace Walpole: Letter to the Countess of Upper Ossory, 17 February 1779

The mob, which everywhere is the majority, will always let itself be led by scoundrels.

Frederick The Great: Letter to Jean Rond d'Alembert, 8 September 1782

Mobs will never do to govern states or command armies.

John Adams: Letter to Benjamin Hitchborn, 27 January 1787

Mobility

Force does not exist for mobility, but mobility for force.

Mahan: Lessons of the War with Spain, 1899

Through mobility we conquer.

Motto, The Cavalry School, Fort Riley, c. 1930

Strange as it may seem, the Air Force, except in the air, is the least mobile of all the Services. A squadron can reach its destination in a few hours, but its establishment, depots, fuel, spare parts, and workshops take many weeks, and even months, to develop.

Winston Churchill: Their Finest Hour, 1949

Pace with variability is the secret of mobility.

B.H. Liddell Hart: Have Armored Forces a Future? 1950

(*See also* Movements.)

Mobilization

To your tents, O Israel.

I Kings, XII, 16

Prepare war, wake up the mighty men.

Joel, III, 9

We should provide in peace what we need in war.

Publilius Syrus: Sententiae, c. 42 B.C.

'Tis time to leave the books in dust,
And oil the unused armor's rust.

Andrew Marvell, 1621–1678

The young men shall fight; the married men shall forge weapons and transport supplies; women will make tents and serve in the hospitals; the children will make up old linen into lint; the old men will have themselves carried into the public squares to rouse the courage of the fighting men, and to preach hatred of kings and the unity of the Republic. The public buildings shall be turned into barracks, the public squares into munitions factories; the earthen floors shall be treated with lye to extract saltpeter. All suitable firearms shall be turned over to the troops; the interior shall be policed with fowling pieces and cold steel. All saddle horses shall be seized for the cavalry; all draft horses not employed in cultivation will draw the artillery and the supply wagons.

Decree by the Committee on Public Safety, French Revolution, 23 August 1793

Leave untended the herd,
The flock without shelter;
Leave the corpse uninterred,
The bride at the altar;
Leave the deer, leave the steer,
Leave nets and barges;
Come with your fighting gear,

Broadswords and targes.

Walter Scott: Pibroch of Donald Dhu, 1816

When a nation is without establishments and a military system, it is very difficult to organize an army.

Napoleon I: Maxims of War, 1831

Lars Porsena of Clusium . . .
Bade his messengers ride forth
To East and West and South and North
To summon his array.

T.B. Macaulay: Lays of Ancient Rome, 1842

Leaped to their feet a thousand men,
Their voices echoing far and near;
"We go, we care not, where or when;
"Our country calls us, we are here!"

The Seventh, verses by unknown author in Harper's Weekly, 27 April 1861 (to the New York 7th Regiment when called to the Colors

We are coming, Father Abraham, three hundred thousand more.

James Sloan Gibbons: Popular song, 16 July 1861, after Lincoln's call for additional volunteers to quell the Southern rebellion.

And there was tumult in the air,
The fife's shrill note, the drum's loud beat,
And through the wide land everywhere,
The answering tread of hurrying feet.

Thomas Buchanan Read, 1853–1916, The Wagoner of the Alleghenies

The flags of war like storm-birds fly,
The charging trumpets blow.

John Greenleaf Whittier, 1807–1892

The Army used to have all the time in the world and no money; now we've got all the money and no time.

George C. Marshall: Remark, January 1942

Monmouth (28 June 1778)

Tell the Philadelphia ladies that the heavenly, sweet, pretty Red Coats—the accomplished Gentlemen of the Guards and Grenadiers have humbled themselves on the plains of Monmouth.

Anthony Wayne: Letter, June 1778

Monroe Doctrine

The American continents . . . are not henceforth to be considered as subjects for future colonization by any European powers.

James Monroe: Message to Congress, 2 December 1823

I called the New World into existence to redress the balance of the Old.

George Canning: To the House of Commons, December 1823 (explaining the background of the Monroe Doctrine)

Monterey, (21 September 1846)

Old Zach's at Monterey.
Bring on yer Santa Anner;
For every time we raise a gun,
Down goes a Mexicaner.

Soldier song during the Mexican War

We were not many—we who stood
Before the iron sleet that day;
Yet many a gallant spirit would
Give half his years, if he but could
Have been with us at Monterey.

Charles Fenno Hoffman, 1806–1884, Monterey

(*See also* Mexican War.)

Morale

You are well aware that it is not numbers or strength that bring victories in war. No, it is when one side goes against the enemy with the gods' gift of a stronger morale that their adversaries, as a rule, cannot withstand them.

Xenophon: Speech to the Greek officers after the defeat of Cyrus at Cunaxa, 401 B.C.

An army is strengthened by labor and enervated by idleness.

Vegetius: De Re Militari, 378

Now when troops gain a favorable situation the coward is brave; if it be lost, the brave become cowards.

Lt Ch'uan, fl. 7th century A.D.

By no means does the outcome of battle depend upon numbers, but upon the united hearts of those who fight.

Kusunoki Masushige, fl. 14th century A.D.

The nature of bad news infects the teller.

Shakespeare: Antony and Cleopatra, i, 2, 1606

I like people who eat well before they fight. It is a good sign.
Maurice de Saxe, 1696–1750

The human heart is the starting point in all matters pertaining to war.
Maurice de Saxe: Mes Rêveries, 1732

A battle is lost less through the loss of men than by discouragement.
Frederick The Great: Instructions for His Generals, xx, 1747

Let us therefore animate and encourage each other, and show the whole world, that a Freeman contending for *Liberty* on his own ground is superior to any slavish mercenary on earth.
George Washington: General Order to the Continental Army, 2 July 1776

Morale makes up three quarters of the game: the relative balance of man-power accounts only for the remaining quarter.
Napoleon I, 1769–1821, Correspondence.

In war the moral is to the material as three to one.
Napoleon I, 1769–1821

In war, everything depends on morale; and morale and public opinion comprise the better part of reality.
Napoleon I, 1769–1821, Pensées

One fights well when his heart is light.
Napoleon I: To General Gaspard Gourgaud, St. Helena, 17 February 1816

A cherished cause and a general who inspires confidence by previous success are powerful means of electrifying an army and are conducive to victory.
Jomini: Précis de l'Art de la Guerre, 1838

It is the morale of armies, as well as of nations, more than anything else, which makes victories and their results decisive.
Jomini: Précis de l'Art de la Guerre, 1838

No system of tactics can lead to victory when the morale of an army is bad.
Jomini: Précis de l'Art de la Guerre, 1838

Universal suffrage, furloughs, and whiskey have ruined us.
Braxton Bragg: After Shiloh, 1862

Combat today requires . . . a moral cohension, a unity more binding than at any other time. If one does not wish the bonds to break, he must make them elastic in order to strengthen them.
Ardant du Picq, 1821–1870, Battle Studies

The men thought that victory was chained to my standard. Men who go into a fight under the influence of such feelings are next to invincible, and are generally victors before it begins.
John S. Mosby: War Reminiscences, vii, 1887

The [French] people had always concentrated on material questions. They thought that the offensive power of the enemy would be broken by the defensive action of new and terrible weapons. In that way they ruined the spirit of their army. That is what chiefly weighed in the scale.
Colmar von der Goltz: of the French defeat after 1870

Battles are beyond all else struggles of morale. Defeat is inevitable as soon as the hope of conquering ceases to exist. Success comes not to him who has suffered the least but to him whose will is firmest and morale strongest.
French Army Field Regulations, 1913

Moral forces may take a back seat at Committees of Imperial Defense or in War Offices; at the front they are put where Joab put Uriah.
Sir Ian Hamilton: The Soul and Body of an Army, x, 1921

The unfailing formula for production of morale is patriotism, self-respect, discipline, and self-confidence within a military unit, joined with fair treatment and merited appreciation from without. It cannot be produced by pampering or coddling an army, and is not necessarily destroyed by hardship, danger, or even calamity . . . It will quickly wither and die if soldiers come to believe themselves the victims of indifference or injustice on the part of their government, or of ignorance, personal ambition, or ineptitude on the part of their leaders.
Douglas MacArthur: Annual Report, Chief of Staff, U.S. Army, 1933

Morale is a state of mind. It is steadfastness and courage and hope. It is confidence and zeal and loyalty. It is elan, esprit de corps and determination. It is staying power, the spirit which endures to the end—the will to win. With it all things are possible, without it everything else, planning, preparation, production, count for naught.

George C. Marshall: Address at Trinity College, Hartford, Connecticut, 15 June 1941

The soldier's heart, the soldier's spirit, the soldier's soul, are everything. Unless the soldier's soul sustains him, he cannot be relied on and will fail himself and his country in the end.

George C. Marshall, 1880–1959

Morale, only morale, individual morale as a foundation under training and discipline, will bring victory.

Sir William Slim: To the officers, 10th Indian Infantry Division, June 1941

Nothing is more dangerous in wartime than to live in the temperamental atmosphere of a Gallup Poll, always feeling one's pulse and taking one's temperature.

Winston Churchill: To the House of Commons, 30 September 1941

Very many factors go into the building-up of sound morale in an army, but one of the greatest is that the men be fully employed at useful and interesting work.

Winston Churchill: The Gathering Storm, 1948

Machines are as nothing without men. Men are as nothing without morale.

E.J. King: Address to the graduating class, U.S. Naval Academy, 19 June 1942

Loss of hope, rather than loss of life, is the factor that really decides wars, battles, and even the smallest combats. The all-time experience of warfare shows that when men reach the point where they see, or feel, that further effort and sacrifice can do no more than delay the end they commonly lose the will to spin it out, and bow to the inevitable.

B.H. Liddell Hart: Defense of the West, 1950

The morale of the soldier is the greatest single factor in war and the best way to achieve a high morale in wartime is by success in battle.

Montgomery of Alamein: Memoirs, vi, 1958

Nothing raises morale better than a dead general.

John Masters: The Road Past Mandalay, 1961

If the history of military organizations proves anything, it is that those units that are told they are second-class will almost inevitably prove that they are second-class.

Brigadier General J.D. Hittle, USMC: in The National Guardsman, July 1962

Morale is neither produced nor destroyed by gadgetry.

Brigadier General S.B. Griffith, USMC: Communist China's Capacity to Make War, January 1965

There will be no liberty on board this ship until morale improves.

*Excerpt from Plan of the Day, USS * * **

I doubt if Mr. McNamara and his crew have any morale setting on their computers.

Rear Admiral Daniel V. Gallery, USN: Eight Bells and All's Well, 1965

(*See also* Esprit, Will to Fight.)

Mountain Warfare

Beyond the Alps lies Italy.

Attributed to Hannibal, 218 B.C.

Those who wage war in mountains should never pass through defiles without first making themselves masters of the heights.

Maurice de Saxe: Mes Rêveries, xxii, 1732

In regard to mountain warfare in general, everything depends on the skill of our subordinate officers and still more on the morale of our soldiers. Here it is not a question of skillful maneuvering, but of warlike spirit and wholehearted devotion to the cause.

Clausewitz: Principles of War, 1812

Mountain operations teach us . . . that in such a country a strong and heroic will is worth more than all the precepts in the world . . . One of the principal rules of this kind of war is, not to risk one's self in the valleys without securing the heights.

Shall I also say that in this kind of war, more than in any other, operations should be directed upon the communications of the enemy? And finally, that good temporary bases or lines of defense at the confluence of the great valleys, covered by strategic reserves, and combined with great mobility and frequent offensive movements, will be the best means of defending the country?

Jomini: Précis de l'Art de la Guerre, 1838

When will blood cease to flow in the mountains? When sugar cane grows in the snows.

Caucasian warriers' proberb

Movement

Aptitude for war is aptitude for movement.

Napoleon I: Maxims of War, 1831

Movement is the safety valve of fear.

B.H. Liddell Hart: Thoughts on War, xvi, 1944

(*See also* Mobility.)

Musket

Would to God this accursed instrument had never been invented . . . So many brave and valiant men would not have met their deaths at the hands very often of the greatest cowards who would not so much as dare look at the man whom they knock down from a distance with their accursed balls.

Marshal Montluc: Commentaires, 1592

Though her sight was not long and her
weight was not small,
Yet her actions were winning, her
language was clear;
And everyone bowed when she opened the
ball
On the arm of some high-gaitered, grim
grenadier.
All Europe admitted the striking success
Of the dances and routs that were given by
Brown Bess.

Rudyard Kipling: Brown Bess, 1911 ("Brown Bess" was the nickname for the standard British Army musket, 1700–1815)

Musketry

This is the music that pleases me.

Victor Emmanuel II of Italy: When he first heard the roar of musketry at Goito, 1848

My opinion is that there ought not to be much firing at all. My idea is that the best mode of fighting is to reserve your fire till the enemy get—or you get them—to close quarters. Then deliver one deadly, deliberate volley—and charge!

Stonewall Jackson, in 1863

Considerable results may be secured by expending a relatively small and easily provided quantity of ammunition provided the fire is well directed.

Ferdinand Foch: Precepts, 1919

Mutiny

. . . a sudden flood of mutiny.

Shakespeare: Julius Caesar, iii, 2, 1599

If he had been deserving of the character you now give him, he would not have been guilty of the crime for which he is condemned.

Lord St. Vincent: Of a seaman convicted of mutiny given a good character in mitigation, 1797

That long official neglect of intolerable grievance, and inexcusable supineness towards measures of progressive improvement had, as they sooner or later infallibly do, at last aroused illegal and exasperated enforcement of redress.

Jedediah Tucker: Memoirs of Earl St. Vincent, 1830 (of the British naval mutinies of 1797)

Why does Colonel Grigsby refer to me to learn how to deal with mutineers? He should shoot them where they stand.

Stonewall Jackson: On receiving a report of refusal of duty by 12-months volunteers, May 1862

Mutiny and revolution are words which do not occur in the vocabulary of a German soldier

General Ludwig Beck, 1880–1944

Mutinies are suppressed in accordance with laws of iron which are eternally the same.

Adolph Hitler: To the Reichstag, July 1934

Napoleon Bonaparte (1769–1821)

. . . this horrid disturber of the peace of mankind.
Lord St. Vincent: Letter, 1812

If there is one soldier among you who wishes to kill his Emperor, he can do it. Here I am! (S'il est parmi vous un soldat qui veuille tuer son Empereur, il peut le faire. Me voila!)
Napoleon I: On his return from Elba, outside Grenoble, 6 March 1814

Bonaparte was a lion in the field only. In civil life, a cold-blooded, calculating unprincipled usurper, without a virtue; no statesman, knowing nothing of commerce, political economy, or civil government, and supplying ignorance by bold presumption.
Thomas Jefferson: Letter to John Adams, July 1814

Had I succeeded, I should have died with the reputation of the greatest man that ever lived. As it is, although I have failed, I shall be considered as an extraordinary man. I have fought fifty pitched battles, almost all of which I have won. I have framed and carried into effect a code of laws that will bear my name to the most distant posterity.
Napoleon I: Letter to Barry E. O'Meara, St. Helena, 3 March 1817

What a romance my life has been.
Napoleon I: On St. Helena, 1817

Thirteen and a half years of success turned Alexander the Great into a kind of madman. Good fortune of exactly the same duration produced the same disorder in Napoleon. The only difference was that the Macedonian hero was lucky enough to die.
Stendhal (Henri Beyle): A Life of Napoleon, xlv, 1818

On May 15th, 1796, General Bonaparte entered Milan at the head of that young army which shortly before had crossed the Bridge at Lodi and taught the world that after all these centuries Caesar and Alexander had a successor.
Stendhal (Henri Beyle): The Charterhouse of Parma, 1840

Tête d'Armée! (Head of the Army!)
Napoleon I: Last words, 5 May 1821

Whose game was empires, and whose stakes were thrones,
Whose table earth—and whose dice were human bones.
Byron: The Age of Bronze, 1823 (of Napoleon)

Bonaparte I never saw; though during the Battle of Waterloo we were once, I understand, within a quarter of a mile of each other. I regret it much; for he was a most extraordinary man.
Wellington: Quoted in Recollections, Samuel Rogers, 1827

Napoleon was the first man of his day on a field of battle, and with French troops.
Wellington, 1769–1852

Bonaparte's whole life, civil, political, and military, was a fraud. There was not a transaction, great or small, in which lying and fraud were not introduced . . . Bonaparte's foreign policy was force and menace, aided by fraud and corruption. If the fraud was discovered, force and menace succeeded.
Wellington, 1769–1852

Although too much of a soldier among sovereigns, no one could claim with better right to be a sovereign among soldiers.
Walter Scott: Life of Napoleon, 1827

His life was the stride of a demi-god, from battle to battle, and from victory to victory.
Goethe: Conversations with Eckermann, 11 March 1828

The French emperor is among conquerors what Voltaire is among writers, a miraculous child. His splendid genius was frequently clouded by fits of humor as absurdly perverse as those of the pet of the nursery, who quarrels with his food, and dashes his playthings to pieces.
T.B. Macaulay: Hallam, September 1828

In my youth we used to march and countermarch all the Summer without gaining or losing a square league, and then we went into winter quarters. And now comes an ignorant, hot-headed young man who flies about from Boulogne to Ulm, and from Ulm to the middle of Moravia, and fights battles in December. The whole system of his tactics is monstrously incorrect.
Ascribed to "an old German officer" by Macaulay, Moore's Life of Byron, 1831

He was sent into this world to teach generals and statesmen what they should avoid. His victories teach what may be accomplished by activity, boldness, and skill; his disasters what might have been avoided by prudence.

Jomini: Prêcis de l'Art de la Guerre, 1838

The man was a Divine Missionary, though unconscious of it; and preached, through the cannon's throat, that great doctrine,—"La carrière ouverte aux talent"—which is our ultimate political evangel.

Thomas Carlyle: On Heroes and Hero-Worship, 1841

Napoleon attempted the impossible, which is beyond even genius.

Ardant du Picq, 1821–1870, Battle Studies

He could use political and social ideas for the purposes of his ambition as dexterously as cannon; but in character he was a Corsican and as savage as any bandit of his isle. If utter selfishness, if the reckless sacrifice of humanity to your own interest and passions be vileness, history has no viler name.

Goldwin Smith: Three English Statesmen, 1867

For Order's cause he labored, as inclined
A soldier's training and his Euclid mind.

George Meredith: Napoleon, 1891

In certain respects, Napoleon was the greatest of all soldiers. He had, to be sure, the history of other great captains to profit by; he had not to invent; he had only to improve. But he did for the military art what constitutes the greatest advance in any art, he reduced it to its most simple, perfect form.

T.A. Dodge: Great Captains, 1895

Napoleon seems to have ended by regarding mankind as a troublesome pack of hounds only worth keeping for the sport of hunting with them.

Geroge Bernard Shaw: The Revolutionist's Handbook, iii, 1903

To husband his troops; to use them judiciously so that the enemy might be attacked at his weakest point with superior forces; to keep control of his men, even when they were scattered, much as a coachman holds the reins, so that they could be concentrated at a moment's notice; to mark down that portion of the opposing army which he aimed at destroying; to discern the critical point where defeat might be turned into rout; to surprise the enemy by the rapidity of his conceptions and operations—those are a few of the essential elements of Napoleon's military genius.

Ferdinand Foch, 1851–1929

National Policy

Where there is no vision, the people perish.

Proverbs, XXIX, 18

Peace is best secured by those who use their strength justly, but whose attitude shows that they have no intention of submitting to wrong.

Address of the Corinthians to the Athenians, 433 B.C.

Whoever is strongest at sea, make him your friend.

Address of the Corcyraeans to the Athenians, 433 B.C.

To an imperial city nothing is inconsistent which is expedient.

Euphemus of Athens: Address to the Camarinaeans, 415 B.C.

Courageous men ought to exchange peace for war as soon as they have been wronged; when they have brought the war to a successful issue, peace may be made with the enemy; but no-one should be uplifted unduly by success in war, nor should any submit to injustice because unwilling to sacrifice the calm delights of peace.

Thucydides: History of the Peloponnesian Wars, c. 404 B.C.

The rulers of the States are the only ones who should have the privilege of lying, either at home or abroad; they may be allowed to lie for the good of the state.

Plato, 428–347 B.C., The Republic, iii

It is no doubt a good thing to conquer on the field of battle, but it needs greater wisdom and greater skill to make use of victory.

Polybius: Histories, x, c. 125 B.C.

Armed forces abroad are of little value unless there is prudent counsel at home. (Parvi enim sunt foris arma, nisi et consilium domi.)

Cicero, 106–43 B.C.

Underwood & Underwood

A. P. WAVELL

1883–1950

"War is a wasteful, boring, and muddled affair."

Let them hate us as long as they fear us. (Oderint dum metuant.)
Caligula, 12–41 A.D.

Great empires are not maintained by timidity.
Tacitus, 55–117 A.D., History

Peace with honor.
Theobald of Champagne: Letter to Louis VI, c. 1125

It is fighting at a great disadvantage to fight those who have nothing to lose.
Francesco Guiciardini: Storia d'Italia, 1564

War ought never to be accepted, until it is offered by the hand of necessity.
Sir Philip Sidney, 1554–1586

No nation need expect to be great unless it makes the study of arms its principal honor and occupation.
Francis Bacon, 1561–1626

'Tis safest making peace with sword in hand.
George Farquhar, 1678–1707, Love and a Bottle, v

A wise government knows how to enforce with temper or to conciliate with dignity.
George Grenville: To the House of Commons on a motion to expel John Wilkes, 1769

Defense is superior to opulence.
Adam Smith: The Wealth of Nations, 1776

If we desire to avoid *insult*, we must be able to *repel it*. If we desire to secure *peace*, it must be known that we are at all times *ready for war*.
George Washington, 1732–1799

To make war with those who trade with us is like setting a bulldog upon a customer at the shop-door.
Thomas Paine: The Age of Reason, 1794

That nation is worthless which does not joyfully stake everything in defense of her honor.
Schiller: The Maid of Orleans, i, 1801

If you wish to avoid foreign collision, you had better abandon the ocean.
Henry Clay: To the House of Representatives, 22 January 1812

Woe to the Monarch who depends
Too much on his red-coated friends.
Thomas Moore, 1779–1852

Wherever the standard of freedom and independence has been or shall be unfurled, there will be America's heart, her benedictions, and her prayers. But she goes not abroad in search of monsters to destroy.
John Quincy Adams: Address in Washington, 4 July 1821

War is not merely a political act, but also a political instrument, a continuation of political relations, a carrying out of the same by other means.
Clausewitz: On War, 1832

A great country cannot wage a little war.
Wellington: To the House of Lords, 16 January 1838

It is a narrow policy to suppose that this country or that is to be marked out as the eternal ally or the perpetual enemy . . . We have no eternal allies, and we have no perpetual enemies. Our interests are eternal and perpetual, and those interests it is our duty to follow.
Lord Palmerston: To the House of Commons, 1848

It is not worthy for a great State to fight for a cause which has nothing to do with its own interest.
Otto von Bismarck: Letter, 1850

The United States is not a nation to which peace is a necessity.
Grover Cleveland: Fourth Annual Message to Congress, 7 December 1896

The United States in her turn may have the rude awakening of those who have abandoned their share in the common birthright of all people, the sea.
Mahan: The Influence of Sea Power Upon History, i, 1890

The question for the United States, as regards the size of its Navy, is not so much what it desires to accomplish as what it is willing or unwilling to concede.
Alfred Thayer Mahan, 1840–1914

While we are conducting war and until its conclusion we must keep all we get; when the war is over we must keep what we want.
William McKinley: Private memorandum, 1898

Speak softly and carry a big stick; you will go far.
Theodore Roosevelt: Address 2 September 1901

The right is more precious than peace.
Woodrow Wilson: War message to Congress, 2 April 1917

In order to assure an adequate national defense, it is necessary—and sufficient—to be in a position, in case of war, to conquer the command of the air.
Giulio Douhet: The Command of the Air, 1921

The high contracting parties solemnly declare in the names of their respective peoples that they condemn recourse to war for the solution of international controversies, and renounce it as an instrument of national policy in their relations with one another.
The Kellogg Peace Pact, signed at Paris, 27 August 1928 (Originally proposed by Aristide Briand, and signed by the United States, Great Britain, Germany, Italy, France, Belgium, Japan, Poland, and Czechoslovakia)

About politics one can make only one completely unquestionable generalization which is that it is quite impossible for statesmen to foresee, for more than a very short time, the results of large-scale political action.
Aldous Huxley: Grey Eminence, x, 1941

The responsibility of the great states is to serve and not to dominate the world.
Harry S. Truman: Message to Congress, 16 April 1945

Freedom of navigation is an important objective of American foreign policy.
Cavendish W. Cannon: Speech while American Ambassador, Belgrade, 18 August 1948 (rejecting Russian-dictated treaty intended to control the Danube)

In War: Resolution
In Defeat: Defiance
In Victory: Magnanimity
In Peace: Good Will.
Winston Churchill: The Gathering Storm (Moral of the Work), 1948

No foreign policy can have validity if there is no adequate force behind it and no national readiness to make the necessary sacrifices to produce that force.
Winston Churchill: The Gathering Storm, xx, 1948

There is no room in war for pique, spite, or rancor.
Winston Churchill: Their Finest Hour, 1949

This strategy would involve us in the wrong war, at the wrong place, at the wrong time, and with the wrong enemy.
Omar N. Bradley: To the Senate Committees on Armed Services and Foreign Relations, regarding the possibility of war with Communist China, 15 May 1951

American armed strength is only as strong as the combat capabilities of its weakest service. Overemphasis on one or the other will obscure our compelling need—not for air-power, sea-power, or land-power—but for American military power commensurate with our tasks in the world.
Omar N. Bradley, 1893–

A doctrine which began by defining war as only a continuation of state policy by other means led to the contradictory end of making policy the slave of strategy.
B.H. Liddell Hart: Strategy, 1954 (of Clausewitz)

A military philosophy and that somewhat more tangible thing—a military policy—are the product of many factors. A philosophy grows from the minds and hearts, social mores and customs, traditions and envoronment of a people. It is the product of national and racial attributes, geography, the nature of a potential enemy threat, standards of living and national traditions, influenced and modified by great military philosophers like Clausewitz and Mahan, and by great national leaders like Napoleon.
Hanson W. Baldwin: In New York Times, 3 November 1957

I would rather stand alone than among those influenced by the carrot or the stick.
John Foster Dulles, 1888–1959

Let every nation know, whether it wishes us well or ill, that we shall pay any price, bear any burden, meet any hardship, support any friend, oppose any foe, to assure the survival and success of liberty.

John F. Kennedy: Inaugural Address, 20 January 1961

(*See also* Policy.)

National Power

In arriving at this decision and resolving not to go to war, the Lacadaemonians were influenced, not so much by the speeches of their allies, as by the fear of the Athenians and their increased power.
Thucydides: History of the Peloponnesian Wars, c. 404 B.C.

For it is not profusion of riches or excess of luxury that can influence our enemies to court or respect us. This can only be effected by fear of our arms.
Vegetius: De Re Militari, i, 378

For empire and greatness it importeth most that a nation do profess arms as their principal honor, study, and occupation.
Francis Bacon, 1561–1626, Essays

Let not thy will roar, when thy power can but whisper.
Thomas Fuller, 1654–1734

Have money and a good army; they insure the glory and safety of a prince.
Frederick William I of Prussia: To his son, later Frederick the Great, 1724

They say they can obtain land and people for the King with the pen; but I say it can be done only with the sword.
Frederick William I of Prussia, 1688–1740

Threats without power are like powder without ball.
Nathan Bailey, d. 1742, Dictionary

. . . the balance of power.
Robert Walpole: To the House of Commons, 13 February 1741

I am tolerably certain that, while the United States of America pursue a just and liberal conduct, *with twenty sail of the line at sea*, no nation on earth will dare to insult them.
Gouverneur Morris: Letter, 1794

To secure a respect to a neutral flag requires a naval force organized and ready to vindicate it from insult or aggression.
George Washington: Farewell Address to Congress, 7 December 1796

Our country demands a most vigorous exertion of her force, directed with judgment.
Nelson: To Sir Hyde Parker, 24 March 1801

Every power has an interest in seeing its neighbors in a state of weakness and decadence.
Stendhal (Henri Beyle): A Life of Napoleon, xxix, 1818

There are no manifestoes like cannon and musketry.
Attributed to Wellington, 1769–1852

Iron weighs at least as much as gold in the scales of military strength.
Jomini: Précis de l'Art de la Guerre, 1838

A great people may be killed, but they cannot be intimidated.
Napoleon I: Political Aphorisms, 1848

Without finances, without assured means of recruitment, and without a fleet, there can be no king.
Napoleon I: Political Aphorisms, 1848

You cannot afford to be a carp in a pond where there are pike about.
Otto von Bismarck, 1815–1898

When it is remembered that the United States, like Great Britain and like Japan, can be approached only by sea, we can scarcely fail to see that upon the sea primarily must be found our power to secure our own borders and to sustain our external policy.
Mahan: Naval Strategy, 1911

We are listened to because we are strong.
Admiral Richard Wainwright, USN: in Scientific American, 9 December 1911

Power emanates from the barrel of a gun.
Mao Tse-tung: On Guerrilla War, 1938

We have tried since the birth of our nation to promote our love of peace by a display of weakness. This course has failed us utterly.

George C. Marshall: Report of the Chief of Staff, U.S. Army, 1945

National strength lies only in the hearts and spirits of men.

S.L.A. Marshall: Men Against Fire, 1947

Diplomacy has rarely been able to gain at the conference table what cannot be gained or held on the battlefield.

General Walter Bedell Smith, USA: on his return from the Geneva Conference on Indo-China. 1954

No country can have great power and a quiet conscience.

Editorial, The Washington Post, 26 April 1965

A mature great power will make measured and limited use of its power. It will eschew the theory of a global and universal duty which not only commits it to unending wars of intervention but intoxicates its thinking with the illusion that it is a crusader for righteousness.

Walter Lippmann: In the Washington Post, 27 April 1965

(*See also* Power.)

National Security

The principal foundations of all states are good laws and good arms; and there cannot be good laws where there are not good arms.

Niccolo Machiavelli: The Prince, xii, 1513

If you believe the doctors, nothing is wholesome; if you believe the theologians, nothing is innocent; if you believe the soldiers, nothing is safe.

Lord Salisbury, 1830–1903

Every danger of a military character to which the United States is exposed can be best met outside her own territory—at sea.

Alfred Thayer Mahan, 1840–1914

No nation ever had an army large enough to guarantee it against attack in time of peace or insure it victory in time of war.

Calvin Coolidge, 1872–1933

Naval Aviation

The Secretary of the Navy has decided that the science of aerial navigation has reached that point where aircraft must form a large part of our naval force for offensive and defensive operations.

Navy Department news release, 10 January 1914

These naval airmen, bold fellows, always on for an adventurous attack . . .

Sir Ian Hamilton: Diary entry, Gallipoli, 8 July 1915

It is impossible to resist an admiral's claim that he must have complete control of, and confidence in, the aircraft of the battle fleet, whether used for reconnaissance, gunfire, or air attack on a hostile fleet . . . The argument that similar conditions obtain in respect of Army cooperation aircraft cannot be countenanced. In one case the aircraft take flight from aerodromes and operate under precisely similar conditions to those of normal independent air force action. Flight from warships and action in connection with naval operations is a totally different matter. One is truly an affair of cooperation only; the other an integral part of modern naval operations.

Winston Churchill: Memorandum for Lord Inskip, 1936

As long as there are aircraft carriers, their aircraft and crews had better be part of the navy.

Sir John Slessor: Strategy for the West, 1954

(*See also* Aerial Combat, Air Power, Aircraft Carrier.)

Naval Officer

He apprehends him fit for a sea officer, by reason he hath been with him a whole year at sea in three engagements, where he saith he behaved himself like a gentleman and an understanding man, a character which I confess wants a good deal of that which must lead me to think a man fit to make a sea officer of: I mean downright diligence, sobriety, and seamanship.

Samuel Pepys: Letter to Captain Rooth, 24 January 1674

Gentlemen shall not be capable of bearing office at sea, except they be tarpaulins too; that is to say, except they are so trained up by a continued habit of living at sea, that they may have a right to be admitted free denizens of Wapping. When a gentleman is preferred at sea, the tarpaulin is very apt to impute it to friend or favour: but if that gentleman hath before his preferment passed through all the steps which lead to it, so that he smelleth as much of pitch and tar, as those that were swaddled in sail-cloth; his having an escutcheon will be far from doing him harm.

Marquess of Halifax: A New Model at Sea, 1694

An upright, God-fearing man, not dainty about his food or drink, robust and alert, with good sea-legs, and in strong voice to give commands to all hands; pleasant and affable in conversation, but imperious in his commands, liberal and courteous to defeated enemies, knowing everything that concerns the handling of the ship.

Samuel de Champlain, 1567–1635, Treatise on Seamanship

A Captain of the Navy ought to be a man of Strong and well connected Sense with a tolerable education, a Gentleman as well as a Seaman both in Theory and Practice.

John Paul Jones: Letter to Joseph Hewes, 19 May 1776

None other than a Gentleman, as well as a Seaman both in Theory and Practice is qualified to support the Character of a Commission Officer in the Navy, nor is any Man fit to Command a Ship of War, who is not also capable of communicating his Ideas on Paper in Language that becomes his Rank.

John Paul Jones, 1747–1792 (This quotation appears to be the basis of the widely quoted, wholly apocryphal passage which next follows.)

It is by no means enough that an officer of the Navy should be a capable mariner. He must be that of course, but also a great deal more. He should be as well a gentleman of liberal education, refined manners, punctilious courtesy, and the nicest sense of personal honor.

Attributed to John Paul Jones, 1747–1792, but almost certainly composed long after his death.

The business of a naval officer is one which above all others, needs daring and decision.

William S. Sims, 1858–1936

(*See also* Officer.)

Naval Operations

Knowledge of naval matters is an art as well as any other and not to be attended to at idle times and on the by . . .

Pericles, fl 460 B.C.

Skill in naval affairs, as in other crafts, is the result of scientific training. It is impossible to acquire this skill unless the matter be treated as of the first importance and all other pursuits are considered to be secondary to it.

Thucydides: History of the Peloponnesian Wars, c. 404 B.C.

Be it wind, be it weet, be it hail, be it sleet,
Our ships must sail the foam.

Ballad of Sir Patrick Spens, 15th century

We are as neare to heaven by sea as by land.

Sir Humphrey Gilbert: During a North Atlantic storm, before the loss of his ship, the Squirrel, *September 1583*

Ye gentlemen of England
That live at home at ease
Ah! little do you think upon
The dangers of the seas.

Song by Martyn Parker, d. 1656

The sea service is not so easily managed as that of the land. There are many more precautions to take and you and I are not capable of judging them.

Duke of Marlborough, 1650–1722, to a fellow Army officer

It follows then as certain as night succeeds day, that without a decisive naval force we can do nothing definitive, and with it everything honorable and glorious.

George Washington: Letter to Marquis de Lafayette, 15 November 1781

Had we taken ten sail, and had allowed the eleventh to escape, when it had been possible to have got at her, I could never have called it well done.

Nelson: Letter after the action of 14 March 1795

A lot of our destroyers and small craft are bumping into one another under the present hard conditions of service. We must be very careful not to damp the ardor of officers in the flotillas by making heavy weather of occasional accidents. They should be encouraged to use their ships with wartime freedom, and should feel they will not be considered guilty of unprofessional conduct, if they have done their best and something or other happens.

Winston Churchill: Note for First Sea Lord, 24 September 1939

You may look at the map and see flags stuck in at different points and consider that the results will be certain, but when you get out on the sea with its vast distances, its storms and mists, and with night coming on, and all the uncertainties which exist, you cannot possibly expect that the kind of conditions which would be appropriate to the movements of armies have any application to the haphazard conditions of war at sea.

Winston Churchill: To the House of Commons, 11 October 1940

On land I am a hero; at sea, I am a coward.

Adolph Hitler: To his Commanders in Chief, 1940

Naval Warfare

With fifty sail of shipping we shall do more good upon their own coast than a great many more will do here at home.

Sir Fancis Drake: During defense planning against the Armada, 1588

My opinion is to go out as soon and as strong as we can and fight the enemy's fleet if they be at sea.

Prince Rupert of the Rhine: To the Earl of Arlington, Dutch Wars, 1672

Naval tactics are based upon conditions the chief causes of which, namely the arms, may change; which in turn causes necessarily a change in the construction of ships, in the manner of handling them, and so finally in the disposition and handling of fleets.

Sébastien-François Bigot de Morogues: Tactique Navale, 1763

Naval tactics, or the art of war at sea, is limited by the possibilities of navigation; and is therefore much less capable of that variety of stratagems which belongs to the hostility of armies.

David Steel: The Elements of Naval Tactics, 1800

Take, sink, burn, or destroy the enemy fleet.

Lord St. Vincent: Orders to Nelson off Toulon, 21 May 1798

Superiority in naval power will henceforth consist in keeping up a proper naval establishment in discipline. The first naval nation to fall will be the one that is first caught napping.

Sir Charles Napier, 1786–1860

The naval strength of the enemy should be the first objective of the forces of the maritime Power both on land and sea.

Colonel G.F.R. Henderson, 1854–1903

The first striking difference between military and naval warfare is that, while—in theory at least—the military forces of a country confine their attacks to the persons and power of their enemy, the naval forces devote themselves primarily to the plunder of his property and commerce.

James A. Farrer: Military Manners and Customs, 1885

The steel decks rock with the lightning
shock, and shake with the great recoil,
And the sea grows red with blood of the
dead and reaches for his spoil—
But not till the foe has gone below or turns
his prow and runs,
Shall the voice of peace bring sweet release to the men behind the guns!

John J. Rooney: The Men Behind the Guns, 1898

In giving up the offensive, the Navy gives up its proper sphere.

Mahan: Naval Strategy, 1911

In war, the proper objective of the Navy is the enemy's navy.

Mahan: Naval Strategy, 1911

Dined with the Admiral . . . A glorious dinner. The sailormen have a real pull over us soldiers in all matters of messing. Linen, plate, glass, bread, meat, wine; of the best, are on the spot, always: even after the enemy is sighted, if they happen to feel a sense of emptiness they have only to go to the cold sideboard.

Sir Ian Hamilton: Diary entry off Gallipoli, 17 June 1915

The whole principle of naval fighting is to be free to go anywhere with every damned thing the Navy possesses.

Sir John Fisher: Memories, 1919

It has been computed that in shore fighting it takes several tons of lead to kill one man: at sea one torpedo can cause the death of many hundreds. On shore the soldier is in almost perpetual discomfort, if not misery—at sea the sailor lives in comparative comfort until the moment comes when his life is required of him.
Admiral of the Fleet Sir Rosslyn Wester-Wemyss: The Navy in the Dardanelles Campaign, 1924

The advantage of sea-power used offensively is that when a fleet sails no one can be sure where it is going to strike.
Winston Churchill: Their Finest Hour, 1949

There is no *blitzkrieg* possible in naval warfare—no lightning flash over the seas, striking down an opponent. Seapower acts more like radium—beneficial to those who use it and are shielded, it destroys the tissues of those who are exposed to it.
B.H. Liddell Hart: Defense of the West, 1950

The seas are no longer a self contained battlefield. Today they are a medium *from which warfare is conducted.* The oceans of the world are the base of operations from which navies project power onto land areas and targets . . . The mission of protecting sea-lanes continues in being, but the Navy's central missions have become to maximize its ability to project power from the sea over the land and to prevent the enemy from doing the same.
Timothy Shea: Project Poseidon, February 1961

(*See also* Naval Operations, Sea Power.)

Navigation

A collision at sea can ruin your entire day.
Attributed to Thucydides, 5th century B.C.

What can be more difficult than to guyde a shyppe engoulfed, when only water and heaven may be seen?
Martin Cortes: Breve Compendio de la Arte de Navegar, 1551 (Richard Eden's translation)

Navigation is that excellent art which demonstrateth . . . how a sufficient ship may bee conducted the shortest good way from place to place.
Captain John Davis: The Seaman's Secret, 1594 (a close paraphrase if not plagiarism of Dr. John Dee's remarks on navigation in his 1570 introduction to Billingsley's Euclid)

You have done your duty in this remonstrance; now obey my orders, and lay me alongside the French Admiral.
Sir Edward Hawke: To his Fleet Navigator after being advised to turn away from the enemy for fear of grounding, Quiberon Bay, 20 November 1759

The winds and waves are always on the side of the ablest navigators.
Edward Gibbon: Decline and Fall of the Roman Empire, lxviii, 1776–1787

I own myself one of those who do not fear the shore, for hardly any great things are done in a small ship by a man that is.
Nelson: To the Admiralty, urging leniency in the case of a captain who had grounded his ship on inshore patrol, November 1804

The most advanced nations are always those who navigate the most.
R.W. Emerson, 1802–1883, Society and Solitude

Our navigator's a jolly tar;
He shot the truck-light for a star,
And wonders where in the hell we are,
In the Armored Cruiser Squadron.
Navy Song, The Armored Cruiser Squadron, c. 1904

Every drunken skipper trusts to Providence. But one of the ways of Providence with drunken skippers is to run them on the rocks.
George Bernard Shaw, Heartbreak House, 1913

Absolute freedom of navigation upon the seas, outside territorial waters, alike in peace and in war, except as the seas may be closed in whole or in part by international action for the enforcement of international covenants.
Woodrow Wilson: To Congress 8 January 1918 (Point II of the Fourteen Points)

Navy

Minos [of Crete] is the first to whom tradition ascribes the possession of a navy.

Thucydides: History of the Peloponnesian Wars, c. 404 B. C.

Their strength lay in the greatness of thier navy, and by that and that alone they gained their empire.

Alcibiades: Of the Athenians, 415 B. C.

King Solomon made a navy of ships in Eziongeber, which is beside Eloth, on the shore of the Red Sea.

I Kings, IX, 26

They that go down to the sea in ships, and occupy their business in great waters; these men see the works of the Lord, and his wonders in the deep.

Psalm CVII

Hearts of oak are our ships,
Jolly tars are our men,
We are always ready, steady, boys, steady,
We'll fight and we'll conquer again and again.

David Garrick: Hearts of Oak, 1759

. . . that they may be a safeguard unto the United States of America, and a security for such as pass on the seas upon their lawful occasions.

Book of Common Prayer: Prayer for the Navy

Of all the public services, that of the Navy is the one in which tampering may be of the greatest danger, which can worst be supplied in an emergency, and of which any failure draws after it the largest and heaviest train of consequences.

Edmund Burke: To the House of Commons, 1769

Without a Respectable Navy—alas America!

John Paul Jones: Letter to Robert Morris, 17 October 1776

Yr Excellency will have observed that whatever efforts are made by the Land Armies, the Navy must have the casting vote in the present contest.

George Washington: Letter to the Comte de Grasse, 28 October 1781

A naval force can never endanger our liberties, nor occasion bloodshed; a land force would do both.

Thomas Jefferson: Letter to James Monroe, 1786

. . . the ruinous folly of a navy.

Thomas Jefferson: During the Tripolitan War, 1805

Of all careers, the navy is the one which offers the most frequent opportunities to junior officers to act on their own.

Napoleon: Political Aphorisms, 1848

Navies are not all for war.

Matthew Fontaine Maury, 1806–1873

Eternal Father! strong to save,
Whose arm hath bound the restless wave,
Who bidd'st the mighty ocean deep
Its own appointed limits keep:
O, hear us when we cry to Thee
For those in peril on the sea!

William Whiting: The Navy Hymn, 1860

Nor must Uncle Sam's web-feet be forgotten. At all the watery margins they have been present. Not only on the deep sea, the broad bay, and the rapid river, but also up the narrow muddy bayou, and wherever the ground was a little damp, they have been and made their tracks.

Abraham Lincoln: Letter to J.C. Conkling, 26 August 1863

Navies do not dispense with fortifications nor with armies, but when wisely handled they may save a country the strain which comes when these have to be called into play.

Mahan: The Influence of Sea Power Upon the French Revolution and Empire, 1892

The necessity of a navy springs from the existence of peaceful shipping.

Alfred Thayer Mahan, 1840–1914

In the brilliant exhibition of enterprise, professional skill, and usual success, by its naval officers and seamen, the country has forgotten the precedent neglect . . . to constitute the Navy as strong in proportion to the means of the country as it was excellent through the spirit and acquirements of its officers.

Mahan: Sea Power in Its Relations to the War of 1812, 1905

If the American nation will speak softly and yet build and keep at a pitch of the highest training a thoroughly efficient Navy, the Monroe Doctrine will go far.

Theodore Roosevelt: Speech, 2 September 1901

The Navy of the United States is the right arm of the United States and is emphatically the peacemaker.

Theodore Roosevelt, 1858–1919

Men go into the Navy . . . thinking they will enjoy it. They do enjoy it for about a year, at least the stupid ones do, riding back and forth quite dully on ships. The bright ones find that they don't like it in half a year, but there's always the thought of that pension if only they stay in. So they stay . . . Gradually they become crazy. Crazier and crazier. Only the Navy has no way of distinguishing between the sane and the insane. Only about 5 percent of the Royal Navy have the sea in their veins. They are the ones who become captains. Thereafter, they are segregated on their bridges. If they are not mad before this, they go mad then. And the maddest of these become admirals.

Attributed to George Bernard Shaw, 1856–1950

The Navy is very old and very wise.

Rudyard Kipling: The Fringes of the Fleet, 1915

We must have a navy so strong and so well proportioned and equipped, so thoroughly ready and prepared, that no enemy can gain command of the sea and effect a landing in force on either our western or our eastern coast.

Republican party platform, 1916

They do not realize that the Army is so absolutely different from the Navy. Every condition in them both is different. The Navy is always at war, because it is always fighting winds and waves and fog. The Navy is ready for an instant blow . . . The ocean is limitless and unobstructed; and the fleet, each ship manned, gunned, and provisioned and fuelled, ready to fight within five minutes.

Sir John Fisher: Memories, 1919

The Navy is a machine invented by geniuses, to be run by idiots.

Herman Wouk, 1915–, The Caine Mutiny

A Navy is organized to gain victory at sea, not to illustrate ideas about human equality.

C. Northcote Parkinson: Comments on "democratization" of Royal Navy officer procurement, in New York Times, 20 August 1961

Tradition, valor, and victory are the Navy's heritage from the past. To these may be added dedication, discipline, and vigilance as the watchwords of the present and future.

All Hands magazine, November 1963

Navy, Royal

It is upon the Navy, under the good Providence of God, that the wealth, safety and strength of the kingdom do chiefly depend.

Preamble to the Articles of War, Charles II, c. 1672

The royal navy of England hath ever been its greatest defence and ornament; its ancient and natural strength; the floating bulwark of the island; an army from which, however strong and powerful, no danger can ever be apprehended to liberty.

William Blackstone: Commentaries on the Laws of England, 1765

Ye mariners of England,
That guard our native seas;
Whose flag has braved a thousand years,
The battle and the breeze!

Thomas Campbell: Ye Mariners of England, 1800

There were gentlemen and there were seamen in the Navy of Charles II. But the seamen were not gentlemen, and the gentlemen were not seamen.

Thomas B. Macaulay: History of England, i, 3, 1848

. . . the old, unskillful, sledge-hammer fashion of the British Navy.

Mahan: Types of Naval Officers, 1901

Traditions of the Royal Navy? I'll give you traditions of the Navy—rum, buggery, and the lash.

Winston Churchill: To the Board of Admiralty, c. 1939

Necessity

Necessity knows no law except to conquer.

Publilius Syrus: Sententiae, 42 B.C.

. . . necessity, which makes even the timid brave.

Sallust: Bellum Catalinae, 44 B.C.

You cannot escape necessities; but you can conquer them.

Seneca, 4 B.C. – 65 A.D., Epistulae ad Lucilium, xxxi

. . . necessity's sharp pinch.

Shakespeare: King Lear, ii, 4, 1605

Necessity knows no law.

Theobald von Bethmann-Hollweg: To the German Reichstag, 4 August 1914 (Cf. Publilius Syrus, ante, 42 B.C.)

Needle Gun

The needle gun, the needle gun,
The death-defying needle gun;
It does knock over men like fun—
What a formidable weapon is the needle gun!

Punch, 1866 (The needle gun was the Prussian breech-loading rifle which won the battle of Königgratz and helped win the Franco-Prussian War.)

Negotiations

Men begin with blows, and when a reverse comes upon them, then have recourse to words.

Address by the Athenian Ambassadors to the Lacadaemonians, 433 B.C.

If without reason one begs for a truce it is assuredly because affairs in his country are in a dangerous state and he wishes to make a plan to gain a respite. Or otherwise, he knows that our situation is susceptible to his plots and he wants to forestall our suspicions by asking for a truce. Then he will take advantage of our unpreparedness.

Ch'ên Hao, fl. 700 A. D.

Our swords shall play the orators for us.

Christopher Marlowe: Tamburlaine, i, 1587

Your Majesty will not hear words, so we must come to the cannon.

Andrea de Loo: To Queen Elizabeth I, 1588 (de Loo was Spanish Ambassador to England.)

When a prince or a state is powerful enough to dictate to his neighbors, the art of negotiation loses its value.

Francois de Callières: The Art of Negotiating with Princes, 1733

Supposing the War to have commenced upon a just Motive; the next Thing to be considered is when a Prince ought in Prudence to receive the Overtures of a Peace: Which I take to be, either when the Enemy is ready to yield the Point originally contended for, or when that Point is found impossible to be ever attained.

Jonathan Swift, "The Conduct of the Allies," 1667–1745

You are now placed in the most happy and enviable situation: no negotiations to retard your operation, and a scene of glory immortal before you.

Lord St. Vincent: Letter to Sir Hyde Parker on the latter's sailing for Denmark, March 1801

. . . a fleet of British ships of war are the best negotiators in Europe.

Nelson: Letter to Lady Hamilton, March 1801

While the negotiation is going on, the Dane should see our flag waving every moment he lifted his head.

Nelson: To Sir Hyde Parker, urging that the British fleet go into Copenhagen, 20 March 1801

It is impossible to frame a treaty of peace in such a manner as to find in it a decision of all questions which can arise between the parties concerned.

Wellington: Despatch, 7 January 1804

Why do they send wild young men to treat for peace with old powers?

Plaint by the Bey of Tunis on acceding to terms imposed by Stephen Decatur, August 1815

It appears to your Majesty's slave that we are very deficient in means, and have not the shells and rockets used by the barbarians. We must, therefore, adopt other methods to stop them, which will be easy, as they have opened negotiations.

Kee-Shen: Ministerial report to the Emperor of China during the Opium War with Great Britain, March 1841

One can never foresee the consequences of political negotiations undertaken under the influence of military eventualities.

Napoleon I: I Political Aphorisms, 1848

Let us never negotiate out of fear. But let us never fear to negotiate.

John F. Kennedy: Inaugural Address, 20 January 1961

Negotiation today means war carried by other means and does not mean bargaining between parties both wanting to reach agreement.
Dean Acheson: Doherty Lecture, University of Virginia, 7 May 1966

(See also Diplomat, Diplomacy.)

Nelson, Horatio (1758–1805)

Before this time tomorrow I shall have gained a Peerage or Westminster Abbey.
Nelson: Before the Battle of the Nile, 1 August 1798

Thank God I have done my duty.
Nelson: Dying words in the cockpit of HMS Victory, *21 Oct. 1805*

There is but one Nelson.
Lord St. Vincent, 1736–1823

For he is England; Admiral,
Till the setting of her sun.
George Meredith, 1828–1909, Trafalgar Day

. . . a genius whose comprehension of rules serves only to guide, not to fetter, his judgment.
Mahan: The Influence of Sea Power Upon the French Revolution and Empire, ii, 1892

Rarely has a man been more favored in the hour of his appearing; never one so fortunate in the moment of his death.
Mahan: Life of Nelson, i, 1897

He brought heroism into the line of duty.
Joseph Conrad, 1857–1924

Neutrality

I shall treat neutrality as equivalent to a declaration of war against me.
Gustavus Adolphus: Proclamation during the Thirty Years War, 1616–1648

The most sincere neutrality is not a sufficient guard against the depredations of nations at war.
George Washington: To Congress, 7 December 1796 (Farewell Address)

Never break the neutrality of any port or place, but never consider as neutral any place from whence an attack is allowed to be made.
Nelson: Letter of instruction, 1804

Neutrals never dominate events. They always sink. Blood alone moves the wheels of history.
Benito Mussolini: Speech in Parma, 13 December 1914

Armed neutrality is ineffectual enough at best.
Woodrow Wilson: To Congress, asking for declaration of war on Germany, 2 April 1917

New Orleans, (8 January 1815)

I will hold New Orleans in spite of Urop and all Hell.
Said to have been written by Andrew Jackson, December 1814

By the Eternal, they shall not sleep on our soil!
Andrew Jackson: Exclamation when apprised of the British landings below New Orleans, 23 December 1814

Next-of-kin

Dear Madam,
I have been shown in the files of the War Department a statement of the Adjutant General of Massachusetts that you are the mother of five sons who have died gloriously on the field of battle. I feel how weak and fruitless must be any words of mine which should attempt to beguile you from the grief of a loss so overwhelming. But I cannot refrain from tendering you the consolation that may be found in the thanks of the Republic they died to save. I pray that our heavenly Father may assuage the anguish of your bereavement, and leave you only the cherished memory of the loved and lost, and the solemn pride that must be yours to have laid so costly a sacrifice upon the altar of freedom.
Abraham Lincoln: Letter to Mrs. Bixbey, of Boston, 21 November 1864

Ney, Michel (1769–1815)

. . . the bravest of the brave.
Napoleon I, 1769–1821

Night Operations

In the daytime the combatants see more clearly; though even then only what is going on immediately around them, and that imperfectly—nothing of the battle as a whole. But in a night engagement like this in which two great armies fought . . . who could be certain of anything?

Thucydides: History of the Peloponnesian Wars, 404 B.C. (of the night attack on Epipolae, 414 B.C.)

I have night's cloak to hide me from their eyes.

Shakespeare: Romeo and Juliet, ii, 2, 1594

In general, I believe that night attacks are only good when you are so weak that you dare not attack the enemy in daylight.

Frederick The Great: Instructions for His Generals, xiii, 1747

. . . two o'clock in the morning courage.

Napoleon I, 1769–1821

No operation of war is more critical than a night march.

Winston Churchill: The River War, xii, 1899

I have never met or heard of troops who can withstand a night attack from the rear.

Bernard Newman: The Cavalry Came Through, 1930

Darkness is a friend to the skilled infantryman.

B.H. Liddell Hart: Thoughts on War, xiv, 1944

Nightingale, Florence (1820–1910)

The wounded they love her as it has been seen,
She's the soldier's preserver, they call her their Queen.
May God give her strength, and her heart never fail,
One of Heaven's best gifts is Miss Nightingale.

The Nightingale in the East, British soldiers' song, Crimean War

A Lady with a Lamp shall stand
In the great history of the land,
A noble type of good,
Heroic womanhood.

Henry Wadsworth Longfellow: Santa Filomena, 1858 (reference is to Florence Nightingale in the Crimean War.)

No-Man's Land

This was a kind of Border, that might be called no Man's Land, being a part of Grand Tartary.

Daniel Defoe: Robinson Crusoe, ii, 1719 (the origin of this phrase)

Noncombatant

But I remember, when the fight was done,
When I was dry with rage and extreme toil,
Breathless and faint, leaning upon my sword,
Came there a certain lord, neat, and trimly dress'd,
Fresh as a bridegroom; and his chin new reap'd
Show'd like a stubble-hand at harvest home;
He was perfumed like a milliner;
And 'twixt his finger and his thumb he held
A pouncet-box, which ever and anon
He gave his nose and took't away again; . . .
And as the soldiers bore dead bodies by,
He call'd them untaught knaves, unmannerly,
To bring a slovenly unhandsome corse
Betwixt the wind and his nobility . . .
To see him shine so brisk, and smell so sweet,
And talk so like a waiting-gentlewoman
Of guns and drums and wounds,—God save the mark!—
And telling me the sovereign'st thing on earth
Was parmaceti for an inward bruise;
And that it was a great pity, so it was,
This villainous saltpetre should be digg'd
Out of the bowels of the harmless earth,
Which many a good tall fellow had destroy'd
So cowardly; and but for these vile guns,
He would himself have been a soldier.

Shakespeare: King Henry IV, i, 3, 1597

It will require the exercise of the full powers of the Federal Government to restrain the fury of the noncombatants.

Winfield Scott: After the attack on Fort Sumter, April 1861

Those who live at home in peace and plenty want the duello part of this war to go on;

but when they have to bear the burden by loss of property and comforts, they will cry for peace.

P.H. Sheridan: Telegram to Major General H.W. Halleck from Kernstown, Virginia, 26 November 1864

... the notable ferocity of noncombatants.

Jean Arthur Rimbaud, 1854–1891, Letter to Izambard

The crisis of confidence always starts among those who do not fight.

General Bernard Serrigny, 1870–1914, Réflexions sur l'Art de la Guerre

War hath no fury like a noncombatant.

C.E. Montague: Disenchantment, xv, 1922

(*See also* Home Front, Manpower)

Noncommissioned Officer

The choice of non-commissioned officers is an object of the greatest importance: The order and discipline of a regiment depends so much upon their behavior, that too much care cannot be taken in preferring none to that trust but those who by their merit and good conduct are entitled to it. Honesty, sobriety, and a remarkable attention to every point of duty, with a neatness in their dress, are indispensable requisites ... nor can a sergeant or corporal be said to be qualified who does not write and read in a tolerable manner.

Baron von Steuben: Regulations for the Order and Discipline of the Troops of the United States, 1779

If a selection of good sergeants and corporals be made by the officer at the head of the regiment, and if that officer will only allow those individuals to do their duty, there is not the least doubt that they will do it.

Diary entry by unknown British sergeant, Peninsular Campaign, c. 1812

The backbone of the Army is the noncommissioned man!

Rudyard Kipling: The 'Eathen, 1896

... A man in khaki kit who could handle men a bit.

Rudyard Kipling: Pharaoh and the Sergeant, 1897

(*See also* Centurion, Drill Instructor, Drum Major, Gunnery Sergeant, Sergeant.)

Nuclear Warfare

... The fortuitous concourse of atoms.

Sir John Russell: Review of Sir Robert Peel's Address, 1835

Some day science may have the existence of mankind in its power and the human race commit suicide by blowing up the world.

Henry Adams: Letter, 1862

It is not probable that war will ever absolutely cease until science discovers some destroying force so simple in its administration, so horrible in its effects, that all art, all gallantry, will be at an end, and battles will be massacres which the feelings of mankind will be unable to endure.

W. Winwood Reade: The Martyrdom of Man, iii, 1872

$E = mc^2$

Albert Einstein: Basic equation from which the atomic bomb was developed, c. 1940

We are not dealing simply with a military or scientific problem but with a problem in statecraft and the ways of the human spirit.

Report on the Control of Atomic Energy, 16 March 1946 (the so-called "Smith Report")

If we embrace this escape from reality, the Myth of the Atomic Bomb, we will drift into the belief that we Americans are safe in the world, safe and secure, because we have this devastating weapon—this and nothing more. We will then tend to relax when we need to be eternally vigilant.

David Lilienthal: Commencement address, Michigan State College, 5 June 1949

The annihilating character of these agencies may bring an utterly unforeseeable security to mankind ... It may be ... that when the advance of destructive weapons enables everyone to kill everybody else no one will want to kill anyone at all. At any rate it seems pretty safe to say that a war which begins by both sides suffering what they dread most—and that is undoubtedly the case now—is less likely to occur than one which dangles the lurid prizes of former days before ambitious eyes.

Winston Churchill: To the House of Commons, 3 November 1953

By carrying destructiveness to a suicidal extreme, atomic power is stimulating and accelerating a reversion to the indirect methods that are the essence of strategy—since they endow war with intelligent properties that raise it above the brute application of force.

B.H. Liddell Hart: Strategy, 1954

A bigger bang for the buck.

Attributed to Adlai Stevenson: in Presidential campaign, 1956

Mass will suffer most from weapons of mass destruction.

K.J. Mackesy: Land Warfare of the Future, 1956

The atom bomb is a paper tiger which the U.S. reactionaries use to scare people. It looks terrible, but in fact isn't. Of course the atom bomb is a weapon of mass slaughter, but the outcome of a war is decided by the people, not by one or two new types of weapons.

Mao Tse-tung, 1893–

Can one guess how great will be the toll of human casualties in a future war? Possibly it would be a third of the 2,700 million inhabitants of the entire world—i.e., only 900 million people. I consider this to be even low if atomic bombs actually fall. Of course it is most terrible. But even half would not be so bad . . . If a half of humanity were destroyed, the other half would remain but imperialism would be destroyed entirely and there would be only Socialism in all the world.

Mao Tse-tung: At Moscow conference, 1957

I have no faith in the so-called controlled use of atomic weapons. There is no dependable distinction between tactical and strategic situations. I would not recommend the use of any atomic weapon, no matter how small, when both sides have the power to destroy the world.

Admiral Charles R. Brown, USN: In Washington, 1958

We intend to have a wider choice than humiliation or all-out nuclear war.

John F. Kennedy: Inaugural Address, 20 January 1961

There is no presumption more terrifying than that of those who would blow up the world on the basis of their personal judgment of a transient situation.

George F. Kennan: After returning from Belgrade, August 1961

The survivors would envy the dead.

Nikita Khruschev, 1962

No annihilation without representation.

Arnold Toynbee: Of the so-called "Nuclear Club," i.e., nations possessing nuclear warfare capabilities, 1964

We have the power to knock any society out of the Twentieth Century.

Robert S. McNamara: During a Pentagon briefing, 1964

There is no illusion more dangerous than the idea that nuclear war can still serve as an instrument of policy, that one can attain political aims by using nuclear weapons and at the same time get off scot-free oneself, or that acceptable forms of nuclear war can be found.

Major General N. A. Talenski: In International Affairs, May 1965

(*See also* Total War.)

Nurses

A soldier of the Legion lay dying in Algiers;
There was lack of woman's nursing, there was lack of woman's tears.

Caroline Elizabeth Norton, 1808–1877, Bingen on the Rhine

The difficulty of finding women equal to a task, after all, full of horrors, and requiring, besides knowledge and goodwill, great energy and great courage, will be great. The task of ruling them and introducing system among them, great; and not the least will be the difficulty of making the whole work smoothly with the medical and military authorities.

Sidney Herbert: Letter to Florence Nightingale, 15 October 1854 (on organization of nurses for the Crimea)

A Lady with a Lamp shall stand
In the great history of the land,
A noble type of good,
Heroic womanhood.

Henry Wadsworth Longfellow: Santa Filomena, 1858 (reference is to Florence Nightingale in the Crimean War)

(*See also* Field Hospital, Health, Nightingale, Wounded.)

Obedience

Willing obedience always beats forced obedience.

Xenophon, 430–350 B.C., Cyropaedia

Tremble and obey!

Emperor K'ang Hsi of China, 1654–1722, Closing phrase of Imperial Rescript

Those who know the least obey the best.

George Farquhar: The Recruiting Officer, iv, 1706

I must, I will be obeyed.

Sir George Rodney: Letter from Gibraltar, 28 January 1780, to the Admiralty

Discipline is summed up in the one word obedience.

Lord St. Vincent, 1735–1823

I do not believe in the proverb that in order to be able to command one must know how to obey . . . Insubordination may only be the evidence of a strong mind.

Napoleon I: Letter from St. Helena, 1817

I profess . . . so much of the Roman principle as to deem it honorable for the general of yesterday to act as a corporal today, if his services can be useful to his country.

Thomas Jefferson, 1743–1826

There is nothing in war which is of greater importance than obedience.

Clausewitz: On War, 1832

Soldiers must obey in all things. They may and do laugh at foolish orders, but they nevertheless obey, not because they are blindly obedient, but because they know that to disobey is to break the backbone of their profession.

Sir Charles Napier, 1782–1853

Theirs not to make reply,
Theirs not to reason why,
Theirs but to do and die.

Alfred Tennyson: The Charge of the Light Brigade, 1854

Obedience being the soul of military organization, I hold it the beginning and end of duty. It is the rein in hand by which the superior does his driving.

Brigadier General C.F. Smith, USA: To Colonel Lew Wallace, September 1861

I want none of this Nelson business in my squadron about not seeing signals.

David G. Farragut: Reprimanding Lieutenant W.S. Schley for failing to obey a signal promptly during the attack on Port Hudson, March 1863

Men must be habituated to obey or they cannot be controlled in battle, and the neglect of the least important order impairs the proper influence of the officer.

R.E. Lee: Circular to the Army of Northern Virginia, 1865

The Army has no right to judge. It has only to obey.

Georges Boulanger, 1837–1891

I am a soldier; it is my duty to obey.

Field Marshal Walther von Brauchitsch, 1881–1948

The spirit of obedience, as distinguished from its letter, consists in faithfully forwarding the general object to which the officer's particular command is contributing.

Mahan: Retrospect and Prospect, 1902

The duty of obedience is not merely military but moral. It is not an arbitrary rule, but one essential and fundamental; the expression of a principle without which military organization would go to pieces, and military success be impossible.

Mahan: Retrospect and Prospect, 1902

The ugly truth is revealed that fear is the foundation of obedience.

Winston Churchill: The River War, ii, 1899

The efficiency of a war administration depends mainly upon whether decisions emanating from the highest approved authority are in fact strictly, faithfully, and punctually obeyed.

Winston Churchill: Their Finest Hour, 1949

The first duty of the soldier is obedience.

Military maxim

(*See also* Discipline, Orders)

Objective, Principle of

If a man does not know to what port he is steering, no wind is favorable.
Seneca, 4 B.C. – 65 A.D.

I have not a thought on any subject separated from the immediate object of my command.
Nelson, 1758–1805

The true method consists in giving each commander of an army corps or division the main direction of his march, and in pointing out the enemy as the objective and victory as the goal.
Clausewitz: Principles of War, 1812

Pursue one great decisive aim with force and determination.
Clausewitz: Principles of War, 1812 (Clausewitz called this "a maxim which should take first place among all causes of victory.")

If our aim is low, while that of the enemy is high, we are bound to get the worst of it.
Clausewitz: On War, 1832

. . . that exclusiveness of purpose, which is the essence of strategy, and which subordinates, adjusts, all other factors and considerations to the one exclusive aim.
Mahan: Naval Strategy, 1911

It may be that the enemy's fleet is still at sea, in which case it is the great objective, now as always.
Mahan: Naval Strategy, 1911

The objective . . . is unquestionably the most important of all the principles of war. It is the connecting link which, alone, can impart coherence to war . . . Without the objective, all other principles are pointless. It gives the commander the "what." The other principles are guides in the "How."
Admiral C.R. Brown, USN: The Principles of War, June 1949

Obstacles

Those expert at preparing defenses consider it fundamental to rely on the strengths of such obstacles as mountains, rivers and foothills. They make it impossible for the enemy to know where to attack. They secretly conceal themselves under the nine-layered ground.
Tu Yu, 735-812

Mountains can be crossed wherever goats cross, and winter freezes most rivers.
Frederick The Great: Instructions for His Generals, xxiv, 1747

Natural hazards, however formidable, are inherently less dangerous and uncertain than fighting hazards.
B.H. Liddell Hart, Strategy, 1954

Odds

CAPTAIN OF THE FLEET— There are eight sail of the line, Sir John.
LORD ST. VINCENT—Very well, Sir.
CAPTAIN OF THE FLEET—There are twenty sail of the line, Sir John.
LORD ST. VINCENT—Very well, Sir.
CAPTAIN OF THE FLEET— There are Twenty-five sail of the line, Sir John.
LORD ST. VINCENT—Very well, Sir.
CAPTAIN OF THE FLEET—There are twenty-seven sail of the line, Sir John
LORD ST. VINCENT—Enough Sir, no more of that. If there are fifty sail I will go through them.
At the battle of Cape St. Vincent, 14 February 1797; Lord St. Vincent (Admiral Sir John Jervis) had 15 ships of the line in his force.

The race is not always to the swift nor the battle to the strong, but that's the way to bet.
Anonymous remark

Offensive

The minds of men are apt to be swayed by what they hear; and they are most afraid of those who commence an attack.
Hermocrates of Syracuse: Speech to the Syracusans, 415 B.C.

The first blow is as much as two.
George Herbert: Outlandish Proverbs, 1640

I should say that in general the first of two army commanders who adopts an offensive attitude almost always reduces his rival to the defensive and makes him proceed in consonance with the movements of the former.
Frederick The Great: Instructions for His Generals, xi, 1747

Underwood & Underwood

MAO TSE-TUNG

1893–

"Political power comes out of the barrel of a gun."

An offensive, daring kind of war will awe the Indians and ruin the French. Blockhouses and a trembling defensive encourage the meanest scoundrels to attack us.

James Wolfe: Letter to Jeffrey Lord Amherst regarding the forthcoming North American campaign, 1758

Pushing on smartly is the road to success.

James Wolfe: Letter to a friend, 1758

The first blow is half the battle.

Oliver Goldsmith: She Stoops to Conquer, ii, 1773

When once the offensive has been assumed, it must be sustained to the last extremity.

Napoleon I: Maxims of War, 1831

The transition from the defensive to the offensive is one of the most delicate operations in war.

Napoleon I: Maxims of War, 1831

The best thing for an army on the defensive is to know how to take the offensive at the proper time, and to take it.

Jomini: Précis de l'Art de la Guerre, 1836

Meantime I desire you to dismiss from your minds certain phrases which I am sorry to find much in vogue amongst you. I hear constantly of taking strong positions and holding them—of lines of retreat, and of bases of supplies. Let us discard such ideas. The strongest position a soldier should desire to occupy is one from which he can most easily advance against the enemy. Let us study the probable lines of retreat of our opponents, and leave our own to take care of themselves.

Major General John Pope USA: General Order to the Officers and Soldiers of the Army of the Potomac, 14 July 1862

To move swiftly, strike vigorously, and secure all the fruits of victory is the secret of successful war.

Stonewall Jackson, 1824–1863

War, once declared, must be waged offensively, aggressively. The enemy must not be fended off, but smitten down. You may then spare him every exaction, relinquish every gain; but till down he must be struck incessantly and remorselessly.

Mahan: The Interest of America in Sea Power, 1896

Hit first! Hit hard! Keep on hitting!

Sir John Fisher: Memories, 1919

Hit hard, hit fast, hit often.

W.F. Halsey: Signal to the Third Fleet, 1944 (cf. Fisher, 1919, ante.)

In war the only sure defense is offense, and the efficiency of the offense depends on the warlike souls of those conducting it.

George S. Patton, Jr.: War As I Knew It, 1947

Since I first joined the Marines, I have advocated aggressiveness in the field and constant offensive action. Hit quickly, hit hard and keep right on hitting. Give the enemy no rest, no opportunity to consolidate his forces and hit back at you. This is the shortest road to victory.

Holland M. Smith: Coral and Brass, i, 1949

The role of an army is to march to the sound of the guns.

Jean Dutourd: Taxis of the Marne, 1957

When the going gets tough, the tough get going.

West Point saying

(*See also* Aggressiveness, Attack, Initiative.)

Officers

You are generals, you are officers and captains. In peace time you got more pay and more respect than they did. Now, in war time, you ought to hold yourselves to be braver than the general mass of men.

Xenophon: To the officers of the Greek army after the defeat of Cyrus at Cunaxa, 401 B.C.

No one can be a good officer who does not undergo more than those he commands.

Xenophon, 430–350 B.C., Cyropaedia

Among those who are placed at the head of armies, there are some who are so deeply immersed in sloth and indolence that they lose all attention both to the safety of their country and their own. Others are immoderately fond of wine, so that their senses are always disordered by it before they sleep. Others abandon themselves to the love of women—a passion so infatuating that those whom it has once possessed will often sacrifice even their honor and their lives to the indulgence of it.

Polybius, 200–118 B.C., Histories

I am a man under authority, having soldiers under me: and I say to this man, Go, and he goeth; and to another, Come, and he cometh.

Matthew, VIII, 9

Soldiers do not like being under the command of one who is not of good birth.

Onosander, fl. 1st century A.D.

He is to take notice of what discords, quarrels, and debates arise among the souldiers of his band; he is to pacify them if it may be; otherwise to commit them; he is to judge and determine such disputes with gravity and good speeches, and where the fault is, to make him acknowledge it and crave pardon of the party he hath abused . . . He is to be careful that every souldier hath a sufficient lodging in garrison, and in the field a hut; he is also to take due care of the sick and maymed, that they perish not for want of means or looking into; he is to take care that the sutlers do not opresse and rack the poor souldiers in their victuals and drinks.

17th century instruction for British army officers

I beseech you to be careful what captains of Horse you choose, what men be mounted: a few honest men are better than numbers . . . If you choose honest godly men to be captains of Horse, honest men will follow them . . . I had rather have a plain russet-coated captain that knows what he fights for, and loves what he knows, than that which you call a gentleman and is nothing else.

Oliver Cromwell: Letter to Sir William Springe, September 1643

Truly the only good officers are the impoverished gentlemen who have nothing but their sword and their cape; but it is essential that they should be able to live on their pay.

Maurice de Saxe: Mes Rêveries, iv, 1732

I myself have undertaken war and have several times seen how a colonel has decided the fate of the state. On such decisive days, one learns to appreciate the value of good officers; then one learns to love these men, seeing with what high-minded contempt of death, with what unshakable strength of mind they oppose the enemy.

Frederick The Great, 1712–1786

When an officer comes on parade, every man in the barrack square should tremble in his shoes.

Frederick The Great, 1712–1786

So be patriots as not to forget we are gentlemen.

Edmund Burke: Thoughts on the Cause of the Present Discontentment, 23 April 1770

An officer is much more respected than any other man who has as little money.

Samuel Johnson: To James Boswell, 3 April 1776

War must be carried on systematically, and to do it, you must have good Officers, there are, in my Judgment, no other possible means to obtain them but by establishing your Army upon a permanent basis; and giving your Officers good pay; this will induce Gentlemen, and Men of Character to engage; and till the bulk of your Officers are composed of such persons as are actuated by Principles of Honor, and a spirit of Enterprize, you have little to expect from them.

George Washington: Letter to the President of Congress, 24 September 1776

An Army formed of good Officers moves like clockwork.

George Washington: Letter to the President of Congress, 24 September 1776

You cannot take too much pains to maintain subordination in your corps. The subalterns of the British army are but too apt to think themselves gentlemen; a mistake which it is your business to rectify. Put them, as often as you can, upon the most disagreeable and ungentlemanly duties.

Francis Grose: Advice to the Officers of the British Army, 1782

Discipline begins in the Wardroom. I dread not the seamen. It is the *indiscreet* conversations of the officers and their *presumptuous* discussions of the orders they receive that produce all our ills.

Lord St. Vincent, 1735–1823

On the conduct of the officers alone depends the restoration of good order, discipline, and subordination in the Navy; the times are unfavorable to it, nevertheless we must do our utmost, and not sacrifice the very vitals of the Service to the miserable popularity of pleasing or indulging this or that officer.

Lord St. Vincent: Letter to Admiral Dickson, 24 January 1802

My brave officers . . . Such a gallant set of fellows! Such a band of brothers! My heart swells at the thought of them!

Horatio Nelson, 1758–1805

As you from this day start the world as a man, I trust that your future conduct in life will prove you both an officer and a gentleman. Recollect that you must be a seaman to be an officer; and also that you cannot be a good officer without being a gentleman.

Attributed to Horatio Nelson, 1758–1805 (advice to a young man just appointed midshipman)

A claim to the position of officer shall from now on be warranted in peace-time by knowledge and education, in time of war by exceptional bravery and quickness of perception . . . All social preference which has hitherto existed is herewith terminated in the military establishment, and everyone, without regard for his background, has the same duties and the same rights.

Decree by the King of Prussia, 6 August 1808

For though, with men of high degree,
The proudest of the proud was he,
Yet, train'd in camps, he knew the art
To win the soldier's hardy heart.
They love a captain to obey
Boisterous as March, yet fresh as May;
With open hand, and brow as free,
Lover of wine and minstrelsy;
Ever the first to scale a tower,
As venturous in a lady's bower . . .

Walter Scott: Marmion, 1808

A military gent, I see . . .

Thackeray: The Newcomes, i, 1, 1854

. . . renunciation of all personal advantage, of all gain, of all comfort—yes! of all desire, if only honor remains! On the other hand, every sacrifice for this, for their King, for their Fatherland, for the honor of Prussian arms! In their hearts, duty and loyalty; for their own lives, no concern! What other class and how many people, above all, in our present age, can pride themselves on such convictions? Model yourselves upon them, posterity, if you would be praised!

Friedrich von der Marwitz: Aus dem Nachtlasse, 1852 (of the officer corps of Frederick the Great)

The officers are young men of the best English families who have left behind them at Eton and Harrow a name for plucky and gentlemanly feeling . . . They do what they are told; lead their men bravely into action; and never think.

Article in The Westminster Review, 1855

Officers of the Army are apt in general to write like kitchen maids.

Lord Palmerston: Letter to Florence Nightingale, 1857

Our army would be invincible if it could be properly organized and officered.

R.E. Lee: Letter to Maj. Gen. John B. Hood, 21 May 1863

The whole present system of the officering and personnel of the Army and Navy of these States, and the spirit and letter of their trebly-aristocratic rules and regulations, is a monstrous exotic, a nuisance and revolt, and belongs here just as much as orders of nobility or the Pope's council of cardinals. I say if the present theory of our Army and Navy is sensible and true, then the rest of America is an unmitigated fraud.

Walt Whitman: Democratic Vistas, 1870

The officer should wear his uniform as the judge his ermine, without a stain.

Rear Admiral John A. Dahlgren, USN: On the night of his death, 12 July 1870

No man can be a great officer who is not infinitely patient of details, for an army is an aggregation of details.

George S. Hillard, 1808–1879, Life and Campaigns of George B. McClellan

This is the officer's part, to make men continue to do things, they know not wherefore; and, when, if choice was offered, they would lie down where they were and be killed.

R.L. Stevenson: Kidnapped, xxii, 1886

He became an officer *and* a gentleman, which is an enviable thing.

Rudyard Kipling: Only a Subaltern, 1888

Gentlemen unafraid . . .

Rudyard Kipling: Barrack-Room Ballads (dedication), 1892

In every military system which has triumphed in modern war the officers have been recognized as the brain of the army, and to prepare them for their trust, governments have spared no pains to give them special education and training.

Emory Upton: The Military Policy of the United States, 1904 (Upton's phrase, "the brain of the army," seems to antedate Spenser Wilkinson's famous title by at least a decade.)

The great thing about an Army officer is that he does what you tell him to do.

Theodore Roosevelt, 1858–1919

The officers they come and they go, but it don't hurt the troop.

Remark by Cavalry first sergeant on completing ten years' service in the same troop, 3d Cavalry, c. 1908

Officers used to serve, not for their starvation pay, but for love of their country and on the off-chance of being able to defend it with their lives . . . They deliberately, indeed joyously, faced up to a life of adventure, roving, action, exile, and poverty because it satisfied and reposed their souls.

Sir Ian Hamilton: The Soul and Body of an Army, x, 1921

An officer and a gentleman is a familiar term to everyone in and out of the Service. Be sure you are both. You cannot be an officer and a gentleman unless you are just, humane, thoroughly trained, unless you have character, a high sense of honor and an unselfish devotion to duty. Be an example of such to everyone.

John W. Weeks: To the graduating class, West Point, June 1922

I divide officers into four classes—the clever, the lazy, the stupid, and the industrious. Each officer possesses at least two of these qualities. Those who are clever and industrious are fitted for the high staff appointments. Use can be made of those who are stupid and lazy. The man who is clever and lazy is fit for the very highest command. He has the temperament and the requisite nerves to deal with all situations. But whoever is stupid and industrious must be removed immediately.

Attributed to General Kurt von Hammerstein, c. 1933

In my experience, based on many years' observation, officers with high athletic qualifications are not usually successful in the higher ranks.

Winston Churchill: Memorandum for Secretary of State for War, 4 February 1941

You are always on parade.

George S. Patton, Jr.: Letter to Cadet George S. Patton, III, USMA 6 June 1944

If love of money were the mainspring of all American action, the officer corps long since would have disintegrated.

The Armed Forces Officer, 1950

The military officer is considered a gentleman, not because Congress wills it, nor because it has been the custom of the people at all times to afford him that courtesy, but specifically because nothing less than a gentleman is truly suited for his particular set of responsibilities.

The Armed Forces Officer, 1950

Service conditions being what they are today, a girl who marries an officer and a gentleman usually has to commit bigamy.

Remark by unknown officer, c. 1960

When I say that officers today must go far beyond the official curriculum, I say it, not because I do not believe in the traditional relationship between the civilian and the military, but you must be more than the servants of national policy. You must be prepared to play a constructive role in the development of national policy.

John F. Kennedy: To the graduating class, U.S. Naval Academy, June 1961

Officers cannot be paid like bus boys, worked like field hands, and released like old, slow halfbacks.

Captain William A. Golden, USN: In U.S. Naval Institute Proceedings, July 1964

It ain't hard to be an officer, but it's damn' hard to be a *good* officer.

Remark by an unknown chief petty officer, USN

(*See also* Action Officer, Adjutant, Admiral, Aide-de-Camp, Brass Hat, Captain, Chief of Staff, Colonel, Commander, General, Naval Officer, Paymaster, Quartermaster, Staff Officer, Staff Secretary.)

Officers' Sons

Your son, my lord, has paid a soldier's debt.

Shakespeare: Macbeth, v, 7, 1605

If e'er my son
Follow the war, tell him it is a school
Where all the principles tending to honor
Are taught, if truly followed.

Philip Massinger, 1583–1640

I trust the other officers of my rank will observe the maxim I do, to prefer the son of a brother officer, when deserving, before any other.

Lord St. Vincent: Letter, 1788

I hope our son will do his duty and make a good soldier.

R.E. Lee: To his wife when R.E. Lee, Jr., enlisted as a private in the Rockbridge Artillery, 15 March 1862

Three generations in a regiment count for less in the eyes of our Army Council than three miserable marks in a miserable competitive examination.

Sir Ian Hamilton: The Soul and Body of an Army, vi, 1921

Sons of heroes are a plague.

Greek Proverb

(*See also* Dependents.)

Official Correspondence

. . . these paper-bullets of the brain.

Shakespeare: Much Ado About Nothing, ii, 3, 1598

You must take pains so to interlard your letters with technical terms, that neither the public, nor the minster to whom they are addressed, will understand them; especially if the transactions you are describing be trivial: it will then give them an air of importance. This is conformable to the maxim in epic and dramatic poetry, of raising the diction at times to cover the poverty of the subject.

Francis Grose: Advice to the Officers of the British Army, 1782

(*See also* Administration.)

Old Soldier

OCTAVIUS—He's a tired and valiant soldier.
ANTONIUS—So is my horse, Octavius.

Shakespeare: Julius Caesar, iv, 1, 1599

Now like a wandring souldier
That has i' th' warres bin maymed,
With the shot of a gunne,
To gallants I runne
And begg: "Sir, help the lamed!
I am a poore old souldier
And better times once viewed,
Though bare now I goe,
Yet many a foe
By me hath been subdued."

The Maunding Souldier, 17th century ballad

A threadbare redcoat, which his taylor duns him for to this day, over which a great broad greasie buff belt, enough to turn anyone's stomach but a disbanded soldier; a perruque ty'd up in a knot to excuse its want of combing, and then because he had been a man-at-armes, he must wear two tuffles of beard forsooth, to lodge a dunghill of snuff upon, to keep his nose in good humour.

Thomas Otway: The Soldier's Fortune, 1680

He is a respectable old man, and has no other failing than that which but too often attends an old soldier.

Colonel Archibald Henderson, USMC: Letter to Lieutenant Colonel Wainwright, 1 July 1837

For my father was a soldier, and, even
when a child,
My heart leaped forth to hear him tell of
struggles fierce and wild.

Caroline Elizabeth Norton, 1808–1877, The Soldier of the Rhine

All the world over, nursing their scars,
Sit the old fighting-men broke in the
wars—

Sit the old fighting-men, surly and grim,
Mocking the lilt of the conquerors' hymn.
Rudyard Kipling: verse from Many Inventions, 1893

I 'eard the feet on the gravel—the feet o' the men what drill—
An' I sez to my flutterin' 'eart-strings, I sez to 'em, "Peace, be still!"
Rudyard Kipling: Back to the Army Again, 1894

You can always tell an old soldier by the inside of his holsters and cartridge boxes. The young ones carry pistols and cartridges; the old ones, grub.
George Bernard Shaw: Arms and the Man, 1894

An old man's soldiering is foulness, and foulness is an old man's love.
Aldous Huxley, 1894–1964, The Devils of Loudun

He had obviously been very man years in [India] and had a face like a bottle of port. He looked as if he lived almost entirely on suction.
Montgomery of Alamein: Memoirs, ii, 1958

Old soldiers never die;
They only fade away!
Soldiers' song

Old soldier, old idiot.
French Proverb

(*See also* Regulars, Soldier, Veteran.)

Opportunity

Opportunity in war is often more to be depended on than courage.
Vegetius: De Re Militari, iii, 378

Four things come not back:
The spoken word; the sped arrow;
Time past; the neglected opportunity.
Omar Ibn, 581–644, Sayings

Strike while the iron is hot.
George Farquhar, The Beaux' Stratagem, iv, 1707

He who seizes the right moment is the right man.
Goethe: Faust, 1808

A sea-officer cannot, like a land-officer, form plans; his object is to embrace the happy moment which now and then offers,—it may be this day, not for a month, and perhaps never.
Nelson: To the British Minister at Genoa, April 1796

In war there is only one favorable moment. Genius seizes it.
Napoleon I: Maxims of War, 1831

(*See also* Chance, Luck.)

Order

Let all things be done decently and in order.
I Corinthians, XIV, 40

Order is Heav'n's first law.
Alexander Pope, 1688–1744, Essay on Man, iv

There must be order! (Ordnung muss sein!)
Marshal Von Hindenburg: During Socialist disorders, 1919

Orders

The man who does something under orders is not unhappy; he is unhappy who does something against his will.
Seneca: Epistulae Morales ad Lucilium, 63 A.D.

When administration and orders are inconsistent, the men's spirits are low, and the officers exceedingly angry.
Chang Yü, fl. 1000

Many generals believe that they have done everything as soon as they have issued orders, and they order a great deal because they find many abuses. This is a false principle; proceeding in this fashion, they will never reestablish discipline in an army in which it has been lost or weakened. Few orders are best, but they should be followed up with care; negligence should be punished without partiality and without distinction of rank or birth; otherwise you will make yourself hated.
Maurice de Saxe: Mes Rêveries, xviii, 1732

Be sure to give out a number of orders. It will at least show the troops you do not forget them. The more trifling they are, the more it shows your attention to the service; and should your orders contradict one another, it will give you an opportunity of altering them, and find subject for fresh regulations.

Francis Grose: Advice to the Officers of the British Army, 1782

The orders I have given are strong, and I know not how my admiral will approve of them, for they are, in a great measure, contrary to those he gave me; but the Service requires strong and vigorous measures to bring the war to a conclusion.

Nelson: Letter to Collingwood, July 1795

I find few think as I do, but to obey orders is all perfection. What would my superiors direct, did they know what is passing under my nose? To serve my King and to destroy the French I consider as the great order of all, from which little ones spring; and if one of those little ones militate against it, I go back to obey the great order.

Nelson: Letter from Palermo, March 1799

I shall endeavor to comply with all their Lordships' directions in such manner as, to the best of my judgment, will answer their intentions in employing me here.

Nelson, 1758–1805

Order, counter-order, disorder.

Helmuth von Moltke ("The Elder"), 1800–1891

Remember, gentlemen, an order that can be misunderstood will be misunderstood.

Helmuth von Moltke ("The Elder"), 1800–1891

An order should contain everything that a commander cannot do by himself, but nothing else.

Helmuth von Moltke ("The Elder"), 1800–1891

I just gave an order—quite a simple matter unless a man's afraid.

Sir Ian Hamilton: The Soul and Body of an Army, xi, 1921

I hope that you will make sure that when you give an order it is obeyed with promptness . . . There is always a danger that anything contrary to Service prejudices will be obstructed and delayed.

Winston Churchill: Note for Secretary of State for War, 8 September 1940

I give orders only when they are necessary. I expect them to be executed at once and to the letter and that no unit under my command shall make changes, still less give orders to the contrary or delay execution through unnecessary red tape.

Erwin Rommel: Letter of instruction to subordinate commanders, 22 April 1944

Promulgation of an order represents not over ten percent of your responsibility. The remaining ninety percent consists in assuring through personal supervision on the ground, by yourself and your staff, proper and vigorous execution.

Geroge S. Patton, Jr., 1885–1945

Operation orders do not win battles without the valor and endurance of the soldiers who carry them out.

Sir A.P. Wavell: Soldiers and Soldiering, 1953

A commander must train his subordinate commanders, and his own staff, to work and act on verbal orders. Those who cannot be trusted to act on clear and concise verbal orders, but want everything in writing, are useless.

Montgomery of Alamein: Memoirs, vi, 1958

Order of Battle

The Spartans are not wont to ask how many the enemy are, but where they are.

Agis II of Sparta, c. 450 B.C.

Evaluation of enemy strength is not an absolute, but a matter of piecing together scraps of reconnaissance and intelligence to form a picture, if possible a picture to fit preconceived theories or suit the demands of a particular strategy. What a staff makes out of the available evidence depends upon the degree of optimism or pessimism prevailing among them, on what they want to believe or fear to believe, and sometimes on the sensitivity or intuition of an individual.

Barbara W. Tuchman: The Guns of August, 1962

(*See also* Intelligence.)

Ordnance

And Uzziah prepared for them throughout all the host shields and spears, and helmets, and habergeons, and bows, and slings to cast stones.

II Chronicles, XXVI, 14

The Ordnance doesn't progress, it rotates.

U.S. Army saying about the Ordnance Department, c. 1930

Organization

Moreover thou shalt provide out of all the people able men, such as fear God, men of truth, hating covetousness; and place such over them, to be rulers of thousands, and rulers of hundreds, rulers of fifties, and rulers of tens: and let them judge the people at all seasons: and it shall be, that every great matter they shall bring unto thee, but every small matter they shall judge: so it shall be easier for thyself, and they shall bear the burden with thee.

Exodus, XVIII, 21–22 (Jethro's advice to Moses)

Generally, management of the many is the same as management of the few. It is a matter of organization.

Sun Tzu, 400–320 B.C., The Art of War

The mass needs, and we give it, leaders . . . We add good arms. We add suitable methods of fighting . . . We also add a rational decentralization . . . We animate with passion . . . An iron discipline . . . secures the greatest unit . . . But it depends also on supervision, the mutual supervision of groups of men who know each other well. A wise organization of comrades in peace who shall be comrades in war . . . And now confidence appears . . . Then we have an army.

Ardant du Picq; 1821–1870, Battle Studies

The primary object of organization is to shield people from unexpected calls upon their powers of adaptability, judgment and decision.

Sir Ian Hamilton: The Soul and Body of an Army, iv, 1921

The military profession organizes men so as to overcome their inherent fears and failings.

Samuel P. Huntington: The Soldier and the State, iii, 1957

Originality

Originality is the most vital of all military virtues.

B.H. Liddell Hart: Thoughts on War, i, 1944

Outposts

The most arduous, while at the same time the most important duties that devolve upon soldiers in the field are those of outposts . . . All concerned should feel that the safety of the army and the honor of the country depend upon their untiring vigilance and activity.

Sir Garnet Wolseley, 1833–1913

Overseas Expeditions

We must take a powerful armament with us from home, in the full knowledge that we are going to a distant land, and that the expedition will be of a kind very different from any which you have hitherto made among your subjects against some enemy in this part of the world . . . Here a friendly country is always near, and you can easily obtain supplies. There you will be dependent on a country which is entirely strange to you, and whence during the four winter months hardly even a message can be sent hither.

Nicias of Athens: Address to the Athenians on the Syracusan expedition, 415 B.C.

When the army marches abroad, the treasury will be emptied at home.

Li Ch'üan, c. 905

(*See also* Expeditionary Service.)

Pacifism

He is a fool who preaches peace in a country that is in the midst of war.
Torquato Tasso: Gerusalemme, v, 1592

Tame the savage spirit of wild war,
That like a lion fostered up at hand,
It may lie gently at the foot of peace.
Shakespeare: All's Well That Ends Well, v, 2, 1602

A people not used to war believeth no enemy dare venture upon him.
Sir Edward Cecil: Commenting on a possible French invasion, 1628

Those who condemn the profession or art of soldiery smell rank of Anabaptism and Quakery.
Sir James Turner: Pallas Armata, 1683

Nought can deform the human race
Like to the armor's iron brace.
William Blake, 1757–1827, Auguries of Innocence

The spirit of this country is totally adverse to a large military force.
Thomas Jefferson: Letter to Chandler Price, 1807

Mankind would greatly gain if nations would refer to the judgment of reason, rather than the decision of the sword, the settlement of disputes.
R.E. Lee, Letter to Colonel Edward Hamley, June 1866

A soldier is an anachronism of which we must get rid.
George Bernard Shaw: The Devil's Disciple, iii, 1897

Professional pacifists, the peace-at-any-price, non-resistance, universal arbitration people . . .
Theodore Roosevelt, 1858–1919, Speech at San Francisco

There is such a thing as a man being too proud to fight.
Woodrow Wilson: Speech, 10 May 1915

I didn't raise my boy to be a soldier.
Allen Bryan: popular song, 1915

He kept us out of war.
Martin H. Glynn: Keynote speech at the Democratic National Convention which nominated Woodrow Wilson, 1916

We are not pacifists. We are against imperialist wars, but it is absurd for the proletariat to oppose revolutionary wars that are indispensable for the victory of Socialism.
V.I. Lenin: Farewell letter to the Swiss workers, 8 April 1917

Pacifism is simply undisguised cowardice.
Adolf Hitler: Speech at Nürnberg, 21 August 1926

War to the hilt between communism and capitalism is inevitable. Today, of course, we are not strong enough to attack. Our time will come in twenty to thirty years. To win, we shall need the element of surprise. The bourgeoisie will have to be put to sleep. So we shall begin by launching the most spectacular peace movement on record. There will be electrifying overtures and unheard-of concessions. The capitalist countries, stupid and decadent, will rejoice to cooperate in their own destruction. They will leap at another chance to be friends. As soon as their guard is down, we shall smash them with our clenched fists.
Attributed to Dmitri Z. Manuilsky: Speech at the Lenin School of Political Warfare, 1931

We are for the abolition of war, we do not want war; but war can only be abolished through war, and to get rid of the gun, we must first grasp it in our own hands.
Mao Tse-tung: Problems of War and Strategy, 1938

Rational pacifism must be based on a new maxim—"If you wish for peace, understand war."
B.H. Liddell Hart: Thoughts on War, i, 1944

War is never prevented by running away from it.
Sir John Slessor: Strategy for the West, 1954

I have never met anyone who wasn't against War. Even Hitler and Mussolini were, according to themselves.
David Low: in New York Times, 10 February 1964

(*See also* Disarmament, Peace.)

Panic

The barbarians were instantly seized with one of those unaccountable panics to which great armies are liable.
Thucydides: History of the Peloponnesian Wars, iv, c. 404 B.C.

Let's fly and save our bacon.
Rabelais: Pantagruel, 1533

The strait pass was dammed
With dead men hurt behind, and cowards living
To die with lengthened shame.
Shakespeare: Cymbeline, v, 3, 1609

Both officers and men must be warned against those sudden panics which often seize the bravest armies when they are not well controlled by discipline, and when they do not recognize that the surest hope of safety lies in order.
Jomini: Précis de l'Art de la Guerre, 1838

The worst enemy a Chief has to face in war is an alarmist.
Sir Ian Hamilton: Gallipoli Diary, I, 1920

(*See also* Fear.)

Parachute

If a man have a tent made of linen of which the apertures have all been stopped up, and it be twelve bracchia across and twelve in depth, he will be able to throw himself down from any great height without sustaining an injury.
Leonardo Da Vinci, 1452–1519, Notebook entry

Partisan Warfare

Woe to the English soldiery,
That little dreads us near!
On them shall come at midnight
A strange and sudden fear;
When, waking to their tents on fire,
They grasp their arms in vain,
And they who stand to face us
Are beat to earth again;
And they who fly in terror deem
A mighty host behind,
And hear the tramp of thousands
Upon the hollow wind.
William Cullen Bryant, 1794–1878, Song of Marion's Men

The more the enemy extends himself, the greater is the effect of an armed populace. Like a slow, gradual fire it destroys the bases of the enemy force.
Clausewitz: On War, 1832

The military value of a partisan's work is not measured by the amount of property destroyed, or the number of men killed or captured, but by the number he keeps watching.
John S. Mosby: War Reminiscences, iv, 1887

To destroy supply trains, to break up the means of conveying intelligence, and thus isolating an army from its base, as well as its different corps from each other, to confuse their plans by capturing despatches, are the objects of partisan war.
John S. Mosby: War Reminiscences, iv, 1887

(*See also* Counter-Insurgency, Guerrilla, Guerrilla Warfare, Insurgency)

Patriotism

He serves me most, who serves his country best.
Homer: The Iliad, x, c. 1000 B.C.

Slain fighting for his country.
Homer: The Iliad, xi, c. 1000 B.C.

The best omen for a man is to fight for his country.
Homer: The Iliad, c. 1000 B.C. (Hector's reply when told by his staff that the omens were unfavorable for fighting that day)

And for our country 'tis a bliss to die.
Homer: The Iliad, xv, c. 1000 B.C.

His sword the brave man draws,
And asks no omen but his country's cause.
Homer: The Iliad, xii, c. 1000 B.C. (V. Hector, Supra.)

This empire has been acquired by men who knew their duty and had the courage to do it, who in the hour of conflict had the fear of dishonor always present to them, and who, if ever they failed in an enterprise, would not allow their virtues to be lost to their country, but freely gave their lives to her as the fairest offering which they could present at her feast.

Pericles: Funeral oration over the Athenian dead, 431 B.C.

Man was not born for himself alone, but for his country.

Plato, 428–347 B.C., Epistle to Archytas

Sweet is the love of one's country.

Cervantes: Don Quixote, ii, 1605

I do love
My country's good with a respect more tender,
More holy, and profound, than mine own life.

Shakespeare: Coriolanus, iii, 3, 1607

Whose bosom beats not in his country's cause?

Alexander Pope: Prologue to Addison's Cato, 1713

What a pity is it
That we can die but once to serve our country!

Joseph Addison: Cato, iv, 1713

He who serves his country well has no need of ancestors.

Voltaire: Merope, i, 1743

Patriotism is the last refuge of a scoundrel.

Samuel Johnson: As recorded by Boswell, 1775

Sink or swim, live or die, survive or perish with my country was my unalterable determination.

John Adams: To Jonathan Sewell, 1774

Think of your forefathers! Think of your posterity!

John Adams: Speech at Quincy, Massachusetts, 22 December, 1802

I only regret that I have but one life to lose for my country.

Nathan Hale: Before being hanged by the British as a spy, Long Island, 22 September 1776

Not the value or command of the whole British Navy would seduce me from the cause of my country.

Captain John Barry, USN, 1745–1803 (Barry was one of the original officers of the Continental Navy.)

I shall always be ready to serve my country.

Gustavus Conyngham: Letter to Benjamin Franklin, 18 November 1779

The patriot's blood's the seed of Freedom's tree.

Thomas Campbell, 1777–1844, Stanzas to the Memory of the Spanish Patriots

It is Time We should establish an American Character—Let that Character be a Love of Country and Jealousy of its honor—This Idea comprehends every Thing that ought to be impressed upon the Minds of all our Citizens, but more especially of those Citizens who are also Seamen and Soldiers.

Benjamin Stoddert: Letter to Captain John Barry, USN, 11 July 1798

Each man must do all in his power for his country.

Isaac Hull: To the ship's company, USS Constitution, *1813*

Breathes there the man with soul so dead,
Who never to himself hath said,
This is my own, my native land!

Walter Scott: Lay of the Last Minstrel, vi, 1805

Let our object be, our country, our whole country, and nothing but our country.

Daniel Webster: Address at the laying of the corner-stone, Bunker Hill Monument, 17 June 1825

Men who when the tempest gathers
Grasp the standards of their fathers
In the thickest fight.

Edward Henry Bickersteth, 1786–1850

Our country: in her intercourse with foreign nations may she always be in the right; but our country, right or wrong!

Stephen Decatur: Toast at a dinner at Norfolk, Virginia, April 1816

He who loves not his country can love nothing.

Byron, 1788–1824, The Two Foscari, iii, 1, 1821

An American kneels only to his god, and faces his enemy.

W.H. Crittenden: On being ordered to kneel down before being shot by the

U. S. Navy

KARL VON CLAUSEWITZ

1780–1831

"The art of war in its highest point of view is policy."

Spaniards as a filibusterer, Havana, 16 August 1851

All I am and all I have is at the service of my country.

Stonewall Jackson: Letter, 1861

Patriotism is the egg from which wars are hatched.

Guy de Maupassant, 1850–1893
My Uncle Sosthenes

A country and government such as ours are worth fighting for, and dying for, if need be.

W.T. Sherman: Memoirs, 1875

In time of war the loudest patriots are the greatest profiteers.

August Bebel: Gegen der Militarismus, 1895

Our country right or wrong. When right, to be kept right; when wrong, to be put right.

Carl Schurz: Address at Chicago, 17 October 1899

I shall always hold myself in readiness to serve my country . . . For forty-eight years my life has been at the call of the flag I love, and it will remain so as long as I live

Robley D. Evans: An Admiral's Log, 1910

Patriotism should be something more than just hating your neighbor as much as you love yourself.

"Saki" (H. H. Munro), 1870–1916

Patriotism is not enough. I must have no hatred or bitterness towards anyone.

Edith Cavell: Before her execution by the Germans, Brussels, 12 October 1915

Patriotism is like a plant whose roots stretch down into race and place subconsciousness; a plant whose best nutriments are blood and tears; a plant which dies down in peace and flowers most brightly in war. Patriotism does not calculate, does not profiteer, does not stop to reason: in an atmosphere of danger the sap begins to stir; it lives; it takes possession of the soul.

Sir Ian Hamilton: The Soul and Body of an Army, x, 1921

Patriotism is not a song in the street and a wreath on a column and a flag flying from a window . . . It is a thing very holy and very terrible, like life itself. It is a burden to be borne, a thing to labor for and to suffer for and to die for; a thing which gives no happiness and no pleasantness—but a hard life and an unknown grave, and the respect and bowed heads of those that follow.

John Masefield, 1878–

I admire men who stand up for their country in defeat, even though I am on the other side.

Winston Churchill: The Gathering Storm, v, 1948

Ask not what your country can do for you—ask what you can do for your country.

John F. Kennedy: Inaugural Address, 20 January 1961

Patriotism is a sense of fear which overcomes a country when war is declared on it.

Major Reginald Hargreaves: Remark to a friend, 30 September 1964

Patton, George Smith, Jr. (1885–1945)

This man would be invaluable in time of war, but is a disturbing element in time of peace.

Major General W.R. Smith, USA: Remark in forwarding the efficiency report of Patton, then a captain, 1927

Patton was an acolyte to Mars.

Colonel J.J. Farley, USA: During conversation in Washington, 17 November 1964

Pay

Come on, brave soldiers; doubt not of the day,
And, that once gotten, doubt not of large pay.

Shakespeare: III King Henry VI, 1590

Without going into detail about the different rates of pay, I shall only say that it should be ample . . . Economy can only be pushed to a certain point. It has limits beyond which it degenerates into parsimony. If your pay and allowances for officers will not support them decently, then you will have only rich men who serve for pleasure or adventure, or indigent wretches devoid of spirit.

Maurice de Saxe: Mes Rêveries, iv, 1732

The soldier should not have any ready money. If he has a few coins in his pocket, he thinks himself too much of a great lord to follow his profession, and he deserts at the opening of the campaign.

Frederick The Great: Instructions for His Generals, xxv, 1747

Our soldiers surely are not luxurious, who live on sixpence a day.

Samuel Johnson: To James Boswell, 13 April 1773

As to pay, I beg leave to assure the Congress that, as no pecuniary consideration could have tempted me to accept this arduous employment at the expense of my domestic ease and happiness, I do not wish to make any profit from it.

George Washington: Letter to the Continental Congress, 16 June 1775

There must be some other stimulus, besides love for their country, to make men fond of the service.

Geroge Washington, 1732–1799

How happy's the soldier who lives on his pay,
And spends half-a-crown out of sixpence a day.

John O'Keeffe, 1747–1833, The Poor Soldier

It is the height of injustice not to pay a veteran more than a recruit.

Napoleon I: Maxims of War, 1831

My present pay is not wholly for present work but is in great part for past services . . . What money will pay Meade for Gettysburg? What Sheridan for Winchester? What Thomas for Chickamauga?

W.T. Sherman: Letter, 1870, on pending legislation to cut Army officers' pay

The nation which forgets its defenders will itself be foregotten.

CalvinCoolidge: Speech, 27 July 1920

If love of money were the mainspring of all American action, the officer corps long since would have disintegrated.

The Armed Forces Officer, 1950

(*See also* Finance, Paymaster.)

Paymaster

He pays you as surely as your feet hit the ground they step on.

Shakespeare: Twelfth Night, iii, 5, 1599

Always grumble and make difficulties when officers go to you for money that is due to them; when you are obliged to pay them endeavor to make it appear granting them a favor, and tell them they are lucky dogs to get it.

Francis Grose: Advice to the Officers of the British Army, 1782

Make your accounts as intricate as you can, and, if possible, unintelligible to everyone but yourself; lest, in case you should be taken prisoner, your papers might give information to the enemy.

Francis Grose: Advice to the Officers of the British Army 1782

(*See also* Pay.)

Peace

They shall beat their swords into plowshares, and their spears into pruning-hooks, nation shall not lift up sword against nation; neither shall they learn war any more.

Isaiah, II, 4

Has not peace honors and glories of her own unattended by the dangers of war?

Hermocrates of Syracuse: Speech to the Sicilian envoys at Gela, 424 B.C.

A bad peace is even worse than war.

Tacitus: Annals, iii, c. 110

The most disadvantageous peace is better than the most just war.

Erasmus: Adagia, 1508

He is a fool who preaches peace in a country that is in the midst of war.

Torquato Tasso: Gerusalemme, v, 1592

Peace is a very apoplexy, lethargy; mulled, deaf, sleepy, insensible; a getter of more bastard children, than war is a destroyer of men.

Shakespeare: Coriolanus, iv, 5, 1607

Peace is tranquil freedom, and is contrary to war, of which it is the end and destruction.

Hugo Grotius: De Jure Belli et Pacis, i, 1625

Peace with honor.
Theobald de Champagne: Letter to Louis the Great, c. 1125 (Also attributed to Sir Kenelm Digby, in a letter to Lord Bristol, 27 May 1625)

No war, or battle's sound
Was heard the world around.
The idle spear and shield were hung up high.
Milton: On the Morning of Christ's Nativity, 1629

He that will have peace, God gives him war.
George Herbert: Outlandish Proverbs, 1640

The first and fundamental law of Nature . . . *to seek peace and ensue it.*
Thomas Hobbes: Leviathan, I, xiv, 1651

Peace hath her victories
No less renowned than war.
Milton: To the Lord Generall Cromwell, 1652

Though peace be made, yet it is interest that keeps peace.
Oliver Cromwell: To Parliament, 4 September 1654

We become reconciled with our enemies because we want to improve our situation, because we are weary of war, or because we fear defeat.
Francois de la Rochefoucauld: Maxims, 1665

War ends in peace, and morning light
Mounts upon midnight's wing.
Michael Wigglesworth: Meat out of the Eater, 1669

Peace itself is war in masquerade.
John Dryden: Absolom and Achitophel, i, 1682

The sword within the scabbard keep,
And let mankind agree.
John Dryden: The Secular Masque, 1700

God send us better times with peace in our borders and war in the enemy's country.
Lord St. Vincent: Letter to Nelson, 22 September 1801

What a beautiful fix we are in now: peace has been declared.
Napoleon I: After the Treaty of Amiens, 27 March 1802

I hope it is practicable, by improving the mind and morals of society, to lessen the disposition to war; but of its abolition I despair.
Thomas Jefferson, 1743–1826

An honorable peace is attainable only by an efficient war.
Henry Clay: To the House of Representatives, 8 January 1813

Peace at any price (Paix à tout prix).
Alphonse de Lamartine, 1790–1869

Peace! and no longer from its brazen portals
The blast of War's great organ shakes the skies!
Henry Wadsworth Longfellow: The Arsenal at Springfield, 1841

War is on its last legs; and a universal peace is as sure as is the prevalence of civilization over barbarism, of liberal governments over feudal forms. The question for us is only how soon?
R.W. Emerson: War, 1849

There is a peace more destructive of the manhood of living man than war is destructive of his material body. Chains are worse than bayonets.
Jerrold W. Douglas, 1803–1857, Peace

Ef you want peace, the thing you've got tu du
Is jes' to show you're up to fightin', tu.
James Russell Lowell, The Biglow Papers, 1848

If peace cannot be maintained with honor, it is no longer peace.
Sir John Russell: Speech at Greenock, 14 September 1853

The war is over. The rebels are our countrymen again.
U.S. Grant: Silencing the cheers of Union troops at Appomattox, 9 April 1865

Let us have peace.
U.S. Grant: Speech accepting the Republican presidential nomination, 29 May 1868

War appears to be as old as mankind, but peace is a modern invention.
Henry Maine: Early History of Institutions, 1875

Eternal peace is a dream, and not even a beautiful one.
Helmuth von Moltke ("The Elder"): Letter to K.K. Bluntschli, 11 December 1880

The legitimate object of war is a more perfect peace.
W. T. Sherman: Epigram 23 February 1882 (inscribed on Sherman's statue, Washington, D. C.)

Thank God for peace! Thank God for peace!
When the great grey ships come in.
Guy Wetmore Carryl, 1873–1904, When the Great Grey Ships Come in

Peace is a goddess only when she comes with sword girt on thigh.
Theodore Roosevelt: Speech at the Naval War College, Newport, Rhode Island, 2 June 1897

Peace, indeed, is not adequate to all progress; there are resistances that can be overcome only by explosion.
Mahan: The Peace Conference, 1899

If I must choose between peace and righteousness, I choose righteousness.
Theodore Roosevelt, 1858–1919, Unwise Peace Treaties

There is a price which is too great to pay for peace, and that price can be put in one word. One cannot pay the price of self-respect.
Woodrow Wilson: Speech at Des Moines, Iowa, 1 February 1916

It must be a peace without victory.
Woodrow Wilson: To the Senate, 22 January 1917

If man does find the solution for world peace it will be the most revolutionary reversal of his record we have ever known.
George C. Marshall: Biennial Report, Chief of Staff, USA, 1 Sept 1945

Could I have but a line a century hence, crediting a contribution to the advance of peace, I would yield every honor which has been accorded by war.
Douglas MacArthur, 1880-1964

And when in his wide courtyards Odysseus had cut down the insolent youths, he hung on high his sated bow and strode to the warm baths to cleanse his blood-stained body. .
Nikos Kazantzakis: The Modern Odyssey, a Sequel, 1958

Peace is the daughter of war.
French Proverb

Peace is produced by war.
Latin Proverb

Eternal peace lasts only until the next war.
Russian Proverb

(*See also* Disarmament, Pacifism.)

Pearl Harbor

The Island of Oahu, with its military depots, both naval and land, its airdromes, water supplies, the city of Honolulu with its wharves and supply points, forms an easy, compact and convenient object for air attack . . . I believe therefore, that should Japan decide upon the reduction and seizure of the Hawaiian Islands . . . attack will be launched on Ford's Island at 7.30 a.m.
William Mitchell: Memorandum for the Chief of Staff, U.S. Army, 1924

Yesterday, December 7, 1941—a date which will live in infamy—the United States of America was suddenly and deliberately attacked by naval and air forces of the Empire of Japan.
Franklin D. Roosevelt: To Congress, 8 December 1941

Throughout the action, there was never the slightest sign of faltering or cowardice. The actions of the officers and men were wholly commendable; there was no panic, no shirking nor flinching, and words fail to describe the truly magnificent display of courage, discipline and devotion to duty of all.
Report by the Executive Officer, USS West Virginia, *after Pearl Harbor, December 1941*

The only thing now to do is to lick hell out of them.
Burton K. Wheeler: Remark after Pearl Harbor, December 1941 (Senator Wheeler had been a leading isolationist)

Remember Pearl Harbor!
World War II slogan

Pearson, Richard (1731–1806)

Should I have the good fortune to fall in with him again, I'll make a Lord of him.
John Paul Jones: On learning that Captain Pearson had been knighted by George III for his defense of HMS Serapis, *taken by USS* Bonhomme Richard, *commanded by Jones, in the battle off Flamborough Head, 23 September 1779*

Pelham, John (1838–1863)

The noble, the chivalric, the *gallant* Pelham, is no more.
J.E.B. Stuart: General Order, 19 March 1863, announcing the death in action of Major Pelham (this is the origin of Pelham's soubriquet, "The Gallant Pelham").

Pen and Sword

I had rather stand the shock of a basilisco than the fury of a merciless pen.
Thomas Browne: Religio Medici, 1634 ("Basilisco" = a brass cannon)

Many wearing rapiers are afraid of goose-quills.
Shakespeare: Hamlet, ii, 1600

How much more cruel the pen may be than the sword.
Robert Burton: Anatomy of Melancholy, 1621

More danger comes by th' quill than by the sword.
Martin Parker: The Poet's Blind Man's Bough, 1641

Among the calamities of war may be justly numbered the diminution of the love of truth by the falsehoods which interest dictates and credulity encourages. A peace will equally leave the warrior and the relater of wars destitute of employment; and I know not whether more is to be dreaded from streets filled with soldiers accustomed to plunder, or from garrets filled with scribblers accustomed to lie.
Samuel Johnson: The Idler, 11 November 1758

A soldier, cried my Uncle Toby, interrupting the corporal, is no more exempt from saying a foolish thing, Trim, than a man of letters—But not so often, an' please your honour, replied the corporal.
Laurence Sterne: Tristram Shandy, viii, 19, 1759

I would rather have written that poem, gentlemen, than take Quebec tomorrow.
James Wolfe: On the eve (12 September 1759) of his death, referring to Gray's Elegy

Four hostile newspapers are more to be feared than a thousand bayonets.
Napoleon I, 1769–1821

The pen is mightier than the sword.
Bulwer-Lytton, 1803–1873, Richelieu, ii

Officers of the Army are apt in general to write like kitchen maids.
Lord Palmerston: Letter to Florence Nightingale, 1857

It is not the business of a Naval officer to write books.
Rear Admiral F. M. Ramsay, USN: Endorsement on an unfavorable fitness report made on Alfred Thayer Mahan, 1893

Our regulations governing the publication by soldiers of their views on military matters are veritable Lettres de Cachet, consigning the intellects of our Services to the Bastille of ignorance.
Sir Ian Hamilton: The Soul and Body of an Army, i, 1921

There is one quality above all which seems to me essential for a good commander, the ability to express himself clearly, confidently, and concisely, in speech and on paper . . . It is a rare quality amongst Army Officers, to which not nearly enough attention is paid in their education. It is one which can be acquired, but seldom is, because it is seldom taught.
Sir A.P. Wavell: Soldiers and Soldiering, 1953

Pentagon, The

This place is a jungle—a jungle.
Robert S. McNamara: Soon after assuming office as Secretary of Defense, 1961

The Pentagon is like a log going down the river with 25,000 ants on it, each thinking he's steering the log.

Henry S. Rowen: in Washington Post, 10 December 1961

Every man who has served with the Navy swells with pride when he hears "Anchors Aweigh," every Marine thrills to the sound of "Halls of Montezuma," Army men cheer at the sound of "The Caissons Go Rolling Along," and the Air Force man responds to "The Wild Blue Yonder"—but did you ever hear anyone shout, and see anyone throw their hats into the air at "Oh, Pentagon, My Pentagon"?

Wilber M. Brucker: Speech at the Las Vegas convention of the Reserve Officers Association, July 1962

The fortunes and affairs of the denizens of the Pentagon are closer and more understandable to the civilian leadership than are the more remote problems and responsibilities of those who man the operating forces.

Admiral Robert B. Carney, USN: Address to the Naval War College, 31 May 1963

Perseverance

Perseverance, dear my lord, keeps honor bright.

Shakespeare: Troilus and Cressida, iii, 3,

Brave admiral, say but one good word:
What shall we do when hope is gone?
The words leapt like a leaping sword:
"Sail on! Sail on! Sail on! Sail on!"

Joaquin Miller, 1841–1913, Columbus

Who hangs on, wins.

German Proverb

Phalanx

In arms the Austrian phalanx stood,
A living wall, a human wood!

James Montgomery, 1771–1854, Make Way for Liberty

Philippine Insurrection (1899–1904)

Damn, damn, damn the Filipinos!
Cross-eyed, kakiak ladrones!
Underneath the starry flag
Civilize 'em with a Krag,
And return us to our own beloved homes.

Soldiers' song, Philippine Insurrection, c. 1899

I'm only a common soldier in the blasted Philippines.
They say I've got brown brothers here, but I dunno what it means.
I like the word fraternity, but still I draw the line.
He may be a brother of Big Bill Taft,
But he ain't no brother of mine!

Robert F. Morrison: In Manila Sunday Times, 1901 (Taft, the first civilian Governor General of the Philippines, had referred to the Filipinos as "little brown brothers.")

Zamboanga, Mindanao,
From the transport you look damn well,
But before I'd serve again in Zamboanga,
I'd rather serve a hitch in Hell.

Soldiers' song, Philippine Insurrection, c. 1902

Oh, I want to know, who's the boss of this show?
Is it me—or Emilio Aguinaldo?

Song, Governor General, 1901 (Aguinaldo was the insurrecto leader, and the song refers to William Howard Taft's difficulties, as Governor General, in quelling the insurrection.)

Physical Fitness

Let us build up physical fitness for the sake of the soul.

Attributed to Plato, 428–347 B.C.

More brawn than brain.

Cornelius Nepos: Epaminondas, v, c. 75 B.C.

What can a soldier do who charges when out of breath?

Vegetius: De Re Militari, iii, 378

The foundation of training depends on the legs and not the arms. All the secret of maneuver and combat is in the legs, and it is to the legs that we should apply ourselves.

Maurice de Saxe: Mes Rêveries, v, 1732

Effeminacy was the chief cause of the ruin of the Roman legions. Those formidable soldiers, who had borne the helmet, buckler and breastplate in the times of the

Scipios under the burning sun of Africa, found them too heavy in the cool climates of Germany and Gaul, and then the Empire was lost.

Jomini: Précis sur l'Art de la Guerre, 1838

I wish to preach, not the doctrine of ignoble ease, but the doctrine of the strenuous life.

Theodore Roosevelt, 1858–1919

Nations have passed away and left no traces,
And history gives the naked cause of it—
One single, simple reason in all cases;
They fell because their peoples were not fit.

Rudyard Kipling: Land and Sea Tales for Scouts and Guides, 1923

Is it really true that a seven-mile cross-country run is enforced upon all in this division, from generals to privates? . . . It looks to me rather excessive. A colonel or a general ought not to exhaust himself in trying to compete with young boys running across country seven miles at a time. The duty of officers is no doubt to keep themselves fit, but still more to think of their men, and to take decisions affecting their safety or comfort. Who is the general of this division, and does he run the seven miles himself? If so, he may be more useful for football than for war. Could Napoleon have run seven miles across country at Austerlitz? Perhaps it was the other fellow he made run. In my experience, based on many years' observation, officers with high athletic qualifications are not usually successful in the higher ranks.

Winston Churchill: Note for the Secretary of State for War, 4 February 1941

A pint of sweat will save a gallon of blood.

George S. Patton, Jr.: Message to troops before landing at Casablanca, 8 November 1942

A man who takes a lot of exercise rarely exercises his mind adequately.

B.H. Liddell Hart: Thoughts on War, xi, 1944

Piecemeal Attacks

Bringing up forces piecemeal . . . amounts to throwing drops of water into a sea.

Ferdinand Foch: Precepts, 1919

Pike-Staff

As plain as a pike-staff.

William Sherlock: 1641–1707, Hatcher of Heresies

Pillage

Know that unless you bring to me the monthly contribution for six months you are to expect an unscantified troop of horse among you, from whom, if you hide yourselves, thay shall fire your houses without mercy, hang up your bodies wherever they find them, and scare your ghosts.

Proclamation by the Royalist Governor of Worcester to a defaulting parish, English Civil War, 1643

Nothing will disorganize an army more or ruin it more completely than pillage.

Napoleon I: Maxims of War, 1831

The only way of pillaging a defeated nation is to cart away any movables that are wanted, and to drive off a portion of its manhood as permanent or temporary slaves.

Winston Churchill: The Gathering Storm, i, 1948

(*See also* Plunder.)

Pilot (Aviator)

Let brisker youths their active nerves prepare,
Fit their light silken wings and skim the buxom air.

Richard Owen Cambridge: Scriblerad, 1751

There won't be any "after the war" for a fighter pilot.

Raoul Lufberry, c. 1917

The engine is the heart of an aeroplane, but the pilot is its soul.

Sir Walter Raleigh: War in the Air, I, 1922

There are old pilots, and bold pilots, but there are no old bold pilots.

Aviators' saying

No pilot is any better than his last landing.

Dictum of "Grampaw Pettibone" in Naval Aviation Newsletter

(*See also* Aerial Combat, Aviation, Flight.)

Pilot (Naval)

. . . Pilots, who have no other thought than to keep the ship clear of danger, and their own silly heads clear of shot.
Nelson: Letter, April 1801 (of his pilots' reluctance to position Nelson's division close aboard the Danish batteries at Copenhagen)

O pilot! 'tis a fearful night,
There's danger on the deep.
Thomas Haynes Bayly, 1797–1839, The Pilot

After you, Pilot.
Captain T.A.M. Craven, USN: To his navigator at the foot of the escape hatch during the sinking of USS Tecumseh, *Mobile Bay, 5 August 1864 (Craven perished with his ship; the navigator survived.)*

(*See also* Navigation, Navigator, Seamanship.)

Planner

When I want any good headwork done, I always choose a man, if suitable otherwise, with a long nose.
Napoleon I, 1769–1821

Planners are always conservative and see all the difficulties, and more can usually be done than they are willing to admit.
Franklin D. Roosevelt, 1882–1945 to General Marshall

Plans

For by wise counsel shalt thou make thy war; and in multitude of counsellors there is safety.
Proverbs, XXIV, 6

It is a bad plan that cannot be altered.
Publilius Syrus, 1st century B.C., Sententiae

The designs of a general should always be impenetrable.
Vegetius: De Re Militari, iii, 378

Where neither the country nor the number of troops you are to act against is known with any precision, a great deal must be left to Fortune . . . The capacity of the Generals may supply the want of Intelligence; but to give them any positive plan or rule of action under such circumstances I apprehend would be absurd.
Lord Ligonier: Letter of Instruction to Sir John Mordaunt for the Rochefort expedition, 1758

Be audacious and cunning in your plans, firm and persevering in their execution, determined to find a glorious end.
Clausewitz: Principles of War, 1812

In forming the plan of a campaign, it is requisite to foresee everything the enemy may do, and be prepared with the necessary means to counteract it.
Napoleon I: Maxims of War, 1831

If I always appear prepared, it is because before entering on an undertaking, I have meditated for long and have foreseen what may occur. It is not genius which reveals to me suddenly and secretly what I should do in circumstances unexpected by others; it is thought and meditation.
Napoleon I, 1769–1821

Nothing succeeds in war except in consequence of a well prepared plan.
Attributed to Napoleon I, 1769–1821

No plan survives contact with the enemy.
Attributed to Helmuth von Moltke ("The Elder"), 1800–1891

In times of peace the general staff should plan for all contingencies of war. Its archives should contain the historical details of the past, and all statistical, geographical, topographical, and strategic treatises and papers for the present and future.
Jomini: Précis de l'Art de la Guerre, 1838

The stroke of genius that turns the fate of a battle? I don't believe in it. A battle is a complicated operation, that you prepare laboriously. If the enemy does this, you say to yourself I will do that. If such and such happens, these are the steps I shall take to meet it. You think out every possible development and decide on the way to deal with the situation created. One of these developments occurs; you put your plan in operation, and everyone says, "What genius . . . " whereas the credit is really due to the labor of preparation.
Ferdinand Foch: Interview, April 1919

The cards in the game of life are the characters of men . . . But when we play the game of death, things are our counters—guns, rivers, shells, bread, roads, forests, ships.

Sir Ian Hamilton: Gallipoli Diary, I, 1920

My war was overthought, because I was not a soldier.

T.E. Lawrence: Seven Pillars of Wisdom, 1926

Successful generals make plans to fit circumstances, but do not try to create circumstances to fit plans.

George S. Patton, Jr.: War As I Knew It, 1947

A good plan violently executed *Now* is better than a perfect plan next week.

George S. Patton, Jr.: War As I Knew It, 1947

An educated guess is just as accurate and far faster than compiled errors.

George S. Patton, Jr.: War As I Knew It, 1947

In total war it is quite impossible to draw any precise line between military and non-military problems.

Winston Churchill: Their Finest Hour, 1949

You don't have to have a piece of paper to know where you are going and to act.

Paul H. Nitze: Address to the Army War College, 27 August 1958.

If, as Clausewitz so justly said, war is a continuation of national policy, so also are war plans.

Barbara W. Tuchman: The Guns of August, vii, 1962

And, as so often happens in war, the least imaginative counsels prevailed.

Brigadier General S.B. Griffith II, USMC: The Battle for Guadalcanal, xix, 1963

(*See also* Programs.)

Plunder

Three hours' plundering is the shortest rule of war. The soldier must have something for his toil and trouble.

Johann Tilly: At the sack of Magdeburg, 20 May 1631

The soldiers of the army have got among them about a million of sterling . . . The night of the battle, instead of being passed in getting rest and food to prepare them for the pursuit of the following day, was passed by the soldiers in looking for plunder. The consequence was, that they were incapable of marching in pursuit of the enemy, and were totally knocked up.

Wellington Letter to Earl Bathurst after Vitoria, 21 June 1813

No reliance can be placed on the conduct of troops in action with the enemy, who have been accustomed to plunder.

Wellington: General Order, 5 March 1814

(*See also* Pillage.)

Police Call

Come get your squeegees, come get your swabs,
Come get your squeegees, come get your swabs,
Come get your squeegees, come get your swabs—
All but Noncoms!

Traditional words of Police Call, 19th and 20th centuries

Policy

In one word, the art of war in its highest point of view is policy.

Clausewitz: On War, 1832

Domestic policy can only defeat us: foreign policy can kill us.

John F. Kennedy: Remark, 1961

Political Objectives

Success in war is determined by the political advantages gained, not victorious battles.

Niccolo Machiavelli, 1469–1527

War is not merely a political act, but also a political instrument, a continuation of political relations, a carrying out of the same by other means.

Clausewitz: On War, i, 1832

Modern wars are not internecine war, in which the killing of the enemy is the object.

The destruction of the enemy, in modern war, and indeed, modern war itself, are means to obtain that object of the belligerent which lies beyond the war.

U.S. War Department General Orders No. 100, 24 April 1863

(*See also* National Policy.)

Politicians

As recruits in these times are not easily got,
And the Marshal must have them—pray, why should we not,
As the last and, I grant it, the worst of our loans to him,
Ship off the Ministry, body and bones to him?
There's not in all England, I'd venture to swear,
Any men we could half so conveniently spare,
And, though they've been helping the French for years past,
We may thus make them useful to England at last.

Thomas Moore: Reinforcements for Lord Wellington, 1813

The politician should fall silent the moment mobilization begins, and not resume his precedence until the strategist has informed the King, after the total defeat of the enemy, that he has completed his task.

Helmuth von Moltke ("The Elder"), 1800–1891

It is a weakness of politicians, who rarely understand their own business, always to aspire to set right the business of others.

Sir John Fortescue: History of the British Army, 1899

As civilization has advanced, the morale of armies has drawn more and more of its strength from conscience, and so, politicians have been more and more forced to cut their best friend, the Devil, when they meet him in public.

Sir Ian Hamilton: The Soul and Body of an Army, x, 1921

Broadly speaking, the career of a politician teaches him to keep a sharp eye on the dramatic effect likely to be produced by his performance upon the mentality of his friends and foes. When he comes upon actual factors like land, sea, munitions, weapons, blood and iron, he is inclined to blink them as details for a soldier.

Sir Ian Hamilton: Listening for the Drum, 1943

In acquiring proficiency in his branch, the politician has many advantages over the soldier; he is always "in the field," while the soldier's opportunities of practicing his trade in peace are few and artificial . . . The politician, who has to persuade and confute, must keep an open and flexible mind, accustomed to criticism and argument; the mind of the soldier, who commands and obeys without question, is apt to be fixed, drilled, and attached to definite rules.

Sir A.P. Wavell: Generals and Generalship, 1939

(*See also* Politico-Military Affairs, Politics.)

Politico-Military Affairs

Politics and arms seem unhappily to be the two professions most natural to man, who must always be either negotiating or fighting.

Voltaire, 1694–1778

The political design is the object, while war is the means.

Clausewitz: On War, i, 1832

Policy is the intelligent faculty, war only the instrument, not the reverse. The subordination of the military view to the political is, therefore, the only thing possible.

Clausewitz: On War, 1832

Our diplomats plunge us forever into misfortune; our generals always save us.

Otto von Bismarck, c. 1850

It is a senseless proceeding to consult the soldiers concerning plans for war in such a way as to permit them to pass purely military judgments on what the ministers have to do; and even more senseless is the demand of theoreticians that the accumulated war material should simply be handed over to the field commander so that he can draw up a purely military plan for the war or for a campaign.

Clausewitz: Krieg und Kriegführung, 1857

When soldiers deal with one another, all goes well; but, as soon as the diplomats

step in, the result is unadulterated stupidity.
Alexander II of Russia: Letter, 1863

Soldier and statesman, rarest unison . . .
James Russell Lowell: Under the Old Elm, 3 July 1875 (of George Washington)

If military-political affairs were only left to a few efficient and reliable officers, the officers would soon agree to everybody's satisfaction.
Helmuth von Moltke ("The Elder"), 1800–1891

Politics uses war for the attainment of its ends; it operates decisively at the beginning and at the end [of the conflict], of course in such manner that it refrains from increasing its demands during the war's duration or from being satisfied with an inadequate success . . . Strategy can only direct its efforts toward the highest goal which the means available make attainable. In this way, it aids politics best, working only for its objectives, but in its operations independent of it.
Helmuth von Moltke ("The Elder"): Ueber Strategie, 1872

In the course of the campaign the balance between the military will and the considerations of diplomacy can only be held by the supreme authority.
Helmuth von Moltke ("The Elder"): Letter to Spenser Wilkinson, 20 January 1890

We are never ready for war, yet we never have a Cabinet who dare tell people the truth.
Sir Garnet Wolseley, 1833–1913, diary entry

Democracy is the best system of government yet devised, but it suffers from one grave defect—it does not encourage those military virtues upon which, in an envious world, it must frequently depend for survival.
Major Guy du Maurier: 1865–1915

That the soldier is but the servant of the statesman, as war is but an instrument of diplomacy, no educated soldier will deny. Politics must always exercise an extreme influence on strategy; but it cannot be gainsaid that interference with the commanders in the field is fraught with the gravest danger.
G.F.R. Henderson: Stonewall Jackson, 1898

War is commonly supposed to be a matter for generals and admirals, in the camp, or at sea. It would be as reasonable to say that a duel is a matter for pistols and swords. Generals with their armies and admirals with their fleets are mere weapons wielded by the hand of the statesman.
Sir John Fortescue: Lecture, 1911

Great political results often flow from correct military action; a fact which no military commander is at liberty to ignore. He may very well not know of those results; it is enough to know that they may happen.
Mahan: Naval Strategy, 1911

I cannot too entirely repudiate any casual word of mine, reflecting the tone which was once so traditional in the Navy . . . that "political questions belong rather to the statesman than to the military man." I find these words in my old lectures, but I very soon learned better.
Mahan: Naval Strategy, 1911

There is no greater fatuity than a political judgment dressed in a military uniform.
Lloyd George, 1863–1945

The decisions of generals are functions of the decisions taken by the statesmen.
Jean Mordacq: Le Ministère Clemenceau, II, 1920

History causes the military problem to become the essence of the political problem.
V.I. Lenin: Speech, 1921

Appearance is not yet reality. The more dominated by military factors a war may seem to be the more political is its actual nature; and this applies equally in reverse.
V.I. Lenin, 1870–1924

The Party commands the gun; the gun will never command the Party.
Mao Tse-tung, 1893–

The Chinese Red Army is an armed body for carrying out the political tasks of the revolution.
Mao Tse-tung: Manifestations of Various Non-Proletarian Ideas in the Party Organization of the Fourth Army, 1929 (Mao was then Political Commissar of the Fourth Army.)

The Bettman Archive

BARON HENRI JOMINI

1779–1869

"War in its ensemble is not a science, but an art."

There are some militarists who say: "We are not interested in politics but only in the profession of arms." It is vital that these simple-minded militarists be made to realize the relationship that exists between politics and military affairs. Military action is a method used to attain a political goal. While military affairs and political affairs are not identical, it is impossible to isolate one from the other.
Mao Tse-tung: On Guerrilla Warfare, 1937

Very near the heart of all foreign affairs is the relationship between policy and military power.
McGeorge Bundy: Preface to collection of Dean Acheson's state papers, 1951

The soldier is the statesman's junior partner.
General Matthew B. Ridgway USA: The Soldier and the Statesman, September 1954

The soldier always knows that everything he does . . . will be scrutinized by two classes of critics—by the Government which employs him and by the enemies of that Government.
Sir William Slim: Unofficial History, 1954

Politics

It would be utterly impossible to maintain discipline if soldiers were allowed to be political partisans, correspondents to newspapers, or members of political clubs. Then, indeed, a standing army would be in truth a curse; then they might bid farewell to civil liberty.
Sir Robert Peel, 1788–1850

I believe you do not mix politics with your profession, in which you are right.
Abraham Lincoln: Letter to Gen Hooker, 26 January 1863

If forced to choose between the penitentiary and the White House for four years, I would say the penitentiary, thank you.
W.T. Sherman: Letter to Major General H.W. Halleck, USA, September 1864

I will not accept if nominated and will not serve if elected.
W.T. Sherman: Telegram to the Republican National Convention, 5 June 1884 (rejecting nomination for the Presidency)

I have no desire for any political office. I am unfitted for it, having neither the education nor the training . . . The Navy is one profession, politics another.
Admiral of the Navy George Dewey, USN: To the press, July 1900

Politics cannot be cut out of the life of a civilian, for then he has no way of expressing his feelings except by a brickbat.
Sir Ian Hamilton: The Soul and Body of an Army, i, 1921

The conversation of military men upon political topics is a rare stimulant for civilians.
Philip Guedalla: Fathers of the Revolution, 1926

I do not approve of this system of encouraging political discussion in the Army among soldiers as such . . . Discussions in which no controversy is desired are a farce. There cannot be controversy without prejudice to discipline. The only sound principle is "No politics in the Army."
Winston Churchill: Note for the Secretary of State for War, 17 October 1941

Power politics is the diplomatic name for the law of the jungle.
Ely Culbertson: Must We Fight Russia? 1946

I have several times had to resist invitations to enter the political field. I do not think that I would make a good politician. War is a pretty rough and dirty game. But politics!
Montgomery of Alamein: Memoirs, xxxiii, 1958

In the United States, we go to considerable trouble to keep soldiers out of politics, and even more to keep politics out of soldiers.
Brigadier General S.B. Griffith II, USMC: Introduction to Mao Tse-tung on Guerrilla Warfare, 1961

The soldier often regards the man of politics as unreliable, inconstant, and greedy for the limelight. Bred on imperatives, the military temperament is astonished by the number of pretenses in which the statesman has to indulge. The terrible simplicities of war contrast strongly to the devious methods demanded by the art of government. The impassioned twists and turns, the dominant concern with the effect to be produced, the appearance of weighing others in terms not of their merit but of their influence—all inevitable characteristics in the civilian whose authority rests upon the popular will—cannot but worry the professional soldier, habituated as he is to a life of hard duties, self-effacement, and respect for services rendered.

Charles de Gaulle, 1890–

If any of us are virtuous 51 percent of the time in life, it's a good record, and in politics an amazing record.

Arthur Sylvester: Remarks at Sigma Delta Chi dinner, New York, 6 December 1962 (at the time of this observation, Mr. Sylvester was Assistant Secretary of Defense for Public Affairs).

(*See also* Politicians, Politico-Military Affairs.)

Positions

War is a business of positions.

Napoleon I, 1769–1821

(*See also* Maneuver, Terrain.)

Power

Power is not revealed by striking hard or often, but by striking true.

Balzac, 1799–1850

Power, in whatever hands, is rarely guilty of too strict limitations on itself.

Edmund Burke: Letter to the Sheriffs of Bristol on the affairs of America, 1777

All power corrupts, and absolute power corrupts absolutely.

Lord Acton, 1834–1902, History of Freedom

Power, force, is a faculty of national life; one of the talents committed to nations by God.

Mahan: The Peace Conference, 1899

Political power emanates from the barrel of a gun.

Mao Tse-tung: On Guerrilla Warfare, 1937

World opinion? I don't believe in world opinion. The only thing that matters is power.

John J. McCloy: During White House conference, September 1961

Few of the important problems of our time have in the final analysis been finally solved by military power alone.

John F. Kennedy: To the graduating class, U. S. Naval Academy, 1 June 1963

Let us have no illusions as we supply the power, the integument of policy, that we are going to be popular in so doing.

Robert M. McClintock: Lecture, Naval War College, 17 September 1964

The problem of power is really the fundamental problem of our time and will remain the basic problem of all future history.

Herbert Rosinski: Power and Human Destiny, 1965

(*See also* National Power.)

Prayer Before Action

Let our axes crush cloth and bones as the jaws of the hyena crush its prey. Make the wounds we give to gape . . . When the wounds of our enemies heal, let lameness remain. Let their stones and arrows fall on us as flowers of the mowa-tree fall in the wind . . . Make their weapons brittle as the long pods of the karta-tree.

Hindu prayer to the war god, Loha Pennu, c. 200 B.C.

O God of battles, steel my soldiers'
hearts;
Possess them not with fear; take from them
now
The sense of reckoning, if the opposed
numbers
Pluck their hearts from them.

Shakespeare: King Henry V, iv, 1, 1598

O Lord! Thou knowest how busy I must be this day: If I forget Thee, do not Thou forget me. March on, boys!

Sir Jacob Astley: Before the battle of Edgehill, 1642

Save and deliver us, we humbly beseech thee, from the hands of our enemies; that we, being armed with thy defense, may be preserved evermore from all perils, to glorify thee, who are the only giver of all victory; through the merits of thy Son, Jesus Christ our Lord. Amen.

The Book of Common Prayer, 1662

Oh God, let me not be disgraced in my old days. Or if Thou wilt not help me, do not help these scoundrels; but leave us to try it ourselves.

Leopold I of Anhalt-Dessau ("Old Dessauer"): Before Kesselsdorf, 14 December 1745 (After this prayer, at the age of seventy, the old field marshal beat the Austrians.)

Dear Lord, on the morrow pray do not let me kill anyone; and, dear Lord, pray do not let anyone kill me.

Prayer by a member of the Clan Fraser on the eve of Culloden, 16 April 1746

There is a time to pray and a time to fight. This is the time to fight.

John Peter Gabriel Mühlenberg: Sermon at Woodstock, Virginia, 1775

May the great God, whom I worship, grant to my country and for the benefit of Europe in general, a great and glorious victory, and may no misconduct in anyone tarnish it, and may humanity after the victory be the predominant feature of the British fleet.

Nelson: Prayer in his diary before Trafalgar, 21 October 1805

Please God—let there be victory, before the Americans arrive.

Sir Douglas Haig: Diary entry, 1917 (attributed)

(*See also* Christian Soldier, God of Battles.)

Preemptive War

If a man does not strike first, he will be the first struck.

Athenogoras of Syracuse: Speech to the Syracusans, 415 B.C.

I know, too, of cases that have occurred in the past when people, sometimes as the result of slanderous information and sometimes merely on the strength of suspicion, have become frightened of each other and then, in their anxiety to strike first before anything is done to them, have done irreparable harm to those who neither intended nor even wanted to do them any harm at all.

Clearchus of Sparta: To Tissaphernes of Persia, after Cunaxa, 401 B.C.

I hold it lawful and Christian policy to prevent a mischief betimes, as to revenge it too late. Only you must resolve, and not delay or dally.

Sir Francis Drake, 1540–1596

. . . The sword drawn to prevent the drawing of swords.

Samuel Purchas: Purchas His Pilgrimage, 1612

Self-defense sometimes dictates aggression. If one people takes advantage of peace to put itself in a position to destroy another, immediate attack on the first is the only means of preventing such destruction.

C.L. Montesquieu, 1689–1755

Offensive war, that is, taking advantage of the present moment, is always imperative when the future holds out a better prospect, not for ourselves, but to our adversary.

Clausewitz: On War, 1832

I would never advise Your Majesty to declare war forthwith, simply because it appeared that our opponent would begin hostilities in the near future. One can never anticipate the ways of divine providence securely enough for that.

Otto von Bismarck: To Wilhelm I of Germany, 1875

The statesman who, knowing his instrument to be ready, and seeing war inevitable, hesitates to strike first is guilty of a crime against his country.

Colmar von der Goltz: Excusing the action of the Boer President Kruger in forcing war with Britain, 1899

When you see a rattlesnake poised to strike, you do not wait until he has struck before you crush him.

Franklin Delano Roosevelt: Fireside Chat, 11 September 1941

Preparedness

Against danger it pays to be prepared.

Aesop: The Wild Boar and the Fox, c. 570 B.C.

Keep the munition, watch the way, make thy loins strong, fortify thy power mightily.
Nahum II, 1

A wise man in time of peace prepares for war.
Horace: Satires, II, c. 30 B.C.

When a strong man armed keepeth his palace, his goods are in peace.
Luke XI, 21–22

In the midst of peace, war is looked on as an eventuality too distant to merit consideration.
Vegetius: De Re Militari, iii, 378

The man who is prepared has his battle half fought.
Cervantes: Don Quixote, ii, 17, 1605

Peace hath so besotted us that as we are altogether ignorant, so are we much the more not sensible of the defect, for we think if we have men and ships our kingdom is safe, as if men were born soldiers.
Sir Edward Cecil: Commenting on a possible French invasion, 1628

One sword keeps another in the sheath.
George Herbert: Jacula Prudentum, 1651

Forewarned, forearmed.
Benjamin Franklin: Poor Richard's Almanac, 1736

In time of peace it is necessary to prepare and always be prepared for war by sea.
Attributed to John Paul Jones, 1747–1792

There is nothing so likely to produce peace as to be well prepared to meet an enemy.
George Washington: Letter to Elbridge Gerry, 29 January 1780

To be prepared for war is one of the most effectual means of preserving peace.
George Washington: First Annual Address to Congress, 8 January 1790

If we desire to avoid *insult,* we must be able to *repel it.* If we desire to secure peace, it must be known that we are at all times *ready for war.*
George Washington, 1732–1799

I wish peace from the *bottom of my soul,* but I desire to see us prepared for war, *in every respect.*
Thomas Truxtun, 1755–1822, letter to Timothy Pickering

The country must have a large and efficient army, one capable of meeting the enemy abroad, or they must expect to meet him at home.
Wellington: Letter, 28 January 1811

Civilized governments ought always to be ready to carry on a war in a short time—they should never be found unprepared.
Jomini: Précis de l'Art de la Guerre, 1838

We ought to ask our country for the largest possible armies that can be raised, as so important a thing as the self-existence of a great nation should not be left to the fickle chances of war.
W.T. Sherman: Letter to General Grant, 20 September 1864

What our sword has won in half a year, our sword must guard for half a century.
Helmuth von Moltke: After the Franco-Prussian War, 1871

Preparations against naval attack and for naval offense is preparedness for anything that is likely to occur.
Mahan: Draft note prior to the Hague Peace Conference, 1899

The real objective of having an Army is to provide for war.
Elihu Root: Annual Report of the Secretary of War, December 1899

Again and again we have owed peace to the fact that we were prepared for war.
Theodore Roosevelt: Lecture at the Naval War College, June 1897

We need to keep in a position of preparedness, especially as regards our Navy, not because we want war, but because we desire to stand with those whose plea for peace is listened to with respectful attention.
Theodore Roosevelt: Speech in New York, 11 November 1902

Preparedness is based on organization. National preparedness means far more than the mere organization of the army and navy.
Major General Leonard Wood, USA,

1860–1927, Our Military History: Its Facts and Fallacies

There is no record in history of a nation that ever gained anything valuable by being unprepared to defend itself.
H.L. Mencken: Prejudices, Series V, 1926

No nation ever had an army large enough to guarantee it against attack in time of peace or insure it victory in time of war.
Attributed to Calvin Coolidge, 1872–1933

The war the Generals always get ready for is the previous one.
H.M. Tomlinson, 1873–, All Our Yesterdays

All forms of war cannot be indiscriminately condemned; so long as there are nations and empires, each prepared callously to exterminate its rival, all alike must be equipped for war.
Sigmund Freud, 1856–1939, letter to Albert Einstein

In the final choice a soldier's pack is not so heavy a burden as a prisoner's chains.
Dwight D. Eisenhower: Inaugural Address, 20 January 1953

A nation cannot in more than the most general sense make itself ready for the distant future; it is the here and now of international relations, by which the future will be shaped, which really count. After all, it is the French Marshal Leboeuf, who is 1870 was "ready to the last gaiter button," who remains a melancholy monument to perfect preparedness; it is the German General Staff, which by 1914 had created a military machine capable of meeting every eventuality, that brought down upon the country the total disaster which they had failed to foresee.
Walter Millis: Arms and Men, 1956

Only when our arms are sufficient beyond doubt can we be certain that they will never be employed.
John F. Kennedy: Inaugural Address, 20 January 1961

We should never separate the idea of peace from the requirement of vigilance in defense.
McGeorge Bundy: Speech, 1963

In time of war an enlightened government must prepare for peace.
Walter Lippmann: In The Washington Post, 18 March 1965

It is a natural trait to prepare for war only when war comes; and when war is over to disarm.
Samuel Eliot Morison: History of the American People, xxv, 1965

Be prepared.
Motto, Boy Scouts

(*See also* Readiness.)

Press-Gangs

It is probable that there may be some disturbance or mob on Thursday next. Will you allow me to suggest to your Lordship whether a proper distribution of half-a-dozen stout, able Press-gangs about Westminster might not be a great utility and perhaps pick up a few stout men should any riot commence.
Letter from agent of Lord North to Lord Sandwich, 1779

(*See also* Conscription, Impressment, Manpower.)

Pride

He that outlives this day, and comes safe home,
Will stand a tip-toe when this day is named.
Shakespeare: King Henry V, iv, 3, 1598

Their military pride promises much, for the first step to make a good soldier is to entertain a consciousness of personal superiority.
Report by an observer of General Anthony Wayne's Pennsylvania Line in Virginia, 1781

Soldiers, we must never be beat—what will they say in England?
Attributed to Wellington at Waterloo, 18 June 1815

The more developed the professional sense of an army the more sensitive and touchy it is to anything which hurts, or seems to hurt, its interests and prerogatives.
Albrecht von Roon: Letter, 29 May 1864

Though pride is not a virtue, it is the parent of many virtues.

John Churton Collins, 1848–1908, Aphorisms

(*See also* Esprit de Corps.)

Princeton (**3 January 1777**)

. . . an old-fashioned Virginia fox-hunt, gentlemen.

George Washington: As he led the Continentals forward onto the British at Princeton

Principle

Where principle is involved, be deaf to expediency.

Matthew Fontaine Maury, 1806–1873

Principles of War

In the art of war there are no fixed rules. These can only be worked out according to circumstances.

Li Chüan, fl. 7th century A.D.

War should be made methodically, for it should have a definite object; and it should be conducted according to the principles and rules of the art.

Napoleon I: Maxims of War, 1831

The principles of war are the same as those of a siege. Fire must be concentrated at one point, and as soon as the breach is made the equilibrium is broken and the rest is nothing.

Napoleon I: Maxims of War, 1831

Keeping your forces united, being vulnerable at no point, moving rapidly on important points—these are the principles which assure victory, and, with fear, resulting from the reputation of your arms, maintains the faithfulness of allies and the obedience of conquered peoples.

Napoleon I: Maxims of War, 1831

There exists a small number of fundamental principles of war, which may not be deviated from without danger, and the application of which, on the contrary, has been in all times crowned with glory.

Jomini: Précis de l'Art de la Guerre, 1838

War is not, as some seem to suppose, a mere game of chance. Its principles constitute one of the most intricate of modern sciences; and the general who understands the art of rightly applying its rules, and possesses the means of carrying out its precepts, may be morally certain of success.

Major General H.W. Halleck, USA: Elements of Military Art and Science, 1846

If men make war in slavish obedience to rules, they will fail.

U.S. Grant: Personal Memoirs, 1885

War acknowledges principles, and even rules, but these are not so much fetters, or bars, which compel its movement aright, as guides which warn us when it is going wrong.

Mahan, 1840–1914

Almost all aspects of the art of war are "theoretical" in time of peace; they only become "practical" when the actual killing begins.

Sir Douglas Haig: Letter to Winston Churchill, 31 January, 1914

I would give you a word of warning on the so-called principles of war, as laid down in *Field Service Regulations.* For heaven's sake, don't treat those as holy writ, like the Ten Commandments, to be learned by heart, and as having by their repetition some magic, like the incantations of savage priests. They are merely a set of common-sense maxims, like "cut your coat according to your cloth," "a rolling stone gathers no moss," "honesty is the best policy," and so forth . . . Clausewitz has a different set, so has Foch, so have other military writers. They are all simply common sense, and are instinctive to the properly trained soldier.

Sir A.P. Wavell: Lecture to the officers, Aldershot Command, c. 1930

. . . to preserve oneself and to annihilate the enemy.

Mao Tse-tung: Strategic Problems of the Anti-Japanese Guerrilla War, 1939

Adherence to one principle frequently demands violation of another. Any leader who adheres inflexibly to one set of commandments is inviting disastrous defeat from a resourceful opponent.

Admiral C.R. Brown, USN: The Principles of War, June 1949

(*See also* Generalship, Strategy, Tactics.)

Prisoner of War

All soldiers taken must be cared for with magnanimity and sincerity so that they may be used by us.
Chang Yü, fl. 1000

You have the captives
Who were the opposites of this day's strife:
We do require them of you.
Shakespeare: King Lear, v, 3, 1605

Information obtained from prisoners should be received with caution, and estimated at its true value. A soldier seldom sees anything beyond his company; and an officer can afford intelligence of little more than the position and movements of the division to which his regiment belongs. On this account the general of any army should never depend upon the information derived from prisoners unless it agrees with the reports received from the advanced guards, in reference to the enemy.
Napoleon I: Maxims of War, 1831

Be'ind the pegged barb-wire strands,
Beneath the tall electric light,
We used to walk in bare-'ead bands,
Explainin' 'ow we lost our fight.
Rudyard Kipling: Half-Ballad of Waterval, 1903

Private

Last night, among his fellow roughs,
He jested, quaffed, and swore;
A drunken private of the Buffs,
Who never looked before.
Today, beneath the foeman's frown,
He stands in Elgin's place,
Ambassador from Britain's crown,
And type of all her race.
Francis Hastings Doyle, 1810–1888, The Private of the Buffs

We were privates, and who is more carefree.
Robert Leckie: Helmet for my Pillow, i, 1957

(*See also* Enlisted Men.)

Privateer

It were well they were restrained by the consent of all princes, since all good men account them but one remove from pirates.
Charles Molloy: De Jure Maritimo, i, 1769

The conduct of all privateering is, as far as I have seen, so near piracy, that I only wonder any civilized nation can allow them. The lawful as well as the unlawful commerce of the neutral flag is subject to every violation and spoliation.
Nelson, 1758–1805, Despatches

Prize-Money

The officer . . . feels a reward in his own bosom, and in his country's thanks. Patriotism and a laudable thirst for renown, will lead *him* to court perils, in defense of his country's rights. These feelings operate upon the sailor also; but to keep up the high tone of his ardor, he must have prize-money in view.
Commodore William Bainbridge, USN: Letter to a friend, 1813

This war is not being conducted for the benefit of officers or to enrich them by the capture of prizes . . . Honor and glory should be the watchword of the Navy, and not profit.
David D. Porter: Endorsement on a report that two captains had failed to inform other ships on station during the chase of a blockade-runner, New Inlet, N.C., 31 October 1864

Profanity

Full of strange oaths and bearded like a pard,
Jealous in honor, sudden and quick in quarrel . . .
Shakespeare: As You Like It, ii, 1599

That in the captain's but a choleric word
Which in the soldier is flat blasphemy.
Shakespeare: Measure for Measure, ii, 2, 1604

. . . Horribly stuff'd with epithets of War.
Shakespeare: Othello, i, 1, 1604

"Our armies swore terribly in Flanders," cried my uncle Toby,—"but nothing to this."

Laurence Sterne: Tristram Shandy, iii, 11, 1760

If any shall be heard to swear, curse, or blaspheme the name of God, the commander is strictly enjoined to punish them for every offense by causing them to wear a wooden collar or some shameful badge, for so long a time as he shall judge proper.
John Adams: Rules and Articles for the Regulation of the Navy of the United Colonies, 28 November 1775

That unmeaning and abominable custom, swearing.
George Washington: General Order against profanity in the Army, 1776

Though you are not to allow swearing in others, it being fobidden by the articles of war, yet by introducing a few oaths occasionally into your discourse, you will give your inferiors some idea of your courage; especially if you should be advanced in years: for then they must think you a daredevil indeed.
Francis Grose: Advice to the Officers of the British Army 1782

Swear at you?—Damn it, Sir, every time I do that, you go a round on the ladder of promotion.
Thomas Truxtun: To Midshipman David Porter, on board USS Constellation, *1799*

Profession of Arms

By the sword shalt thou live.
Genesis, XXVII, 40

The Lord is a man of war.
Exodus, XV, 3

He who makes war his profession cannot be otherwise than vicious. War makes thieves, and peace brings them to the gallows.
Attributed to Niccolo Machiavelli, 1469–1527

War being an occupation by which a man cannot support himself with honor at all times, it ought not to be followed as a business by any but princes or governors of commonwealths; and if they are wise men they will not suffer any of their subjects or citizens to make that their only profession.
Niccolo Machiavelli, 1469–1527

If e'er my son
Follow the war, tell him it is a school
Where all the principles tending to honor
Are taught, if truly followed.
Philip Massinger, 1583–1640

I hate that heady and adventurous crew . . .
That by death only seek to get a living,
Make scars their beauty and count the loss of limbs
The commendation of a proper man.
Nero, i, 1624 (author unknown)

Although custom and example render the profession of arms the noblest of all, I, for my own part, who only regard it as a philosopher, value it at its proper worth, and, indeed, find it very difficult to give it a place among the honorable professions, seeing that idleness and licentiousness are the two principle motives which now attract most men to it.
Descartes, 1596–1650

The military pedant . . . always talks in a camp, and is storming towns, making lodgments and fighting battles from one end of the year to the other. Everything he speaks smells of gunpowder; if you take away his artillery from him he has not a word to say for himself.
Joseph Addison: The Spectator, 30 June 1711

A soldier's time is passed in distress and danger or in idleness and corruption.
Samuel Johnson: To James Boswell, 10 April 1778

Every man thinks meanly of himself for not having been a soldier.
Samuel Johnson: To James Boswell, 10 April 1778

Military talent is greatly overrated by the world, because the means by which it shows itself are connected with brute force and the most terrible results; and men's faculties are dazzled and beaten down by a thunder and lightning so formidable to their very existence. If playing a game of chess involved the blowing up of gunpowder and the hazard of laying waste a city, men whould have the same grand idea of chess.
Leigh Hunt: The Companion, xiii, 1828

The fewer the employments followed by a nation, the more that of arms predominates.
Clausewitz: On War, 1832

But really, the Service makes brutes of us all.

Frederick Marryat: Peter Simple, 1834

. . . the glories and the servitudes of military life (Les gloires et les servitudes de la vie militaire).

Ardant du Picq, 1821–1870, Battle Studies (cf. de Vigny, 16, ante)

Soldiering, my dear madam, is the coward's art of attacking mercilessly when you are strong, and keeping out of harm's way when you are weak.

George Bernard Shaw: Arms and the Man, ii, 1894

In no event will there be money in it; but there may always be honor and quietness of mind and worthy occupation—which are far better guarantees of happiness.

Mahan: The Navy as a Career, 1895

The profession has its utility, but I should be sorry to see any friend of mine belong to it.

Sir Walter Elliot, 1803–1887

A soldier's life is a hard one, interspersed with some real dangers.

André Maurois: Les Silences de Colonel Bramble, 1917

Thank God this is over and we can get back to real soldiering again.

Remark by an unknown Regular after the Armistice, 1918

The professional military mind is by necessity an inferior and unimaginative mind; no man of high intellectual quality would willingly imprison his gifts in such a calling.

H.G. Wells: Outline of History, xl, 1920

Such professions as the soldier and the lawyer . . . give ample opportunity for crimes but not much for mere illusions . . . If you have lost a battle you cannot believe you have won it; if your client is hanged you cannot pretend you have got him off.

G.K. Chesterton, 1874–1936

For a man with a few outside interests to occupy his spare time—and, of course, a feeling for the Service, I can imagine no better profession than that of arms.

John W. Thomason, Jr.: Letter to a young Marine officer, 1939

Where in contemporary life will you find a profession in which a man can be fired from his job "for conduct unbecoming an officer and a gentleman"?

Hoffman Nickerson, Officer and Gentleman, 1888-

I find I have like all of the soldiers of different races who have fought with me and most of those who have fought against me. This is not strange, for there is a freemasonry among fighting soldiers that helps them to understand one another even if they are enemies.

Sir William Slim, Unofficial History, ix, 1959

The military profession is more than an occupation; it is a style of life.

Morris Janowitz: The Professional Soldier, ix, 1960

What you have chosen to do for your country by devoting your life to the service of your country is the greatest contribution that any man could make.

John F. Kennedy: To the graduating class, U.S. Naval Academy, 6 June 1961

When there is a visible enemy to fight in open combat . . . many serve, all applaud, and the tide of patriotism runs high. But when there is a long, slow struggle, with no immediate visible foe, your choice will seem hard indeed.

John F. Kennedy: Address to the graduating Class, U.S. Naval Academy, 6 June 1961

I'm inclined to think that a military background wouldn't hurt anyone.

William Faulkner: At West Point, 20 April 1962

A nation at war or planning against war rightfully expects the professional soldier to protect it from its own excesses, as well as from the enemy.

Statement by unknown author, c. 1963

(*See also* Military Mind, Regular, Soldier.)

Professional Study

War is a matter of vital importance to the state, the province of life or death, the road to survival or ruin. It is mandatory that it be thoroughly studied.

Sun Tzu, 400–320 B.C., The Art of War, i

The Lacadaemonians made war their chief study. They are affirmed to be the first who reasoned on the events of battles and committed their observations thereon to writing with such success as to reduce the military art, before considered totally dependent on courage or fortune, to certain rules and fixed principles.

Vegetius: De Re Militari, iii, 378

The courage of a soldier is heightened by his knowledge of his profession.

Vegetius: De Re Militari, i, 378

Reading and Discourse are requisite to make a Souldier perfect in the Art Military, how great soever his practical knowledge may be.

George Monk (Duke of Albemarle): Observations Upon Military & Political Affairs, 1671

War is not an affair of chance. A great deal of knowledge, study, and meditation is necessary to conduct it well.

Frederick The Great: Instructions for His Generals, xi 1747

The warrior who cultivates his mind polishes his arms.

Chevalier de Boufflers, 1738–1815

Dry books of tactics are beneath the notice of a man of genius, and it is a known fact, that every British officer is inspired with a perfect knowledge of his duty, the moment he gets his commission.

Francis Grose: Advice to the Officers of the British Army, 1782

Officers can never act with confidence until they are masters of their profession.

Henry Knox: 1750–1806

Read and re-read the campaigns of Alexander, Hannibal, Caesar, Gustavus Adolphus, Turenne, Eugene, and Frederick. Make them your models. This is the only way to become a great captain and to master the secrets of the art of war.

Napoleon I: Maxims of War, 1831

War is an art, to attain perfection in which, much time and experience, particularly for the officers, are necessary.

John C. Calhoun, 1782–1850

An ignorant officer is a murderer. All brave men confide in the knowledge he is supposed to possess; and when the death-trial comes their generous blood flows in vain. Merciful God! How can an ignorant man charge himself with so much bloodshed? I have studied war long, earnestly and deeply, but yet I tremble at my own derelictions.

Sir Charles Napier, 1782–1853

I say that those who take up the science of war must not fail to master the Classics.

Yoshida Shōin, 1830–1859

In our days we no longer believe in what Chatham called "Heaven-born generals." It is agreed that modern warfare is the offspring of science and civilization,—that it has its rules and its principles, which it is necessary to thoroughly master before being worthy to command, and that it is wiser to profit by such lessons of history, as taught in the work before us than to purchase experience by the blood of battlefields.

Colonel George Washington Cullum, USA: In United States Service Magazine, II, 1864

The instruments of battle are valuable only if one knows how to use them.

Ardant du Picq, 1821–1870, Battle Studies

Neither gallantry nor heroism will avail much without professional training.

Sir Edward Cardwell, 1813–1886

Conscientious study will not perhaps make them great, but it will make them respectable; and when responsibility of command comes, they will not disgrace their flag, injure their cause, nor murder their men.

Lieutenant General Richard Taylor, CSA: Destruction and Reconstruction, 1879

It is criminal to hand over in action the lives of gallant soldiers to men who are deplorably ignorant of the elements of their profession.

Sir Garnet Wolseley: After the Egyptian campaigns, 1882–1885

No study is possible on the battlefield; one does there simply what one can in order to apply what one knows. Therefore, in order to do even a little, one has already to know a great deal and know it well.

Ferdinand Foch: Principles of War, 1919

Though the military art is essentially a practical one, the opportunities of practicing it are rare. Even the largest-scale peace maneuvers are only a feeble shadow of the real thing. So that a soldier desirous of acquiring skill in handling troops is forced to theoretical study of Great Captains.

Sir A.P. Wavell: Lecture to officers, Aldershot Command, c. 1930

For my strategy, I could find no teachers in the field: but behind me there were some years of military reading, and even in the little that I have written about it, you may be able to trace the allusions and quotations, the conscious analogies.

T.E. Lawrence, 1888–1935

Generalship, at least in my case, came not by instinct, unsought, but by understanding, hard study, and brain concentration. Had it come easy to me, I should not have done it as well.

T.E. Lawrence, 1888–1935

If your book could persuade some of our new soldiers to read and mark and learn things outside drill manuals and tactical diagrams, it would do a good work.

T.E. Lawrence: Letter to Liddell Hart, 26 June 1933

A study of the laws of war is necessary as we require to apply them to war. To learn this is no easy matter and to apply them in practice is even harder; some officers are excellent at paper exercises and theoretical discussions in the war colleges, but when it comes to battle there are those that win and those that lose.

Mao Tse-tung: On the Study of War, 1936

War is the highest form of struggle between nations, and thus the study of military matters brooks not a moment's delay, and must be learned not only by our commanders, but also by members of the Party.

Mao Tse-tung: On the Study of War, 1936

It should be the duty of every soldier to reflect on the experiences of the past, in the endeavor to discover improvements, in his particular sphere of action, which are practicable in the immediate future.

B.H. Liddell Hart: Thoughts on War, i, 1944

Don't ignore the yesterdays of war in your study of today and tomorrow.

Douglas Southall Freemen: Lecture at Marine Corps Schools, Quantico, Virginia, 1950

(*See also* History, Military Mind.)

Programs

It is folly to raise a single company, squadron or battery before it is known exactly what place it is to take in some definite organization authorized for some definite purpose.

Sir Ian Hamilton: The Soul and Body of an Army, iv, 1921

A staff officer renders no service to the country who aims at ideal standards, and thereafter simply adds and multiplies until impossible totals are reached.

Winston Churchill: Note to the War Cabinet, 15 October 1940

The Army used to have all the time in the world and no money; now we've got all the money and no time.

George C. Marshall: Remark, January 1942

Promotion

For promotion cometh neither from the east, nor from the west, nor yet from the south. And why? God is the judge; he putteth down one, and setteth up another.

Psalm 75, 7–8

Comrades, you have lost a good captain to make a bad general.

Saturninus, c. 100 B. C.

. . . these times
Where none will sweat but for promotion.

Shakespeare: As You Like It, ii, 3, 1599

Who does i' the wars more than his captain can
Becomes his captain's captain.

Shakespeare: Antony and Cleopatra, iii, 1, 1606

I have seen some extremely good colonels become very bad generals.

Maurice de Saxe: Mes Rêveries, 1732

Promotion with my Lords is like kissing—it goes by favor.

18th century saying among officers of the Royal Navy

Above all, be careful never to promote an intelligent officer; a brave, chuckle-headed fellow will do full as well to execute your orders. An officer, that has an iota of knowledge above the common run, you must consider as your personal enemy.

Francis Grose: Advice to the Officers of the British Army, 1782

If ever you wish to rise a step above your present degree, you must learn that maxim of the art of war, of currying favor with your superiors; and you must not only cringe to the commander-in-chief himself, but you must take especial care to keep in with his favorites, and dance attendance on his secretary.

Francis Grose: Advice to the Officers of the British Army, 1782

The promotion to the Flag has happily removed a number of officers from the command of ships-of-the-line, who at no period of their lives were capable of commanding them; and I am sorry to have occasion to observe, that the present state of the upper part of the list of Captains is not much better than it stood before.

Lord St. Vincent: Letter to Earl Spencer, 21 March 1779

And, verily, it gives
A precedent of hope, a spur of action
To the whole corps, if once in their remembrance
An old, deserving soldier makes his way.

Friedrich Schiller: Wallenstein, i, 1, 1798

Every French soldier carries a marshal's baton in his knapsack.

Attributed to Louis XVIII of France, 1755–1824

I will never agree to the promotion of an officer who, for ten years, has not been under fire.

Napoleon I, 1769–1821

As to rewards and promotion, it is essential to respect long service and at the same time open a way for merit.

Jomini: Précis de l'Art de la Guerre, 1838

If the skill of a general is one of the surest elements of victory, it will readily be seen that the judicious selection of generals is one of the most delicate points in the science of government and one of the most essential parts of the military policy of a state. Unfortunately, this choice is influenced by so many petty passions that chance, rank, age, favor, part spirit, or jealousy will have as much to do with it as the public interest and justice.

Jomini: Précis de l'Art de la Guerre, 1838

In the Army there's sobriety, Promotion's very slow . . .

Benny Havens, Oh! West Point song, 1838 (attributed to "Lieutenant O'Brien, 8th Infantry")

I have now obtained what I have been looking for all my life—a flag—and having obtained it, all that is necessary to complete the scene is victory. If I die in the attempt, it will be only what every officer has to expect.

David G. Farragut: To his wife, 1862

Forward! if any man is killed, I'll make him a corporal.

Captain Adna R. Chaffee, USA; To his troop, 6th Cavalry, during the Kiowa-Comanche campaign, 1874

If you wish in the world to advance
On your merits you're bound to enhance,
You must stir it and stump it
And blow your own trumpet
Or, trust me, you haven't a chance.

W.S. Gilbert: Ruddigiore, 1887

More like a mother she were—
Showed me the way to promotion an' pay,
An' I learned about women from 'er!

Rudyard Kipling: The Ladies, 1892

Originality never yet led to preferment.

Sir John Fisher: Memories, 1919

One should only give jobs to people who are looking for at least one promotion.

Benito Mussolini: Remark to Count Ciano, 8 August 1940

We are now at war, fighting for our lives, and we cannot afford to confine Army appointments to persons who have excited no hostile comment in their career.

Winston Churchill: Note to the Chief of the Imperial General Staff, 19 October 1940

Smooth answers smooth the path to promotion.

B.H. Liddell Hart: Thoughts on War, v, 1944

The value of "tact" can be over-emphasized in selecting officers for command: positive personality will evoke a greater response than negative pleasantness.

B.H. Liddell Hart: Thoughts on War, xi, 1944

More and more does the "System" tend to promote to *control*, men who have shown themselves efficient *cogs* in the machine . . . There are few commanders in our higher commands. And even these, since their chins usually outweigh their foreheads are themselves outweighed by the majority—of commanders who are essentially staff officers.

B.H. Liddell Hart: Thoughts on War, xi, 1944

The man who never does more than supinely pass on the opinion of his seniors is brought to the top, while the really valuable man, the man who accepts nothing ready-made but has an opinion of his own, gets put on the shelf.

Erwin Rommel: The Rommel Papers, xviii, 1953

An extensive use of weedkiller is needed in the *senior* ranks after a war; this will enable the first class young officers who have emerged during the war to be moved up.

Montgomery of Alamein: Memoirs, ii, 1958

We make generals today on the basis of their ability to write a damned letter. Those kinds of men can't get us ready for war.

Lewis B. Puller: Marine, 1962

By good fortune in the game of military snakes and ladders, I found myself a general.

Sir William Slim: Unofficial History, vii, 1959

(*See also* Rank, Seniority.)

Promptness

Prompness contributes a great deal to success in marches and even more in battles.

Frederick The Great: Instructions for His Generals, xxiii, 1747

I have always been a quarter of an hour before my time, and it has made a man of me.

Attributed to Nelson, 1758–1805

Promptness is the greatest of military virtues, evincing, as it does, zeal, energy, and discipline. The success of arms depends more upon celerity than any one thing else.

D. H. Hill: Circular, Chattanooga, 7 September 1963

It is always too little or too late—or both.

David Lloyd George: After the fall of Finland, 1940

(*See also* Expedition.)

Propaganda

. . . stuffing the ears of men with false reports.

Shakespeare, 1564–1616

In a time of war the nation is always of one mind, eager to hear something good of themselves, and ill of the enemy. At this time the task of news-writers is easy; they have nothing to do but tell that the battle is expected, and afterwards that a battle has been fought, in which we and our friends, whether conquering or conquered, did all, and our enemies did nothing.

Samuel Johnson: The Idler, 11 November 1758

Vilify! Vilify! Some of it will always stick.

Beaumarchais, 1732–1799

Great captains have always published statements for the benefit of the enemy, that their own troops were very strong in numbers; while to their own people, the enemy was represented as very inferior.

Napoleon I, 1769–1821, Pensees

The appalling thing about war is that it kills all love of truth.

George Brandes: Letter to Georges Clemenceau, March 1915

Propaganda, as inverted patriotism, draws nourishment from the sins of the enemy. If there are no sins, invent them! The aim is to make the enemy appear so

great a monster that he forfeits the rights of a human being. He cannot bring a libel action, so there is no need to stick at trifles.
Sir Ian Hamilton: The Soul and Body of an Army, x, 1921

The printing press is the greatest weapon in the armory of the modern commander.
T.E. Lawrence, 1888–1935

. . . That branch of the art of lying which consists in very nearly deceiving your friends without quite deceiving your enemies.
Aphorism, source unknown, quoted by Dean Acheson in The Reporter, 12 August 1965

(*See Also* Psychological Warfare, Public Information.)

Property

He guards his own property who wishes the common property to be safe.
Publilius Syrus: Sententiae, 1st century B.C.

If it's small enough to pick up, turn it in; if you can't move it, paint it.
Soldiers' saying

A good property sergeant is never caught short.
Marine Corps saying

Prudence

In the moment of victory, tighten your helmet-strap.
Lieutenant Tadayoshi Sakurai: Human Bullets, 1907

Victory loves prudence. (Amat victoria curam.)
Latin Proverb

Prussia

The world does not rest more surely upon the shoulders of Atlas than does Prussia upon her Army.
Frederick The Great: After Hohenfriedberg, June 1745

War is the national industry of Prussia.
Mirabeau: De la Monarchie Prussienne, 1788

Prussia was hatched from a cannon ball.
Attributed to Napoleon I, 1769–1821

Prussia, with all the veils that hide the thing, is a military organization led by a military corporation.
Ardant du Picq, 1821–1870, Battle Studies

Their natural gait is the goose-step.
H.L. Mencken: Prejudices, Series II, 1920

Psychological Warfare

To seduce the enemy's soldiers from their allegiance and encourage them to surrender is of especial service, for an adversary is more hurt by desertion than by slaughter.
Vegetius: De Re Militari, iii, 378

Give the enemy young boys and women to infatuate him, and jades and silks to excite his ambitions.
Chen Hâo, fl. 700

Heart is that by which the general masters. Now order and confusion, bravery and cowardice, are qualities dominated by the heart. Therefore the expert at controlling his enemy frustrates him and then moves against him. He aggravates him to confuse him and harasses him to make him fearful. He thus robs the enemy of his heart and of his ability to plan.
Chang Yü, c. 1000

It is your attitude, and the suspicion that you are maturing the boldest designs against him, that imposes on your enemy.
Frederick The Great: Instructions for His Generals, ix, 1747

As the excited passions of hostile people are of themselves a powerful enemy, both the general and his government should use their best efforts to allay them.
Jomini: Précis de l'Art de la Guerre, 1838

Get 'em skeered, and keep the skeer on 'em.
Attributed to Nathan Bedford Forrest, 1821–1877

Communists . . . always seek to make use of their enemy and in the meantime take care not to be used by him.
Chiang Kai-Shek, 1887– , Soviet Russia in China

The Red Army fights not merely for the sake of fighting but in order to conduct propaganda among the masses, arm them, and help them to establish revolutionary political power.

Mao Tse-tung: Manifestations of Various Non-Proletarian Ideas in the Party Organization of the Fourth Army, 1929 (Mao was then Political Commissar of the Fourth Army).

The place of artillery preparation for frontal attack by the infantry in trench warfare will in future be taken by revolutionary propaganda, to break down the enemy psychologically before the armies begin to function at all.

Adolf Hitler, 1889–1945

Mental confusion, contradiction of feeling, indecisiveness, panic: these are our weapons.

Adolf Hitler: To Hermann Rauschning, 1939

The mind of the enemy and the will of his leaders is a target of far more importance than the bodies of his troops.

Brigadier General S.B. Griffith II, USMC: Introduction to Mao Tse-tung on Guerrilla Warfare, 1961

(*See also* Propaganda)

Public Information

In war, truth is the first casualty.

Aeschylus, 525–456 B. C.

It is the merit of a general
To impart good news, and to conceal the bad.

Sophocles, 496–406 B.C.

Four hostile newpapers are more to be feared than a thousand bayonets.

Napoleon I, 1769–1821

"False as a bulletin" became a proverb in Napoleon's time.

Thomas Carlyle: Heroes and Hero-Worship, vi, 1840

News of battle!—news of battle!
Hark! 'tis ringing down the street.

William Edmondstoune Aytoun: Edinburgh after Flodden, 1849

Not more than two newspapers will be published in Savannah; their editors and publishers will be held to the strictest accountability, and will be punished severely, in person and property, for any libelous publication, mischievous matter, premature news, exaggerated statements, or any comments whatever upon the acts of the constituted authorities.

W.T. Sherman: Special Field Order on occupying Savannah, 26 December 1864

Anti-press, anti-publicity, Official Secrets Acts have kept step with loud professions of belief in open diplomacy and the wisdom of the sovereign people.

Sir Ian Hamilton: The Soul and Body of an Army, i, 1921

Publicity, like war, prevents the process of civilization.

Gertrude Stein, 1874–1946

Information is power.

Arthur Sylvester: Remarks at Sigma Delta Chi dinner, New York, 1962

News flowing from actions taken by the Government is part of the weaponry.

Attributed to Arthur Sylvester, Assistant Secretary of Defense (Public Affairs) during the Cuban crisis, 6 December 1962 (this is the so-called "news managment" dictum which contributed to Mr. Sylvester's unpopularity with Washington correspondents.)

(*See also* Propaganda, Psychological Warfare, Public Relations.)

Public Relations

Many wearing rapiers are afraid of goose-quills.

Shakespeare: Hamlet, ii, 2, 1600

Don't tell them anything. When it's over, tell them who won.

Ernest J. King: When asked to state a public relations policy for the Navy, c. 1942

(*See also* Public Information.)

Pursuit

I have pursued mine enemies, and destroyed them; and turned not again until I had consumed them.

II Samuel, XXII, 38

A pursuit gives even cowards confidence.
Xenophon: Speech to the Greek army at Calpe, in the Euxine, 400 B.C.

To a flying enemy, a silver bridge.
Cervantes: Don Quixote, ii, 1605

Once the enemy has taken flight they can be pursued with no better weapons than air-filled bladders. But if the officer you have ordered in pursuit prides himself on the regularity of his formations and the precautions of his march . . . there is no use in having sent him. He must attack, push, and pursue without cease.
Maurice de Saxe: Mes Rêveries, xxxii, 1732

Next to victory, the act of pursuit is the most important in war.
Clausewitz: Principles of War, 1812

When you strike the enemy and overcome him, never give up the pursuit as long as your men have strength to follow; for an enemy routed, if hotly pursued, becomes panic-stricken, and can be destroyed by half their number.
Stonewall Jackson, 1824–1863

Strenuous, unrelaxing pursuit is therefore as imperative after a battle as is courage during it.
Mahan: Naval Strategy, 1911

Pursuit will hold the first place in your thoughts. It is at the moment when the victor is most exhausted that the greatest forfeit can be exacted from the vanquished.
Winston Churchill: To Lord Wavell, 13 December 1940, after his victory in the Libyan desert

While coolness in disaster is the supreme proof of a commander's courage, energy in pursuit is the surest test of his strength of will.
Sir A.P. Wavell: Unpublished "Recollections," 1946

A rapid advance is paradise for the tactician but hell for the quartermaster.
Military maxim

A stern chase is a long chase.
Naval maxim

Pushbutton War

You press the button, and we'll do the rest.
Advertising slogan for the first Kodak cameras, c. 1888

There is another alarming peril found in a modern fallacy that computers, or economics, or numbers of weapons win wars. Alone, they do not . . . Our nation will defy every lesson of history if we think that stockpiles of weapons or the decision of computers win wars. Man, his wits, and his will are still the key to war and peace, victory and defeat.
Admiral George W. Anderson, USN: In U. S. Naval Institute Proceedings, July 1964

Quartermaster

Nobody ever heard of a quartermaster in history.

Nathanael Greene: To George Washington, demurring at being detailed Quartermaster-General of the Continental Army, March 1778

The standing maxim of your office is to receive whatever is offered you, or you can get hold of, but not to part with anything you can keep.

Francis Grose: Advice to the Officers of the British Army, 1782

If [the general] allows himself to be guided by the supply officers, he will never move, and his expeditions will fail.

Napoleon I, 1769–1821

If quartermasters and civilian officials are left to take their own time over the organization of supplies, everything is bound to be very slow. Quartermasters often tend to work by theory and base all their calculations on precedent, being satisfied if their performance comes up to the standard which this sets. This can lead to frightful disasters when there is a man on the other side who carries out his plans with greater drive and thus greater speed.

Erwin Rommel, 1891–1944

(*See also* Logistics, Property, Supply.)

Quebec (13 September 1759)

Let not my brave soldiers see me drop—The day is ours—Keep it.

James Wolfe: On being mortally wounded during the assault on Quebec, 13 September 1759

What, do they run already? Then I die happy.

James Wolfe: Last words, on the Plains of Abraham, Quebec, 13 September 1759

There would appear in this celebrated campaign fully as much guid luck as guid guiding.

Sir Thomas Selkirk: Diary entry, 1804

Quisling

There was also a Major Quisling, who with a handful of young men had aped and reproduced in Norway on an insignificant scale the Fascist movement.

Winston Churchill: The Gathering Storm, 1948 (the origin of the term, "Quisling")

(*See also* Fifth Column.)

Raiding

If we are to have any campaign in 1941, it must be amphibious in its character, and there will certainly be many opportunities for minor operations, all of which will depend on surprise landings of lightly equipped, nimble forces accustomed to work like packs of hounds instead of being moved about in the ponderous manner which is appropriate to the regular formations.

Winston Churchill: Note for Secretary of State for War, 25 August 1940

Rallying Cry

Go yourselves, every man of you, and stand in the ranks and either a victory beyond all victories in its glory awaits you, or failing you shall fail greatly, and worthy of your past.

Demosthenes: Third Philippic, 341 B.C.

Come, my knights, and let us all go and die there! This is the day!

Grand Master de la Vallette: During the great siege of Malta, 1565

If we are marked to die, we are enow
To do our country loss; and if to live,
The fewer men, the greater share of honor.

Shakespeare: King Henry V, 1598

Offscouring of Scoundrels, would ye live forever!

Attributed to Frederick The Great, 1712–1786

Come on, boys, if the day is long enough, we'll have them in hell before tonight.

Benedict Arnold: Leading the decisive assault at Saratoga, 7 October 1777

If I advance, follow me! If I retreat, kill me! If I die, avenge me!

Comte de la Rochejaquelin: To his volunteers in the revolt of la Vendée, 1793

Fifty-Seventh, die hard!

By a British colonel (Inglis) to the 57th Foot (West Middlesex Regiment) at Albuera, 16 May 1811

Some of us must die; cross yourselves and march forward.

W.S. Rosecrans: To his troops at Murfreesboro, Tennessee, 31 December 1862

Men, let us die like soldiers.

Lieutenant C.B. McLellan, USA: To his troop of the 6th Cavalry on charging a Confederate position, Sailor's Creek, Virginia, 6 April 1865

(*See also* Charge.)

Rank

Let rule entrusted be
To him who treats his rank
As if it were his soul.

Lao Tze: The Way of Life, 6th century B.C.

Where is the injustice, if I or anyone who feels his own superiority to another refuses to be on a level with him?

Alcibiades: Speech to the Athenians, 415 B.C.

Render therefore to all their dues; tribute to whom tribute is due; custom to whom custom; fear to whom fear; honor to whom honor.

Romans XIII, 7

Titles do not reflect honor on men, but rather men on their titles.

Niccolo Machiavelli: Discorsi, iii, 38, 1531

An' two men ride a horse, one must ride behind.

Shakespeare: Much Ado About Nothing, ii, 5, 1598

Take but degree away, untune that string,
And hark, what discord follows!

Shakespeare: Troilus and Cressida, i, 3, 1601

The general's disdain'd
By him one step below; he by the next;
That next by him beneath; so every step,
Exampled by the first pace that is sick
Of his superior, grows to an envious fever.

Shakespeare: Troilus and Cressida, i, 1601

That in the captain's but a choleric word
Which in the soldier is flat blasphemy.

Shakespeare: Measure for Measure, ii, 2, 1604

What is commendable in an officer may be in the highest degree reprehensible in a private man.

Francis Grose: Advice to the Officers of the British Army, 1782

I think with the Romans of old, that the general of today should be a common soldier tomorrow, if necessary.

Thomas Jefferson: Letter to James Madison, 1797

Rank is a great beautifier.

Bulwer-Lytton: The Lady of Lyons, ii, 1, 1838

I think rank of but trivial importance so that it is sufficient for the individual to exercise command.

R.E. Lee, 1807–1870

When differences arise between officers of the government, the ranking officer must be obeyed.

Abraham Lincoln: Letter to Mr. Moulton, 31 December 1863

The barrier of rank is the highest of all barriers in the way of access to the truth.

B.H. Liddell Hart: Thoughts on War, xi, 1944

Every officer has his ceiling in rank, beyond which he should not be allowed to rise—particularly in war-time.

Montgomery of Alamein: Memoirs, vi, 1958

R.H.I.P. (Rank hath its privileges.)

U. S. military saying

(*See also* Promotion, Seniority.)

Rapidity

There has never been a protracted war from which a country has benefited.

Sun Tzu, 400–320 B.C., The Art of War, ii

No great success can be hoped for in war in which rapid movements do not enter as an element. Even the very elements of Nature seem to array themselves against the slow and over-prudent general.

Dennis Hart Mahan, 1802–1871 (Professor Mahan was the father of Alfred Thayer Mahan.)

(*See also* Expedition, Speed.)

Rashness

Beware of rashness, but with energy and sleepless vigilance go forward and give us victories.

Abraham Lincoln: Letter to Major General Joseph Hooker, 26 January 1863

Rashness in war is prudence.

Sir John Fisher: Memories, 1919

(*See also* Audacity, Boldness.)

Rations

I know not which way to deal with the mariners to make them rest contented with sour beer.

Lord Howard of Effingham: While concentrating forces against the Armada, May 1588

Give them great meals of beef, and iron and steel, they will eat like wolves and fight like devils.

Shakespeare: King Henry V, iii, 7, 1598

Soldiers' stomachs always serve them well.

Shakespeare: I King Henry VI, ii, 3, 1591

A poor sailor, in double danger both of the fight and of shipwreck, by day parched with the heat of the sun, by night nipt and whipt with blustering tempests; and when he is wet, cold, and hungry, should not the poor soul have a can of beer to refresh him, but he must say "*Mors est in olla*" when he drinks it, or a cake of bread, but he must think he is set to a penance when he eats it.

Nathaniel Knott: An Advice of a Seaman, 1634 ("Mors est in olla"= "It is death to drink.")

No soldier can fight properly unless he is properly fed on beef and beer.

Duke of Marlborough, 1650–1722, Sayings

I hope your excellency is doing all in Your power to supply your half Starved Fellow Citizens. Flour, Rum, and Droves of Bullocks, should without Delay be forwarded to this Army or the Southern Department will soon want one to defend it.

Horatio Gates: Letter to Governor Thomas Jefferson of Virginia, 1780

You need not mind, whether the provisions issued to the soldiers be good or bad. If it

Brown Brothers

NAPOLEON I

1769–1821

"In war the moral is to the material as three to one."

were always good, they would get too much attached to eating to be good soldiers . . . If the soldiers complain of the bread, taste it, and say, better men have eat much worse. Talk of the *bompernickle,* or black rye bread of the Germans and swear you have seen the time you would have jumped at it.

Francis Grose: Advice to the Officers of the British Army, 1782

An army marches on its stomach.

Attributed to Napoleon I, 1769–1821

(*See also* Subsistence.)

Readiness

It is a doctrine of war not to assume the enemy will not come, but rather to rely on one's readiness to meet him; not to presume that he will not attack, but rather to make one's self invincible.

Sun Tzu, 400–320 B.C., The Art of War, viii

Achilles, though invulnerable, never went into battle but completely armed.

Lord Chesterfield: Letter, 15 January 1753

An army should be ready, every day, every night, and at all times of the day and night, to give all the resistance of which it is capable . . . The soldier should always be furnished completely with arms and ammunition; the infantry should never be without its artillery, its cavalry, and its generals; and the different divisions of the army should be constantly ready to support, to be supported, and to protect themselves.

Napoleon I: Maxims of War, 1831

I think the necessity of being *ready* increases—Look to it.

Abraham Lincoln: Letter (in entirety above) to Governor Curtin of Pennsylvania, 8 April 1861

The picket boats must always be kept in readiness at night, with their torpedoes ready for instant service, and if an ironclad should come down they must destroy her even if they are all sunk. For this purpose you must select men of nerve to command them, who will undertake anything, no matter how desperate.

David D. Porter: Order to the James River Division, North Atlantic Squadron, 2 December 1864

There has been a constant struggle on the part of the military element to keep the end—fighting, or readiness to fight—superior to mere administrative considerations . . . the military man, having to do the fighting, considers that the chief necessity; the administrator equally naturally tends to think the smooth running of the machine the most admirable quality.

Mahan: Naval Administration and Warfare, 1903

. . . the one thing needed, namely, to be ready to the utmost on the day of battle.

Mahan: Naval Strategy, 1911

We are ready now, sir, that is, as soon as we finish refueling.

Commander J.K. Taussig, USN: On being asked by Admiral Bayly, the British commander at Queenstown, Ireland, when the newly arrived U.S. destroyer division would be ready for service, 4 May 1917

I am sure you will be on your guard against the capital fault of letting diplomacy get ahead of naval preparedness.

Winston Churchill: Letter to Sir Samuel Hoare, 25 August 1935

It is a law of life that has yet to be broken that a nation can only earn the right to live soft by being prepared to die hard in defense of its living.

Sir Archibald Wavell: Other Men's Flowers, 1945

Being ready is not what matters. What matters is winning after you get there.

Lieutenant General V.H. Krulak, USMC: To a Marine unit leaving for Vietnam, April 1965

First to Fight.

U.S. Marine Corps slogan first popular in World War I

Semper paratus (Always ready).

Motto of the U.S. Coast Guard

(*See also* Preparedness.)

Rear Guard

Rear guards are the safety of armies and often they carry victory with them.

Frederick The Great: Instructions for His Generals, vi, 1747

Rebel

Well, if ever I love another country, damn *me*!

Remark by a discouraged Confederate soldier to Major General Leonidas Polk, 1863

I am a good old rebel—
Yes; that's just what I am—
And for this land of freedom
I do not give a damn.
I'm glad I fit agin 'em,
An I only wish we'd won;
And I don't ax no pardon
For anything I've done.

Innes Randolph, 1837–1887, A Good Old Rebel

(*See also* Confederacy.)

Rebellion

The rebellions of the belly are the worst.

Sir Francis Bacon, 1561–1626

A little rebellion now and then is a good thing.

Thomas Jefferson: Letter to James Madison, 30 January 1787

Soldiers only make rising and riots; they are generals and colonels who make rebellions.

Horace Walpole: Letter to Mrs. Carter, 25 July 1789

He who fights against his country is a child who would kill his own mother.

Napoleon I: Refusing a pardon for a French emigré taken in arms against the French government (in Political Aphorisms, 1848)

. . . Rebellion!
How many a spirit born to bless,
Hath sunk beneath that withering name,
Whom but a Day's—an Hour's—success
Had wafted to eternal fame.

Thomas Moore, 1779–1852

The only justification of rebellion is success.

Thomas B. Reed: To the House of Representatives during debate on legislation to reimburse William and Mary College for Civil War damages, 12 April 1878

It doesn't take a majority to make a rebellion; it takes only a few determined leaders and a sound cause.

H.L. Mencken, 1880–1956, Prejudices

There is one thing of which every rebellion is mortally afraid—treachery . . . One well informed traitor will spoil a national rising.

T.E. Lawrence: Letter to Colonel A.P. Wavell, 21 May 1923

Rebellion must have an unassailable base, something guarded not merely from attack, but from the fear of it: such a base as the Arab Revolt had in the Red Sea ports, the desert, or in the minds of men converted to its creed.

T.E. Lawrence, 1888–1935

(*See also* Insurgency, Revolt, Revolution.)

Receiving Ships

So infamously rotten and corrupt, as to have sown the seeds of all the theft, false musters, and general departures from the regulations of the service; and the men in them are idle and profligate.

Lord St. Vincent: Letter to First Lord of the Admiralty, September 1797 (Jervis referred to "Guard-Ships," the 18th century equivalent of receiving ships.)

Reconnaissance

A man trusts his ears less than his eyes.

Herodotus, 484–424 B.C., Clio, i, 8

Those who do not know the conditions of mountains and forests, hazardous defiles, marshes and swamps, cannot conduct the march of an army.

Sun Tzu, 400–320 B.C., The Art of War

Agitate the enemy and ascertain the pattern of his movement. Determine his dispositions and so ascertain the field of battle. Probe him and learn where his strength is abundant and where deficient.

Sun Tzu, 400–320 B.C., The Art of War, vi

When it is desired to apply oneself to this essential part of war, the most detailed and exact maps of the country that can be found are taken and examined and re-examined frequently. If it is not in time of war, the places are visited, camps are chosen, roads are examined, the mayors of the villages, the butchers, and the farmers are talked to. One becomes familiar with the footpaths, the depth of the woods, their nature, the depth of the rivers, the marshes that can be crossed and those which cannot . . . The road is chosen for such and such a march, the number of columns in which the march can be made estimated and all strong camping places on the route are examined.

Frederick The Great: Instructions for His Generals, vi, 1747

. . . skillfully reconnoitering defiles and fords, providing himself with trusty guides, interrogating the village priest and the chief of relays, quickly establishing relations with the inhabitants, seeking out spies, seizing letters.

Napoleon I: Maxims of War, 1831

Time spent on reconnaissance is seldom wasted.

British Army Field Service Regulations, 1912

In order to conquer that unknown which follows us until the very point of going into action, there is only one means, which consists in looking out until the last moment, even on the battlefield, for *information.*

Ferdinand Foch: Precepts, 1919

Take your time. Stay away from the easy going. Never go the same way twice.

Gunnery Sergeant Charles C. Arndt, USMC: Guadalcanal, 1942 (GySgt Arndt's rules for successful reconnaissance)

You never can do too much reconnaissance.

George S. Patton, Jr.: War As I Knew It, 1947

Recruits

We must not be faint-hearted, nor behave as if we were mere novices in the art of war, who when defeated in their first battle are full of cowardly apprehensions and continually retain the impress of their disaster.

Nicias of Athens: Speech to the Athenian troops, 414 B.C.

An army raised without proper regard to the choice of its recruits was never yet made good by length of time.

Vegetius: De Re Militari, i, 378

The young soldier should have a lively eye and carry his head erect; his chest should be broad, his shoulders muscular and brawny, his fingers long, his arms strong, his waist small, his build compact, his legs and feet wiry rather than fleshy. When all these qualities are found in a recruit, a little height may be dispensed with.

Vegetius: Re Re Militari, i, 378

He stirs, he warms—
The warlike Youth—He listens to the Charms
Of Plunder, fine lac'd coats, and glitt'ring Arms.

George Farquhar: The Recruiting Officer, 1706

Men just dragged from the tender Scenes of domestick life; unaccustomed to the din of Arms; totally unacquainted with every kind of Military skill, . . . when opposed to Troops regularly train'd, disciplined, and appointed, superior in knowledge, and superior in Arms, makes them timid and ready to fly from their own shadows.

George Washington: Letter to the President of Congress, 24 September 1776

Oh, why the deuce should I repine,
And be an ill foreboder?
I'm twenty-three and five feet nine—
I'll go and be a sodger.

Robert Burns: Extempore, 1784

There is not so helpless and pitiable an object in the world as a landsman beginning a sailor's life.

R.H. Dana, Jr.: Two Years Before the Mast, i, 1840

Listen, young heroes! your country is calling!
Time strikes the hour for the brave and the true!
Now, while the foremost are fighting and falling,
Fill up the ranks that have opened for you!

Oliver Wendell Holmes, Sr.: Poem, 1862

When the 'arf-made recruity goes out to the East,
'E acts like a babe an' 'e drinks like a beast.

Rudyard Kipling: The Young British Soldier, 1890

Unconscious of that dignity which belongs to his calling, the raw recruit readily gives in to any scheme urged by the designing.

Colonel C. Field, RMLI: Britain's Sea Soldiers, xiii, 1924

There are those who say, "I am a farmer," or, "I am a student"; "I can discuss literature but not the military arts." This is incorrect. There is no profound difference between the farmer and the soldier. You must have courage. You simply leave your farms and become soldiers . . . When you take your arms in hand, you become soldiers; when you are organized, you become military units.

Mao Tse-tung: On Guerrilla Warfare, 1937

The best soldier comes from the plow.

Spanish Proverb

(*See also* Recruit Training, Recruiting.)

Recruit Training

These men, as soon as enlisted, should be taught to work on entrenchments, to march in ranks, to carry heavy burdens, and to bear the sun and dust. Their meals should be coarse and moderate; they should be accustomed to lie sometimes in the open air and sometimes in tents. After this they should be instructed in the use of their arms. And if any long expedition is planned, they should be encamped as far as possible from the temptations of the city.

Vegetius: De Re Militari, i, 378

The first thing to be taken care of in the disciplining of men, is to dress them, to teach them the air of a soldier, and to drive out the clown.

Lieutenant John MacIntire, RM: A Military Treatise on the Discipline of the Marine Forces When at Sea, 1763

Body and spirit I surrendered whole
To harsh Instructors—and received a soul.

Rudyard Kipling: The Wonder (Epitaphs of the War), 1919

Feed 'em up and give 'em hell. Teach 'em where they are. Make 'em so mad they'll eat steel rather than get another dressing from you. Make 'em hard but don't break 'em.

Laurence Stallings: What Price Glory? 1926

The ore, the furnace, and the hammer are all that is needed for a sword.

Indian Proverb

(*See also* Drill Instructor, Recruit, Recruiting.)

Recruiter

The greatest rakes are the best recruiters.

Lieutenant Blackadder: Diary entry, February 1705

What! No bastards! and so many Recruiting Officers in Town! I thought 'twas a Maxim among them, to leave as many Recruits in the Country as they carry'd out.

George Farquhar: The Recruiting Officer, 1706

If your Worship pleases to cast up the whole Sum, viz., Canting, Lying, Impudence, Pimping, Bullying, Swearing, Whoring, Drinking, and a Halberd, you will find the sum total amount to a Recruiting Serjeant.

George Farquhar: The Recruiting Officer, 1706 (the halberd was the normal side-arm and badge of rank of an 18th century NCO).

. . . the vile crimps who recruit for the foreign regiments of Spain.

Lord St. Vincent: Letter to the Secretary of the British Legation, Tunis, July 1796

(*See also* Recruit, Recruiting.)

Recruiting

Will you tell me, Master Shallow, how to choose a man? Care I for the limb, the thewes, the stature, bulk, and big assemblance of a man? Give me the spirit, Master Shallow.

Shakespeare: II King Henry IV, iii, 2, 1597

. . . The Recruiting Trade, with all its Train
Of endless Plague, Fatigue, and endless Pain.

George Farquhar: The Recruiting Officer, 1706

Troops are raised by enlistment with a fixed term, without a fixed term, by compulsion sometimes, and most frequently by tricky devices.

Maurice de Saxe: Mes Rêveries, i, 1732

I heartily wish we could raise men as fast as you equip ships.

Lord St. Vincent: Letter to Lord Keith, 5 April 1803

In respect to recruiting the army, my own opinion is, that the government have never taken an enlarged view of the subject. It is expected that people will become soldiers in the line, and leave their families to starve . . . What is the consequence? That none but the worst description of men enter the regular service.

Wellington: Letter, 28 January 1811

People talk of [soldiers] enlisting from their fine military feeling—all stuff—no such thing. Some of our men enlist from having got bastard children—some for minor offenses—many more for drink; but you can hardly conceive such a set brought together, and it really is wonderful that we should have made them the fine fellows that they are.

Wellington, 1769–1852

It has been said by officers enthusiastic in their profession that there are three causes which make a soldier enlist, viz., being out of work, in a state of intoxication, or, jilted by his sweetheart. Yet the incentives to enlistment, which we desire to multiply, can hardly be put by Englishmen of the nineteenth century in this form, viz., more poverty, more drink, more faithless sweethearts.

Florence Nightingale: Notes on Matters Affecting the Health, Efficiency, and Hospital Administration of the British Army, 1858

(*See also* Recruit, Manpower.)

Reforms

There is a time for all things: there is even a time for change; and that is when it can no longer be resisted.

Duke of Cambridge, 1819–1904

Any change, even for the better, is to be deprecated.

Military maxim

Regiment

Remember your regiment and follow your officers.

Captain Charles A. May, USA: To his squadron of the 2d Dragoons before charging the Mexicans at Resaca de la Palma, 9 May 1846

The regiment is the family. The Colonel, as the father, should have a personal acquaintance with every officer and man, and should instill a feeling of pride and affection for himself, so that his officers and men would naturally look to him for personal advice and instruction.

W.T. Sherman: Memoirs, 1875

Ho! get away you bullock-man, you've
'eard the bugle blowed,
There's a regiment a-comin' down the
Grand Trunk Road.

Rudyard Kipling: Route Marchin' 1892

Regular

No music with him but the drum and fife.

Shakespeare: Much Ado About Nothing, ii, 3, 1598

I am a soldier,
A name, that, in my thoughts, becomes
me best.

Shakespeare: King Henry V, iii, 3, 1598

If they be well ordered and kept in by the rules of good discipline, they fear not the face or the force of the stoutest foe . . . And though at first they be stoutly resisted, yet they will as resolutely undertake the action the second time, though it is to meet death itself in the face.

Donald Lupton: A Warre-like Treatise of the Pike, 1642

I have eaten Smoke from the Mouth of a Cannon, Sir; don't think I fear Powder, for I live upon't.

George Farquhar: The Recruiting Officer, 1706

The man who devotes himself to war should regard it as a religious order into which he enters. He should have nothing, know no concern other than his troops, and should hold himself honored in his profession.

Maurice de Saxe: Mes Rêveries, iv, 1732

. . . those fellows who are contented to swallow gun-powder because they have nothing else to eat.

Henry Fielding: Tom Jones, 1749

Regular troops alone are equal to the exigencies of modern war, as well for defense as offense, and when a substitute is attempted it must prove illusory and ruinous.

George Washington, 1732–1799

When the perfect order and exact discipline which are essential to regular troops are contemplated, and with what ease and precision they execute the difficult maneuvers indispensable to the success of offensive or defensive operations, the conviction cannot be resisted that such troops will always have a decided advantage over more numerous forces composed of uninstructed militia or undisciplined recruits.

Alexander Hamilton, 1755–1804

The steady operations of war against a regular and disciplined army can only be successfully conducted by a force of the same kind.

Alexander Hamilton: The Federalist, No. xxv, 1787

None but the worst description of men enter the regular service.

Wellington: Letter, 28 January 1811

It is a subject of joy that we have so few of the desperate characters which compose modern regular armies.

Thomas Jefferson: Letter to James Monroe, 1813

By God, these are Regulars!

Major General Phineas Riall: To his staff at the battle of Chippewa as Winfield Scott's brigade attacked the British line, 5 July 1814

The man for hire, the soldier, is a poor braggart, victim, and executioner, a scapegoat daily sacrificed to his people and for his people, who make sport of him; he is a martyr ferocious and humble at the same time.

Alfred de Vigny: Servitudes et Grandeurs Militaires, 1835

You are paid for doing this kind of work.

Colonel Shriver, Maryland militia: To Colonel R.E. Lee, USA, declining Lee's offer to let the militia storm John Brown at Harper's Ferry, 18 October 1859

There was Sergeant John McCaffery and
Captain Donahue,
Oh, they made us march and toe the line
in gallant Company Q.
Oh the drums would roll. Upon my soul,
this is the style we'd go:
Forty miles a day on beans and hay in the
Regular Army O!

Edward Harrigan: The Regular Army O!, minstrel song, c. 1875

It's Tommy this, an' Tommy that, an'
"Chuck him out, the brute!"
But it's "Savior of 'is country" when the
guns begin to shoot.

Rudyard Kipling: Tommy, 1890

A man who has been doing nothing but killing people for years! What does he care? What does any soldier care?

G.B. Shaw: Arms and the Man, iii, 1894

I'm a professional soldier! I fight when I have to, and am very glad to get out of it when I haven't to.

G.B. Shaw: Arms and the Man, iii, 1894

I do not love my country's foes,
Nor call 'em angels; still
What *is* the sense of 'atin' those
'Oom you are paid to kill?

Rudyard Kipling: Piet, 1903

Regular troops, engaged for the war, are the only safe reliance of a government, and are in every point of view, the best and most economical.

Emory Upton: The Military Policy of the United States, 1904

General, these are American regulars. In a hundred and fifty years they have never been beaten. They will hold.

Colonel Preston Brown, USA: To General Degoutte, French Army, when the latter questioned the ability of the infantry and Marines of the 2d U.S. Division to stop the German advance on Paris, June 1918.

They suffer only from the regular soldiers' fault; there are too few of them.

Sir Ian Hamilton: Gallipoli Diary, I, 1920

These, in the day when heaven was falling,
The hour when earth's foundations fled,
Followed their mercenary calling
And took their wages and are dead.

Their shoulders held the sky suspended;
They stood, and earth's foundations stay;
What God abandoned, these defended,
And saved the sum of things for pay.

A.E. Housman: Epitaph on an Army of Mercenaries, 1922 (to the British Expeditionary Force, 1914)

They were the old breed of American regular, regarding the service as home and war as an occupation.

John W. Thomason, Jr.: Fix Bayonets! 1926 (of the 4th Marine Brigade, 1917)

Military men, retarded by their splendid calling, often spend a lifetime in their adolescence, thus retaining a Puck-like quality which frequently obstructs their careers as statesmen.

Philip Guedalla: Fathers of the Revolution, 1926

The regular officer has the traditions of forty generations of serving soldiers behind him, and to him the old weapons are the most honored.

T.E. Lawrence, 1888–1935

. . . singularly free from the distaste for novel devices so often found in professional soldiers.

Winston Churchill: Their Finest Hour, 1949

I longed for more Regular troops with which to rebuild and expand the Army. Wars are not won by heroic militia.

Winston Churchill: Their Finest Hour, 1949

War is the professional soldier's time of opportunity.

B.H. Liddell Hart: Defense of the West, 1950

. . . a cool, experienced soldier, who would keep his head and, hot though it was, his temper, and display, if needed, a certain rugged diplomacy of a type not always too evident in our foreign relations.

Sir William Slim: Unofficial History, ix, 1959

(*See also* Profession of Arms, Soldier.)

Regulations

As far as propriety, laws, and decrees are concerned, the army has its own code, which it ordinarily follows. If these are made identical with those used in governing a state, the officers will be bewildered.

Tu Mu, 803–852, Wei Liao Tzu

Nobody in the British Army ever reads a regulation or an order as if it were to be a guide for his conduct, or in any other manner than as an amusing novel.

Wellington, 1769–1852

As the preacher knows his Bible, as the lawyer knows his statutes, every general should know the regulations and articles of war.

Brigadier General C.F. Smith: To Colonel Lew Wallace, September 1861

The genius of war abhors uniformity, and tramples upon forms and regulations.

Alexander Kinglake: Invasion of the Crimea, ii, 1863

To insure Peace of Mind, ignore the Rules and Regulations.

George Ade: Forty Modern Fables, 1901

Regulations are all very well for drill but in the hour of danger they are no more use . . . You must learn to think.

Ferdinand Foch, 1851–1929

One of the first regulations might be to think.

MacKenzie Hill (pseud. Major General Orlando Ward, USA): In Field Artillery Journal, c. 1938

Regulations were made to be broken.

Soldiers' saying

Reinforcements

Reinforcements are always more formidable to an enemy than the troops with which he is already engaged.

Brasidas of Sparta: Speech to the Lacadaemonian troops, Battle of Amphipolis, 422 B.C.

If a little help reaches you in the action itself, it determines the turn of fortune for you. The enemy is discouraged and

his excited imagination sees the help as being at least twice as strong as it really is.
Frederick The Great: Instructions for His Generals, xx 1747

One fresh man in action is worth ten fatigued men.
John Stark: On being urged to hurry his people into position, Bunker Hill, 17 June 1775

A seasonable reinforcement renders the success of a battle certain, because the enemy will always imagine it stronger than it really is, and lose courage accordingly.
Napoleon I: Maxims of War, 1831

Relief from Command

In war you must either trust your general or sack him.
Sir John Dill: Letter to Sir A.P. Wavell, 1941

(*See also* Change of Command.)

Religious War

When ye encounter the unbelievers, strike off their heads, until ye have made a great slaughter among them. Verily, if God pleased, He could take vengeance upon them without your assistance, but He commandeth you to fight His battles.
The Koran, XLVII

Before the discovery of gunpowder, both the Indies were in the jaws of Hellish Satan and in the very darkest obscurity, more like cattle or wild beasts in customs and beliefs than like reasonable creatures of the Great God. Gunnery has been the only means by which the command of Christ could be performed.
Michael Mieth, fl. 1683

There has never been a kingdom given to so many civil wars as that of Christ.
Montesquieu, 1689–1755

Defoe says that there were 100,000 country fellows in his time ready to fight to the death against popery, without knowing whether popery was a man or a horse.
William Hazlitt, 1778–1830

Wars of opinion, as they have been the most destructive, are also the most disgraceful of conflicts; being appeals from right to might, and from argument to artillery.
C.C. Colton: Lacon, 1820

Civilized men have done their fiercest fighting for doctrines.
William Graham Sumner, 1840–1910, Essays

And the Spear was a Desert Physician
who cured not a few of ambition,
And drave not a few to perdition with
medicine bitter and strong;
And the shield was a grief to the fool and
as bright as a desolate pool,
And as straight as the rock of Stamboul
when our cavalry thundered along,
For the coward was drowned with the
brave when our battle sheered up like
a wave,
And the dead to the desert we gave, and
the glory to God in our song.
James Elroy Flecker, 1884–1915, War Song of the Saracen

Although the force of fanatical passion is far greater than that exerted by any philosophical belief, its sanction is just the same. It gives men something which they think is sublime to fight for, and this serves them as an excuse for wars which it is desirable to begin for totally different reasons. Fanatacism is not a cause of war. It is the means which helps savage peoples to fight.
Winston Churchill: The River War, i, 1899

The logical end of a war of creeds is the final destruction of one.
T.E. Lawrence, 1888–1935

Replacements

I don't know what effect these men will have on the enemy, but, by God, they frighten *me*.
Wellington: Of a draft of troops sent to him in Spain, 1809

The fighting troops are not being replaced effectively, although masses of drafts are sent to the technical and administrative services, who were originally on the most lavish scale, and who have since hardly suffered at all by the fire of the enemy. The first duty of the War Office is to keep up the rifle infantry strength.
Winston Churchill: Note to the Chief

of the Imperial General Staff, 3 May 1943

Replacements are spare parts—supplies. They must be asked for in time by the front line, and the need for them must be anticipated in the rear.
George S. Patton, Jr.: War As I Knew It, 1947

Reports

You should have a clever secretary to write your dispatches, in case you should not be so well qualified yourself. This gentleman may often serve to get you out of a scrape.
Francis Grose: Advice to the Officers of the British Army, 1782

Reports are not self executive.
Florence Nightingale: Marginal comment on a document, 1857

Must the operational reports from the Middle East be of their present inordinate length and detail? . . . I suggest that the average weekly wordage of these routine telegrams should be calculated for the last two months and Air Marshal Longmore asked to reduce them to, say, one-third their present length.
Winston Churchill: Note for the Chief of Air Staff, 12 January 1941

. . . that useless type of report that obscures rather than clarifies.
Correlli Barnett: The Swordbearers, vi, 1963

(*See also* Action Report, Administration, Official Correspondence.)

Reprimand

Admonish your friends in private; praise them in public.
Publilius Syrus: Sententiae, c. 50 B.C.

Rebukes ought not to have a grain of salt more than of sugar.
Thomas Fuller: Gnomologia, 1732

Always use the most opprobrious epithets in reprimanding the soldiers, particularly men of good character: for these men will not in the least be hurt, as they will be concious that they do not deserve them.
Francis Grose: Advice to the Officers of the British Army 1782

. . . reprehension, which the pride of military character cannot digest.
Lord St. Vincent: Letter to the Admiralty, June 1800

If I had been censured every time I have run my ship, or fleets under my command, into great danger, I should long ago have been *out* of the Service, and never *in* the House of Peers.
Nelson: Letter to the Admiralty, March 1805

Never give a man a dollar's worth of blame without a dime's worth of praise.
Colonel L.P. Hunt, USMC: When on duty in Washington, 1937

Long ago I had learned that in conversation with an irate senior, a junior should confine himself to the three remarks, "Yes, sir," "No, sir," and "Sorry, Sir." Repeated in the proper sequence, they will get him through the most difficult interview with the minimum discomfort.
Sir William Slim: Unofficial History, v, 1959

Reprisal

Reprisals are a sorry recourse.
Napoleon I: Political Aphorisms, 1848

In districts and neighborhoods where the army is unmolested, no destruction of property should be permitted; but should guerrillas or bushwhackers molest our march, or should the inhabitants burn bridges, obstruct roads, or otherwise manifest hostility, then army commanders should order and enforce a devastation more or less relentless.
W.T. Sherman: Special Field Orders No. 120, 9 November 1864 (order for the march from Atlanta to the sea)

I will soon commence on Loudon County, and let them know there is a God in Israel.
P.H. Sheridan: Telegram to Major General H.W. Halleck, 26 November 1864

The vengeful passions are uppermost in the hour of victory.
B.H. Liddell Hart: Defense of the West, 1950

The Italian Embassy, Washington, D. C.

GIULIO DOUHET

1869–1930

"Aerial warfare will be the most important element in future wars."

Research and Development

There's a way to do it better—Find it.
Thomas A. Edison: To a research associate, c. 1919

The best scale for an experiment is 12 inches to a foot.
Sir John Fisher: Memories, 1919

Having precise ideas often leads to a man doing nothing.
Paul Valéry, 1871–1945

Inventions do not make their first bow to armies on the battlefield. They have been in the air for some time; hawked about the ante-chambers of the men of the hour; spat upon by common sense; cold-shouldered by interests vested in what exists; held up by stale functionaries to whom the sin against the Holy Ghost is "to make a precedent."
Sir Ian Hamilton: The Soul and Body of an Army, xi, 1921

The [tank] had been conceived, but before it could be born and waddle across no-man's land to browse upon the barbed wire of the Germans, it had first to get through the barbed wire of the bureaucrats whereon fluttered still the poor rags once worn by dead inventors.
Sir Ian Hamilton: The Soul and Body of an Army, xi, 1921

The first stage is the formulation of a felt want by the fighting Service. Once this is clearly defined in terms of simple reality it is nearly always possible for the scientific experts to find a solution.
Winston Churchill: Note for Director of Scientific Research, 16 October 1939

It is by devising new weapons, and above all by scientific leadership, that we shall best cope with the enemy's superior strength.
Winston Churchill: Memorandum for the War Cabinet, 3 September 1940

This helplessness of the art of war, we can now see, was due to the slowness with which armies changed away from an old conception of warfare to a new one. All the technical means for ending this helplessness were present early in the war; the gasoline engine, the caterpillar tractor, the idea of an armored vehicle capable of crossing trenches and standing machine-gun fire, the aeroplane and the light machine gun were all available. What was not available was the idea of war as a changing art or science affected by every change in the techniques of production and transport, and inevitably out of date if these techniques were not employed to the full.
Tom Wintringham: The Story of Weapons and Tactics, 1943

Military history is filled with the record of military improvements that have been resisted by those who would have profited richly from them.
B.H. Liddell Hart: Thoughts on War, xi, 1944

There are many sciences with which war is concerned, but war is not such a science itself, and any forecast for the indefinite future presupposes a certitude that is not possible.
James V. Forrestal: To the Air Policy Commission, 3 December 1947

The machine-gun and the tank would still remain blueprint dreams if their development had awaited the specifications of clear-cut military requirements.
Hanson W. Baldwin: "Slow-Down in the Pentagon," Foreign Affairs, January 1965

Reserve

The great secret of battle is to have a reserve. I always had.
Wellington, 1769–1852

Fatigue the opponent, if possible, with few forces and conserve a decisive mass for the critical moment. Once this decisive mass has been thrown in, it must be used with the greatest audacity.
Clausewitz: Principles of War, 1812

The reserve is a *club*, prepared, organized, reserved, carefully maintained with a view to carrying out the one act of battle from which a result is expected—the decisive attack.
Ferdinand Foch: Precepts, 1919

There is always the possibility of accident, of some flaw in materials, present in the general's mind: and the reserve is unconsciously held to meet it.
T.E. Lawrence, 1888–1935

It is in the use and withholding of their reserves that the great Commanders have generally excelled. After all, when once the last reserve has been thrown in, the Commander's part is played . . . The event must be left to pluck and to the fighting troops.
Winston Churchill: Painting as a Pastime, 1932

[Hitler] had created a spider's web of communications, *but he forgot the spider.*
Winston Churchill: Their Finest Hour, 1949 (of Hitler's failure to provide adequate reserves for the defense of Europe)

Reservists

Reserves are so much eyewash, and take in only short-sighted mathematicians who equate the value of armies with their numerical strength, without considering their moral value.
General Pierre Cherfils, b. 1849

He's an absent-minded beggar, but he heard his country's call,
And his reg'ment didn't need to send to find him!
Rudyard Kipling: The Absent-Minded Beggar, 1899 (dedicated to reservists called up for the Boer War)

Reservists are nothing. (Les réserves, c'est zéro!)
French Army saying, pre-World War I

The reservist is twice the citizen.
Attributed to Winston Churchill, 1874–1965

(*See also* Manpower, Militia.)

Reserve Officers Training Corps

We must . . . make military instruction a regular part of collegiate instruction. We can never be safe till this is done.
Thomas Jefferson: Letter to James Monroe, 1813

Resistance

And when the moment came, they were minded to resist and suffer, rather than to fly and save their lives; they ran away from the word of dishonor, but on the field of battle their feet stood fast, and in an instant, at the height of their fortune, they passed away from the scene, not of their fear but of their glory.
Pericles: Funeral oration over the Athenian dead, 431 B.C.

A soldier told Pelopidas, "We are fallen among the enemies." Said he, "How are we fallen among them more than they among us?"
Plutarch, 46–120 A.D., Apothegms of Kings and Great Commanders

There is one certain means by which I can be sure never to see my country's ruin: I will die in the last ditch.
William III (Prince of Orange), 1650–1702

If you cannot fight the enemy, carry me on deck and I will.
John Barry: During the action between USS Alliance *and HMS* Atalanta, *29 May 1781*

If I were an American, as I am an Englishman, while a foreign troop was landed in my country, I would never lay down my arms—never, never, never!
William Pitt: To the House of Commons, 18 November 1777

The French people do not make peace with an enemy who occupies its territory.
Constitution of 1793, Art. 4, Ch. xxv

Armies do not always suffice to save a nation; while a country defended by its people is ever invincible.
Napoleon I: Political Aphorisms, 1848

Never say die.
Charles Dickens: Pickwick Papers, ii, 1836

A nation, fighting for its liberty, ought not to adhere rigidly to the accepted rules of warfare. Mass uprisings, revolutionary methods, guerrilla bands everywhere; such are the only means by which a small nation can hope to maintain itself against an adversary superior in numbers and equipment. By their use a weaker force

can overcome its stronger and better organized opponents.
Karl Marx: Letter, 1849

If a general and his men fear death and are apprehensive over possible defeat, then they will unavoidably suffer defeat and death. But if they make up their minds, from the general down to the last foot-soldier, not to think of living but of standing in one place and facing death together, then, though they may have no other thought than meeting death, they will instead hold on to life and gain victory.
Yoshida Shōin, 1830–1859

They shall not pass. (On ne passe pas.)
Henri Philippe Petain: To General de Castelnau at Verdun, 26 February 1916 (This became the watchword of Verdun's defenders, and is also attributed to Nivelle who ended his 23 June 1916 Order of the Day with Petain's phrase.)

I shall fight before Paris, I shall fight in Paris, I shall fight behind Paris. (Je me bats devant Paris, je me bats à Paris, je me bats derrière Paris.)
Georges Clemenceau: To the Chamber of Deputies, 4 June 1918 (based on Foch's 26 March 1918 exclamation to Haig: "I would fight in front of Amiens. I would fight in Amiens. I would fight behind Amiens.")

If it is thought best for France in her agony that her Army should capitulate, let there be no hesitation on our account, because whatever you may do, we shall fight on for ever and ever and ever.
Winston Churchill: Message to Paul Reynaud, June 1940

We shall not flag or fail. We shall go on to the end, we shall fight in France, we shall fight in the seas and oceans, we shall fight with growing confidence and growing strength in the air, we shall defend our island, whatever the cost may be, we shall fight on the beaches, we shall fight on the landing grounds, we shall fight in the fields and in the streets, we shall fight in the hills; we shall never surrender.
Winston Churchill: To the House of Commons after Dunkirk, 4 June 1940

Atomic war is bad enough; biological warfare would be worse; but there is something that is worse than either. The French can tell you what it is; or the Czechs, or the Greeks, or the Norwegians, or the Filipinos; it is subjection to an alien oppressor.
Elmer Davis: No World, If Necessary, 30 March 1946

Resolution

Watch ye, stand fast in the faith, quit you like men, be strong.
I Corinthians, XVI, 13

Be bloody, bold, and resolute.
Shakespeare: Macbeth, iv, 1, 1605

If we are mark'd to die, we are enow
To do our country loss; and if to live,
The fewer men, the greater share of honor.
Shakespeare: King Henry V, iv, 3, 1598

I have not the particular shining bauble or feather in my cap for crowds to gaze at or kneel to, but I have power and resolution for foes to tremble at.
Oliver Cromwell, 1599–1658

If our number is small, our hearts are great.
Sir Henry Morgan: To his men before attacking Porto Bello, 1667

Stand your ground men. Don't fire unless fired upon; but if they mean to have a war, let it begin here.
Captain Jonas Parker: To the Minute Men, Lexington Green, before the redcoats opened fire, 19 April 1775

Since his Majesty thinks that nothing but audacity and resolve are needed to succeed in the naval officer's calling, I shall leave nothing to be desired.
Admiral Pierre Villeneuve: On receiving Napoleon's instructions prior to Trafalgar, 1805

Come one, come all! This rock shall fly
From its firm base as soon as I.
Walter Scott: The Lady of the Lake, 1810

Great extremities require extraordinary resolution. The more obstinate the resistance of an army, the greater the chances of success. How many seeming impossibilities have been accomplished by men whose only resolve was death!
Napoleon I: Maxims of War, 1831

Never say die as long as there's a shot in the locker.
Naval saying, 18th century

This place should be defended with the spirit which actuated the defenders of Thermopylae, and if left to myself such is my determination.
Stonewall Jackson: Letter to R.E. Lee concerning Maryland Heights, June 1861

We will sink with our colors flying.
Lieutenant George U. Morris, USN: Reply, as executive officer of the sinking frigate, USS Cumberland, *when called upon to strike his colors to the rebel ironclad, CSS* Virginia, *Hampton Roads, 8 March 1862*

We will beat the enemy or sink at our post.
Lieutenant William B. Cushing, USN: To Rear Admiral S.P. Lee, off Wilmington, North Carolina, 14 April 1863

With equal or even inferior power . . . he will win who has the resolution to advance.
Ardant du Picq, 1821–1870, Battle Studies

And cowards' funerals, when they come,
Are not wept so well at home,
Therefore, though the best is bad,
Stand and do the best, my lad.
A.E. Housman: A Shropshire Lad (The Day of Battle), 1896

Retreat hell! We just got here.
Attributed to Captain Lloyd S. Williams, USMC: Belleau Wood, 5 June 1918

I shall return.
Douglas MacArthur: On departing Corregidor, 11 March 1942

If I go on, I shall die; if I stay behind, I shall be dishonored; it is better to go on.
Ashanti war song

A stout heart breaks bad luck.
Spanish Proverb
(*See also* Will to Fight.)

Responsibility

Responsibility is the test of a man's courage.
Lord St. Vincent, 1735–1833

Never mind, General, all this has been *my* fault; it is *I* that have lost this fight, and you must help me out of it in the best way you can.
R.E. Lee: Note to Major General Cadmus M. Wilcox, CSA, 3 July 1863, at Gettysburg

I neither ask nor desire to know anything of your plans. Take the responsibility and act, and call on me for assistance.
Abraham Lincoln: To U.S. Grant on his appointment to command the Union Armies, 1864

. . . responsibility, the best of educators.
Mahan: Life of Nelson, vi, 1897

The degree of criticism varies in direct proportion to the distance from the point of responsibility.
Brigadier General W. M. Fondren, USA: At Quarry Heights, C.Z., 1962

Rest and Recreation

Men would rather have their fill of sleep, love, singing and dancing, than of war.
Homer: Iliad, xiii, c, 1000 B.C.

I would give all my fame for a pot of ale and safety.
Shakespeare: King Henry V, iii, 1598

I know not how, but martial men are given to love. I think it is, but as they are given to wine; for perils commonly ask to be paid in pleasures.
Francis Bacon: Essays, x, 1625

Drinking is the soldier's pleasure.
John Dryden: Alexander's Feast, 1697

* * * wants to go to Lisbon, and I have told him he may stay there 48 hours which is as long as any reasonable man can wish to stay in bed with the same woman.
Wellington: Letter from Portugal, 1811

Rest is sweet after strife.
Owen Meredith, 1831–1891, Lucile, i

Retirement

I have done the state some service, and they know't.
Shakespeare: Othello, v, 1604

My sword I give to him that shall succeed me in my pilgrimage, and my courage and skill to him that can get it. My marks and scars I carry with me, to be a witness for me, that I have fought his battles who now will be my reward.
John Bunyan: The Pilgrim's Progress, II, 1678

Fought all his battles o'er again;
And thrice he routed all his foes, and thrice he slew the slain.
John Dryden: Alexander's Feast, 1697

With some Regret I quit the active Field,
Where Glory full Reward for Life does yield.
George Farquhar: The Recruiting Officer, 1706

So, safe on shore the pensioned sailor lies,
And all the malice of the storm defies.
William Somerville: The Author, an Old Man, 1750 (to his arm-chair)

Exert your talents and distinguish yourself, and don't think of retiring from the world until the world will be sorry that you retire. I hate a fellow whom pride or cowardice or laziness drives into a corner, and who does nothing when he is there but sit and growl. Let him come out as I do, and bark.
Attributed to Samuel Johnson, 1709–1784

The time factor . . . rules the profession of arms. There is perhaps none where the dicta of the man in office are accepted with such an uncritical deference, or where the termination of an active career brings a quicker descent into careless disregard. Little wonder that many are so affected by the sudden transition as to cling pathetically to the trimmings of the past.
B.H. Liddell Hart: Thoughts on War, vi, 1944

Young soldiers, old beggars.
German Proverb

Retreat

To them that fleeth cometh neither power nor glory.
Homer: Iliad, xv, c. 1000 B.C.

When soldiers run away in war they never blame themselves: they blame their general or their fellow-soldiers.
Demosthenes; Third Olynthiac, 348 B.C.

The man who runs away will fight again.
Menander, 342–292 B.C.

I took to my heels as fast as I could.
Terence, 190–150 B.C., Eunuchus

A fine retreat is as good as a gallant attack.
Baltasar Gracian: The Art of Worldly Wisdom, xxxviii, 1647

In all the trade of war no feat
Is nobler than a brave retreat.
Samuel Butler: Hudibras, I, iii, 1663

I would by no means retreat while any hope of success remained.
John Paul Jones: Report to the American Commissioners in Paris, April 1778

There are few generals that have run oftener, or more lustily than I have done. But I have taken care not to run too far, and commonly have run as fast forward as backward, to convince the Enemy that we were like a Crab, that could run either way.
Nathanael Greene: To Henry Knox, 18 July 1781

If an army throws away all its cannon, equipments, and baggage, and everything which can strengthen it, and can enable it to act together as a body; and abandons all those who are entitled to its protection, but add to its weight and impede its progress; it must be able to march by roads through which it cannot be followed, with any prospect of being overtaken, by an army which had not made the same sacrifice.
Wellington: While pursuing Soult in Portugal, May 1804

In a retreat, besides the honor of the army, the loss of life is often greater than in two battles.
Napoleon I: Maxims of War, 1831

However skillful the maneuvers in a retreat, it will always weaken the morale of an army, because in losing the chances of success these last are transferred to the enemy. Besides, retreats always cost more men and materiel than the most bloody engagements; with this difference, that in a battle the enemy's loss is nearly equal to

your own—whereas in a retreat the loss is on your side only.
Napoleon I: Maxims of War, 1831

My wounded are behind me, and I will never pass them alive.
Zachary Taylor: To General John E. Woll when retreat was suggested at Buena Vista, 22 February 1847

If we are surrounded we must cut our way out as we cut our way in.
U.S. Grant: On being told that he was cut off from his transports on the river, Belmont, Missouri, 7 November 1861

The Army voted for peace with its legs.
V.I. Lenin: Of retreats by the Czarist armies, 1917

Retreat hell! We just got here.
Attributed to Captain Lloyd S. Williams, USMC: Belleau Wood, 5 June 1918

He who leaves a fight, loses it.
French Proverb

Reveille

Trumpet, blow loud,
Send thy brass voice through all these lazy tents.
Shakespeare: Troilus and Cressida, i, 3, 1601

Oh! How I hate to get up in the morning,
Oh! How I'd love to remain in bed;
For the hardest blow of all
Is to hear the bugle call,
"You've got to get up, you've got to get up,
"You've got to get up this morning!"
Irving Berlin: Song, Oh! How I hate to Get Up in the Morning, 1918

I can't get 'em up, I can't get 'em up, I can't get 'em up in the morning;
I can't get'em up, I can't get 'em up, I can't get 'em up at all.
Corporal's worse than privates;
Sergeant's worse than corporals;
Lieutenant's worse than sergeants;
And Captain's worst of all!
Traditional words of Reveille

Revolt

Whoever conquers a free town and does not demolish it commits a great error, and may expect to be ruined himself: because whenever the citizens are disposed to revolt, they betake themselves of course to that blessed name of liberty, and the laws of their ancestors, which no length of time nor kind usage whatever will be able to eradicate.
Niccolo Machiavelli: The Prince, v, 1513

The spectacle of a spontaneous rising of a nation is rarely seen. Though there be in it something grand and noble which commands our admiration, the consequences are so terrible that, for the sake of humanity, we ought to hope never to see it.
Jomini: Précis de l'Art de la Guerre, 1838

There could be no rest-houses for revolt, no dividend of joy to be paid out. Its spirit was accretive, to endure as far as the senses would endure and to use each such advance as the base for further adventure, deeper privation, further pain.
T.E. Lawrence, 1888–1935

A time comes when there is demoralization above, a growing revolt below; the morale of the army is undermined. The old structure of society is tottering. There are actual insurrections; the army wavers. Panic seizes the rulers, a general uprising begins.
M. J. Olgin, 1874–1937, Why Communism?

(*See also* Revolution, Rebellion.)

Revolution

The people arose as one man.
Judges, xx, 8

A people's voice is dangerous when charged with wrath.
Aeschylus: Agamemnon, 458 B.C.

To make a revolution, you have more need of a given amount of stupidity on one side than a given dose of light on the other.
Antonine de Rivarol, 1753–1801

Sire, it is not a revolt; it is a revolution.
Duc de la Rochfoucault-Liancourt: To Louis XVI, on the storming of La Bastille, 14 July 1789

War to the castles, peace to the cottages.

Nicolas Chamfort: Motto for the French Revolution, 1790

Revolutions are not to be made with rose-water.
Byron, 1788–1824, Letters

. . . A war of all peoples against all kings.
Decree by the French Directory, 1793

The men who lose their heads most easily, and who generally show themselves weakest on days of revolution are the military; accustomed as they are to have an organized force facing them and an obedient force in their hands, they readily become confused before the tumultuous uproar of a crowd and in the presence of hesitation and occasional connivance of their own men.
Alexis de Tocqueville, 1805–1859

Revolution is the locomotive of history.
Karl Marx,1818–1883

In time of revolution, with perseverance and courage, a soldier should think nothing impossible.
Napoleon I: Political Aphorisms, 1848

I know, and all the world knows, that revolutions never go backward.
William H. Seward: Speech at Rochester, N.Y., October 1858 (also attributed to Wendell Phillips, Boston, 17 February 1861)

A disorganized army and a complete breakdown of discipline has been the condition as well as the result of every victorious revolution.
Friedrich Engels: Letter to Karl Marx, 26 September 1851

Every society is pregnant with revolution, and force is the handmaiden of revolution.
Karl Marx, 1818–1883

The basic question of every revolution is the question of power in the state.
V.I. Lenin, 1870–1924

It is impossible to predict the time and progress of revolution. It is governed by its own more or less mysterious laws. But when it comes it moves irresistibly.
V.I. Lenin: Speech in the Polytechnic Museum, Moscow, 23 August 1918

Every great revolution brings ruin to the old army.
Leon Trotzsky: To Professor Milyukov, 1921

The purity of a revolution can last a fortnight.
Jean Cocteau, 1889–

Revolution is a transfer of property form class to class.
Leon Samson: The New Humanism, 1930

The central task and the highest form of a revolution is to seize political power by armed force, to settle problems by war.
MaoTse-tung: Problems of War and Strategy, 1954

A revolutionary war is never confined within the bounds of military action. Because its purpose is to destroy an existing society and its institutions and to replace them with a completely new state structure, any revolutionary war is a unity of which the constitutent parts, in varying importance, are military, political, economic, social, and psychological.
Brigadier General S.B. Griffith II, USMC: Introduction to Mao Tse-tung on Guerrilla Warfare, 1961

Revolutions rarely compromise; compromises are made only to further the strategic design. Negotiation, then, is undertaken for the dual purpose of gaining time to buttress a position and to wear down, frustrate, and harass the opponent. Few, if any, essential concessions are to be expected from the revolutionary side.
Brigadier General S.B. Griffith II, USMC: Introduction to Mao Tse-Tung on Guerrilla Warfare, 1961

Revolutions are always the work of a conscious minority.
Walter Lippman: Washington Post, 12 April 1966

(*See also* Rebellion, Revolt.)

Riflemen

Brave Rifles . . . !
Winfield Scott (To the Regiment of Mounted Rifles after the capture of Mexico City, 14 September 1847; the phrase was adopted at the regimental motto.)

Form, Form, Riflemen Form!
Ready, be ready to meet the storm!
Riflemen, Riflemen, Riflemen form!
Tennyson: Riflemen Form!, 1859

. . . Quick-footed, quick-minded and, as far as possible, light-hearted.
Sir A.P. Wavell: The Training of the Army for War, February 1933 (Wavell's attributes for the ideal infantry rifleman)

Risk

What though success will not attend on all?
Who bravely dares, must sometimes risk a fall.
Tobias Smollet, 1721–1771, Advice

In war, something must always be allowed to chance and fortune, seeing that it is in its nature hazardous and an option of difficulties; the greatness of an object should come under consideration as opposed to the impediments that be in the way.
Attributed to James Wolfe, 1727–1759

You have done your duty in this remonstrance; now obey my order, and lay me alongside the French Admiral.
Sir Edward Hawke: To his Fleet Navigator after being advised to turn away from the enemy for fear of grounding, Quiberon Bay, 20 November 1759

We should on all occasions avoid a general Action, or put anything to Risque, unless compelled by a necessity, into which we ought never to be drawn.
George Washington: To the President of Congress, September 1776

It is true I must run great risk; no gallant action was ever accomplished without danger.
John Paul Jones: Letter to the American Commissioners in Paris, 1778

If I find the French convoy in any place where there is a probability of attacking them, you may depend they shall either be taken or destroyed at the risk of my squadron . . . which is built to be risked on proper occasions.
Nelson: To an Austrian Army liaison officer in Italy, April 1796

I was aware that I was risking infinitely too much, but something must be risked for the honor of the Service.
Sir John Moore: Letter from Spain, December 1808

First reckon, then risk.
Helmuth von Moltke ("The Elder"), 1800–1891

Boats so delicate [destroyers] which, to be handled effectively must be handled with great daring, necessarily run great risks, and their commanders must, of course, realize that a prerequisite to successfully handling them is the willingness to run such risks. That they will observe proper precautions is, of course, required, but it is more important that our officers should handle these boats with dash and daring than that the boats should be kept unscratched.
William S. Sims, 1858–1936, endorsement (pre-World War I) on correspondence regarding destroyers

Anyone can see the risk from air attack which we run . . . This risk will have to be faced. Warships are meant to go under fire.
Winston Churchill: Note for First Sea Lord, 15 July 1940

The habit of gambling contrary to reasonable calculations is a military vice which, as the pages of history reveal, has ruined more armies than any other cause.
B.H. Liddell Hart: Thoughts on War, v, 1944

One does not send soldiers into battles telling them there is no risk. That is a fraud, the same kind of deception as telling one's children that cod-liver oil has a delicious taste.
Jean Dutourd: Taxis of the Marne, 1957

He who risks nothing gets nothing. (Qui ne risque rien, n'a rien.)
French Proverb

Rivalry

Rivalry is good for mortals.
Hesiod, c. 735 B.C., Works and Days

Troops should not be encouraged to foster a spirit of jealousy and unjust detraction towards other arms of the service, where

all are mutually dependent and mutually interested, with functions differing in character but not in importance.
J.E.B. Stuart: After Second Bull Run, 1862

There's no love lost between soldiers and sailors.
English Proverb

River Crossings

With regard to forcing a passage across rivers, I believe it is hardly possible to prevent it. A river crossing ordinarily is supported by such massive artillery fire that it is impossible to prevent an advance force from crossing, entrenching, and throwing up works to over the bridgehead.
Maurice de Saxe: Mes Rêveries, xxiii, 1732

The defense of a river crossing is the worst of all assignments especially if the front that you are to defend is long; in this case defense is impracticable.
Frederick the Great: Instructions for His Generals, xxii, 1747

The passage of great rivers in the presence of the enemy is one of the most delicate operations in war.
Attributed to Frederick the Great, 1712–1786

Let us cross over the river, and rest under the shade of the trees.
Stonewall Jackson: Last words, 10 May 1863

And crimson-dyed was the river's flood,
For the foe had crossed from the other side,
That day, in the face of a murderous fire
That swept them down in its terrible ire;
And their life-blood went to color the tide.
Nathaniel Graham Shepherd: Calling the Roll, 1835–1869

The qualities most required for river crossing, stability and cheerfulness.
Sir William Slim: Unofficial History, v, 1959

River Warfare

Whenever an army operates along a river, subsistence becomes easier.
Frederick The Great: Instructions for His Generals, iv, 1747

Fighting is nothing to the evil of the river—getting on shore, running afoul of one another, losing anchors, etc.
David G. Farragut: Letter off Vicksburg, 1862

The Father of Waters again goes unvexed to the sea.
Abraham Lincoln: Letter to J.C. Conkling, 26 August 1863 (after the fall of Vicksburg)

It was a slow, dirty, sand-bar kind of a war.
Lieutenant John C. Roberts, USN: In U. S. Naval Institute Proceedings, March 1965

Rockets

. . . Rockets or squibs, which, being thrown up into the air, will cast forth flames of fire, and in coming down towards the ground will shew like stars falling from heaven.
Niccolo Fontana Tartaglia, 1499–1559, Colloquies Concerning the Art of Shooting in Great and Small Pieces of Artillery

And the rockets' red glare, the bombs
bursting in air,
Gave proof through the night that our flag
was still there.
Francis Scott Key: The Star-Spangled Banner, 1814

Rogues' March

Poor old soldier,
Poor old soldier,
Tarred and feathered and sent to hell
Because he wouldn't soldier well.
Soldiers' words to the Rogues' March, 19th century

Rommel, Erwin (1891-1944)

. . . A splendid military gambler, dominating his problems of supply and scornful of opposition.
Winston Churchill, 1874–1965

Rubicon

The die is cast. (Alea iacta est.)
Julius Casesar: On leading his troops across the Rubicon River and thus commencing civil war, 49 B.C.

The Rubicon is a rivulet to cross: to re-cross, it is a mighty river.
A. R. M. Carr, fl. 1959

Ruse-de-Guerre

Ruses are of great usefulness. They are detours which often lead more surely to the objective than the wide road which goes straight ahead.
Frederick The Great: Instructions for His Generals, ix, 1747

We are bred up to feel it a disgrace even to succeed by falsehood; the word spy conveys something as repulsive as slave; we will keep hammering along with the conviction that honesty is the best policy, and that truth always wins in the long run. These pretty little sentiments do well for a child's copy-book, but a man who acts on them had better sheathe has sword forever.
Sir Garnet Wolseley: Soldier's Pocket-Book, 1869

All is fair in love and war.
English Proverb

Russia

There is only one defense against the Russians and that is a very hot climate.
Stendhal (Henri Beyle): A Life of Napoleon, lvi, 1818

The Russian Army is a wall which, however far it may retreat, you will always find in front of you.
Jomini: To Marshal Canrobert, of the projected Crimean campaign, 1854

I cannot forecast to you the action of Russia. It is a riddle wrapped in a mystery inside an enigma; but perhaps there is a key. That key is Russian national interest.
Winston Churchill: Radio broadcast, 1 October 1939

It is clear that this process of encroachment and consolidation by which Russia has grown in the last five hundred years from the Duchy of Muscovy to a vast empire has got to be stopped.
Dean Acheson, 1893–

How can one expect any sort of respect for normal international agreements from a regime that in the thirty-seven years since the Revolution has shot as spies and traitors, amongst others, all the members of their first Inner Cabinet and all members of the party Politburo as constituted after Lenin's death except Stalin, forty-three out of fifty-three Secretaries of the Central Organization of the Party, seventy out of the eighty members of the Soviety War Council, three out of every five marshals and about 60 percent of the generals of the Soviet Army?
Sir John Slessor: Strategy for the West, 1954

Russians, in the knowledge of inexhaustible supplies of manpower, are accustomed to accepting gigantic fatalities with comparative calm.
Barbara W. Tuchman: The Guns of August, 1962

Safety

The desire for safety stands againt every great and noble enterprise.
Tacitus: Annals, c. 110

I would give all my fame for a pot of ale and safety.
Shakespeare: King Henry V, iii, 2, 1598

The time for taking all measures for a ship's safety is while still able to do so.
Fleet Admiral Chester W. Nimitz, in a letter to the Pacific Fleet, 18 February 1945.

. . . the woolly theme of being safe everywhere.
Winston Churchill: Their Finest Hour, 1949

Safety lies forward.
Military maxim

Sailor

We be three poor mariners
Newly come from the seas;
We spend our lives in jeopardy,
While others live at ease.
The Mariner's Glee, 1609

They'll tell thee, sailors, when away,
In every port a mistress find.
John Gay, 1685–1732, Sweet William's Farewell

There's a Providence sits up aloft,
To keep watch for the life of poor Jack!
Charles Dibdin, 1745–1814, Poor Jack

No man will be a sailor who has the contrivance to get himself in jail, for being in a ship is being in jail, with a chance of being drowned . . . A man in jail has more room, better food, and commonly better company.
Samuel Johnson: To James Boswell, 16 March 1759

How little do the landsmen know
Of what we sailors feel,
When waves do mount and winds do blow!
But we have hearts of steel.
The Sailor's Resolution, 18th century

Fill up the mighty sparkling Bowl,
That I, a true & loyal Soul,
May drink and sing without controul,
To support my pleasure.
Thus may each jolly Sailor live
When Fears & Dangers o'er,
For past Misfortunes never Grieve,
When He's arriv'd on Shore.
Sailors' drinking song, 18th century

Nothing is so unmanageable as a sailor, except by his own officers.
Letter by unknown British Army officer, Walcheren Expedition, 1809

It is part of a sailor's life to die well.
Stephen Decatur, 1779–1820, of James Lawrence, 1813

And what do you think they got for their dinner?
'Twas water soup, but slightly thinner.
And what do you think they got for their suppers?
Belaying pin soup and a roll in the scuppers.
Blow, Boys, Blow, chanty, c. 1812

The sailor in a squadron fights only once in every campaign; the soldier fights every day. The sailor, whatever may be the fatigues and dangers on the sea, undergoes fewer of these than the soldier. He never suffers from hunger or thirst; he has always with him his quarters, his kitchen, his hospital and his pharmacy.
Napoleon I: Maxims of War, 1831

Remember still the gallant tar, who roams
Through rocks and gulfs, the ocean's gloomy vast;
To quell your foes, and guard your peaceful homes,
Who bides the battle's shock and tempest's blasts.
Tars of Columbia, song, early 19th century

Jack dances and sings and is always content;
In his vows to his lass he'll ne'er fail her;
His anchor's a-trip when his money's all spent—
And this is the life of a sailor.
The True English Sailor, 19th century song

Six days shalt thou labor and do all thou art able,
And on the seventh—holystone the decks and scrape the cable.
Seamen's traditional "Philadelphia Catechism," c. 1830

Sailors are too sanguine to despair, even at the last moment.
Frederick Marryat: Peter Simple, 1834

He was a true sailor, every finger a fish-hook.
R.H. Dana, Jr.: Two Years Before the Mast, xxv, 1840

The wonder is always new that any sane man can be a sailor.
R.W. Emerson, 1802–1883, English Traits

Sailors never should be shy.
W.S. Gilbert: HMS Pinafore, i, 1878

Why, Jack's the king of all,
For they all love Jack!
Frederic Edward Weatherly, 1848–1929, They All Love Jack

"Have you any news of my boy jack?"
Not this tide.
"When d'you think that he'll come back?"
Not with this wind blowing and this tide . . .
"Oh, dear, what comfort can I find?"
None this tide,
Nor any tide,
Except that he did not shame his kind—
Not even with that wind blowing, and that tide.
Rudyard Kipling: My Boy Jack, 1916

And I'd rather be bride to a lad gone down
Than a widow to one safe home.
Edna St. Vincent Millay, 1892–1951, Keen

There are three things a sailor doesn't need—a car, civilian clothes, and a wife.
Remark by a chief petty officer, c. 1965.

Sailors, with their built-in sense of order, service, and discipline, should really be running the world.
Nicholas Monsarratl: In The New Yorker, 2 May 1966

True sailors die on the turn of the tide, going out with the ebb.
Sailors' saying

(*See also* Seaman.)

Sanitation

If the facilities for washing were as great as those for drink, our Indian Army would be the cleanest body of men in the world.
Florence Nightingale: Observations on the Evidence Contained in the Stational Reports Submitted to the Royal Commission on the Sanitary State of the Army in India, 1863

(*See also* Health)

Saxe, Maurice de (1696–1750)

When *le Marêchal* Saxe and the proud Pompadour
Were driving out gaily in gilt coach and four,
Frelon spied the pair—"Oh, see them," he cried,
"The sword of our King, and its sheath, side by side."
Verses by unknown author, c. 1735

Frederick the Great asked me the names of the people who were present. I told him the names of a number of princes of distinguished blood who were entering upon a military career, some of whom showed great promise. "Yes," he said, "but I think an empire needs a certain amount of cross breeding. I am all in favor of bastards. Look, for example, at Marshal Saxe."
Prince de Ligne, 1735–1814, Memoirs

. . . That connoisseur of the art of war.
B.H. Liddell Hart: Thoughts on War, xiv, 1944

Scorched Earth

Until we can repopulate Georgia, it is useless for us to occupy it; but the utter destruction of its roads, houses and people will cripple their military resources. I can make this march, and make Georgia howl.
W.T. Sherman: Telegram to U.S. Grant from Atlanta, 9 September 1864

In case of a forced retreat of Red Army units, all rolling stock must be evacuated; to the enemy must not be left a single engine, a single railway car, not a single pound of grain or a gallon of fuel . . . In occupied regions conditions must be made unbearable for the enemy and all his accomplices. They must be hounded and annihilated at every step and all their measures frustrated.

Joseph V. Stalin: Address to the Russian people, 3 July 1941

(*See also* Horrors of War.)

Scott, Winfield (1786–1866)

The General, of course, stands out prominently, and does not hide his light under a bushel, but he appears the bold, sagacious, truthful man he is.

R.E. Lee: Comment on Scott's autobiography when published, 1864

Sea

. . . Over the wine-dark sea.

Homer: The Iliad, c. 1000 B.C.

The Sea! The Sea!

Xenophon, 430–355 B.C., Anabasis, IV, vii

The sea is His and He made it.

Psalm XCV

It is pleasant, when the sea is high and the winds are dashing the waves about, to watch from shore the struggles of another. (Suave, mari magno turbantibus aequora ventis/E terra magnum alterius spectare laborem.)

Lucretius, 99–55 B.C., De Rerum Natura

Beyond all things is the ocean.

Seneca, 4 B.C. – 65 A.D.

Now would I give a thousand furlongs of sea for an acre of barren ground.

Shakespeare: The Tempest, i, 1, 1611

Now for the Services of the Sea, they are innumerable . . . It is an open field for Merchandise in Peace, a pitched field for the most dreaded fights of Warre.

Samuel Purchas: Purchas, His Pilgrimage, 1625

Neither nature nor art has partitioned the sea into empires. The ocean and its treasures are the common property of all men. Upon this deep and strong foundation do I build, and with this cogent and irresistible argument do I fortify our rights & liberties.

John Adams, 1735–1826

The sea is valor's charter,
A nation's wealthiest mine.

Robert Treat Paine, 1773–1811

I love the sea as I do my own soul.

Heinrich Heine, 1797–1856, Travel Pictures

Roll on, thou deep and dark blue ocean—roll!
Ten thousand fleets sweep over thee in vain;
Man marks the earth with ruin—his control
Stops with the shore.

Byron: Childe Harold's Pilgrimage, 1816

There is witchery in the sea, its songs and stories, and in the mere sight of a ship, and the sailor's dress, expecially to a young mind, which has done more to man navies, and fill merchantmen, than all the press gangs of Europe.

R.H. Dana, Jr.: Two Years Before the Mast, 1840

The first and most obvious light in which the sea presents itself from the political and social viewpoint is that of a great highway; or better, perhaps, of a wide common, over which all men may pass in all directions . . .

Mahan: The Influence of Sea Power Upon History, i, 1890

Who hath desired the Sea?—the sight of salt water unbounded—
The heave and the halt and the hurl and the crash of the comber windward-hounded?

Rudyard Kipling: The Sea and the Hills, 1901

And now the old ships and their men are gone; the new ships and the new men, many of them bearing the old auspicious names, have taken up their watch on the stern and impartial sea, which offers no opportunities but to those who know how to grasp them with a ready hand and undaunted heart.

Joseph Conrad, 1857–1924

Sea Lawyer

I will have no Body in my Company that

The Bettman Archive

MAURICE DE SAXE

1696–1750

"It is not big armies that win battles; it is the good ones."

can write; a Fellow that can write, can draw Petitions.
George Farquhar: The Recruiting Officer, 1706

. . . Those men of Belial known as sea-lawyers.
Sir Ian Hamilton: The Soul and Body of an Army, vi, 1921

(*See also* Law Specialist, Uniform Code of Military Justice.)

Sea Power

There is nothing in the world more soft and weak than water, yet for attacking things that are firm and strong, nothing surpasses it.
Lao Tze, fl. 6th century B.C.

He who commands the sea has command of everything.
Themistocles, 514–449 B.C.

Whoever is strongest at sea, make him your friend.
Address of the Corcyraeans to the Athenians, 433 B.C.

He who rules on the sea will very shortly rule on the land also.
Khair-Ed-Din (Barbarossa), d. 1546

For foure things our Noble sheweth to me, King, shippe, and sword, and power of the sea.
Richard Hakluyt: Voyages, 1589

Let us be back'd with God and with the seas
Which he hath given for fence impregnable,
And with their helps only defend ourselves:
In them and in ourselves our safety lies.
Shakespeare: III King Henry VI, iv, 3, 1591

To be master of the sea is an abridgement of monarchy . . . There be many examples where sea-fights have been final to the war . . . But this much is certain; that he that commands the sea is at great liberty, and may take as much and as little of the war as he will. Whereas these, that be strongest by land, are many times nevertheless in great straits.
Francis Bacon: Of the True Greatness of Kingdoms and Estates, 1597

Whosoever commands the sea commands the trade; whosoever commands the trade of the world commands the riches of the world, and consequently the world itself.
Sir Walter Raleigh: Historie of the Worlde, 1616

The dominion of the sea . . . is the best security of the land. The wooden walls are the best walls of this kingdom.
Thomas Coventry: To the House of Commons, 17 June 1635

A man-of-war is the best ambassador.
Oliver Cromwell, 1599–1658

To the question, What shall we do to be saved in this world? There is no answer but this, Look to your moat.
Marquess of Halifax: A New Model at Sea, 1694

Like other amphibious animals we must come occasionally on shore: but the water is more properly our element, and in it . . . as we find our greatest security, so we exert our greatest force.
Lord Bolingbroke: Idea of a Patriot King, 1749

Vain are their threats, their armies all are vain;
They rule the balanced world, who rule the Main.
David Mallet, 1705–1765

For who are so free as the sons of the wave?
David Garrick: Hearts of Oak, 1759

The royal navy of England hath ever been its greatest defence and ornament; it is its ancient and natural strength; the floating bulwark of the island.
William Blackstone: Commentaries on the Laws of England, 1765

The trident of Neptune is the scepter of the world.
A.M. Le Mierre: Commerce, c. 1775

Without a Respectable Navy—alas, America!
John Paul Jones: Letter to Robert Morris, 17 October 1776

Under all circumstances, a decisive naval superiority is to be considered a fundamental principle, and the basis upon which all hope of success must ulitmately depend.
George Washington: Letter, 1780

. . . the expected Naval superiority which was the pivot upon which everything turned.

George Washington: Letter to Benjamin Franklin, 20 December 1780

Without a decisive naval force we can do nothing definitive, and with it everything honorable and glorious.

George Washington: To the Marquis de la Fayette, 15 November 1781

Naval power is the natural defense of the United States.

John Adams, 1735–1826

Britannia needs no bulwarks,
No towers along the steep;
Her march is o'er the mountain waves
Her home is on the deep.

Thomas Campbell: Ye Mariners of England, 1800

In the wreck of the continent, and the disappointment of our hopes there, what has been the security of this country but its naval preponderance?

William Pitt: To the House of Commons, 2 February 1801

I do not say the Frenchman will not come; I only say he will not come by sea.

Lord St. Vincent: At a parley in London while First Lord of the Admiralty, 1803

Wherever wood can swim, there I am sure to find this flag of England.

Napoleon I: At Rochefort, July 1815

Had I been master of the sea, I should have been lord of the Orient.

Napoleon I, 1769–1821

It is not the taking of individual ships or convoys, be they few or many, that strikes down the money power of a nation; it is the possession of that overbearing power on the sea which drives the enemy's flag from it, or allows it to appear only as the fugitive; and by controlling the great common, closes the highways by which commerce moves to and from the enemy's shores. This overbearing power can only be exercised by great navies.

Mahan: The Influence of Sea Power Upon History, 1890

The world has never seen a more impressive demonstration of the influence of sea power upon its history. Those far distant, storm-beaten ships, upon which the Grand Army never looked, stood between it and the dominion of the world.

Mahan: The Influence of Sea Power Upon the French Revolution and Empire, II, 1892 (on the Royal Navy during the Napoleonic Wars)

If blood be the price of admiralty,
Lord God, we ha' paid in full!

Rudyard Kipling: The Song of the English, 1893

The British Army should be a projectile to be fired by the British Navy.

Sir Edward Grey, 1862–1933

We speak glibly of "sea power" and forget that its true value lies in its influence on the operations of armies. For a defensive war a navy may suffice alone, but how fruitless, how costly, and how long drawn out a war must be, that for lack of an adequate army is condemned to the defensive, is the great lesson we have to learn.

Sir Julian Corbett: The Successors of Drake, 1900

When we speak of command of the seas, it does not mean command of every part of the sea at the same moment, or at every moment. It only means that we can make our will prevail ultimately in any part of the seas which may be selected for operations, and thus indirectly make our will prevail in every part of the seas.

Winston Churchill: To the House of Commons, 11 October 1940

My military education and experience in the First World War has all been based on roads, rivers, and railroads. During the past two years, however, I have been acquiring an education based on oceans and I've had to learn all over again.

George C. Marshall: Remark, late 1943

Command of the sea is the indispensable basis of security, but whether the instrument that commands swims, floats, or flies is a mere matter of detail.

Sir Herbert Richmond: Statesmen and Sea Power, 1946

To control the sea, the Navy must be capable of destroying the source of weapons which threaten ships and operations at sea—submarine bases, air bases,

missile bases, and any other bases from which control of the sea can be challenged.
Arleigh A. Burke: In U.S. Naval Institute Proceedings, 1962

The pathway of man's journey through the ages is littered with the wreckage of nations, which, in their hour of glory, forgot their dependence on the sea.
Brigadier General J.D. Hittle USMC: Speech in Philadelphia 28 October 1961

It is from the sea that we will realize our ultimate victory.
Herbert C. Bonner, June 1962

He who controls communications by sea controls his fate; the master of the seas is master of the situation.
Barbara W. Tuchman: The Guns of August, 1962

You can liken sea power to the human hand where with the greatest delicacy, sensitivity, and perception it can be used to solve the combination of the lock on the safe, or the skilled hand of the surgeon who, with the greatest delicacy, can excise the cancerous growth from the human body, or again it can be clenched in a fist representing the brute force philosophy of an all-out nuclear exchange.
Rear Admiral J.S. Mc Cain, Jr.: Testimony to Congress, 12 April 1962

Control of the seas means security. Control of the seas means peace. Control of the seas can mean victory. The United States must control the sea if it is to protect our security.
John F. Kennedy: To all hands, USS Kitty Hawk, *June 1963*

Ours is a maritime nation, requiring the most powerful navies to protect our free rights to the farthest reaches of the seas.
Lyndon B. Johnson: To the Navy League, 27 October 1964

Merely to face upon the oceans is not, *ipso facto,* to have maritime power, but only presents an opportunity. The test is what is done with the opportunity.
Paul H. Nitze: "Trends in the Use of the Sea," Marine Corps Gazette, March 1965

No major war has been fought that has not been, at some stage, critically dependent on the sea.
Paul H. Nitze: In Navy magazine, April 1965

Ex scientia, tridens. (From knowledge, sea power.)
Motto of the U.S. Naval Academy

(*See also* Amphibious Warfare, Fleet, National Policy, Naval Warfare, Navy.)

Seabees

Every time I pass a bulldozer, I want to stop and kiss it.
W.F. Halsey: On Guam, 1945

Never hit a Seabee. He may be your grandfather.
World War II saying

Seaman

Ships are to little purpose without skillful Sea Men.
Richard Hakluyt: Voyages, 1589

All you that would be seamen must bear a valiant heart.
Martyn Parker, d. 1656

Of sea-captains, young or old, and the
mates, and of all intrepid sailors,
Of the few, very choice, taciturn, whom
fate can never surprise nor death
dismay . . . .
Walt Whitman, 1819–1892, Song for All Seas, All Ships

He scorns to sleep 'neath the smoking
rafter
He plows with his ship the raging deep,
Though the storm king roars and the winds
howl after,
To him it is only a thing of laughter,
The sea king loves it better than sleep.
Henry Wadsworth Longfellow, 1807–1882

(*See also* Sailor.)

Seamanship

The mistaking of a rope, by an unskilful person, either in fight or upon a lee shore, may be the loss of all.
Reasons Against the Proposition of Lessening the Number of Men Aboard the King's Ships, author unknown, c. 1619

I have as much pleasure in running into port in a gale of wind as ever a boy did in a feat of skill.
David G. Farragut, 1801–1870

To insure safety at sea, the best that science can devise and that naval organization can provide must be regarded only as an aid, and never as a substitute for good seamanship, self-reliance, and sense of ultimate responsibility which are the first requisites in a seaman and naval officer.
C.W. Nimitz: Letter to U.S. Pacific Fleet, 13 February 1945

Second-in-Command

I have always felt the inutility and inconvenience of the office of second-in-command. It has a great and high sounding title, without duties or responsibilities of any description; at the same time it gives pretensions . . . Every officer in an army should have some duty to perform . . . The second-in-command has none that any body can define; excepting to give opinions for which he is in no manner responsible.
Wellington: Letter to Marshal Beresford, 2 December 1812

Secrecy

O divine art of subtlety and secrecy! Through you we learn to be invisible, through you inaudible; and hence hold the enemy's fate in our hands.
Sun Tzu, 400-320 B.C., The Art of War, xiii

The business asketh silent secrecy.
Shakesperare: II King Henry IV, ii, 1597

Secrecy is one of the first requisites in a commander. In order, therefore, to get a name for this great military virtue, you must always be silent and sullen, particularly at your own table; and I would advise you to secure your secrets the more effectually, by depositing them in the safest place you can think of; as, for instance, in the breast of your wife or mistress.
Francis Grose: Advice to the Officers of the British Army, 1782

If a subaltern should only venture to ask you what it is o'clock? you must not inform him, in order to show that you are fit to be entrusted with secrets.
Francis Grose: Advice to the Officers of the British Army, 1782

Frederick the Great was right when he said that if his night-cap knew what was in his head he would throw it into the fire.
Jomini: Précis de l'Art de la Guerre, 1838

No serving soldier can tell his fellow-countryman anything about an Army which is not (1) quite commonplace; (2) an expression of the views of the Authorities of the moment.
Sir Ian Hamilton: The Soul and Body of an Army, i, 1921

War and truth have a fundamental incompatibility. The devotion to secrecy in the interests of the military machine largely explains why, throughout history, its operations commonly appear in retrospect the most uncertain and least efficient of human activities.
B.H. Liddell Hart: Thoughts on War, i, 1944

Security

When invading an enemy's country, men should always be confident in spirit, but they should fear, too, and take measures of precaution; and thus they will be at once most valorous in attack and impregnable in defense.
Archidamus of Sparta: Speech to the Lacadaemonian expeditionary forces departing against Athens, 431 B.C.

He passes through life most securely who has least reason to reproach himself with complaisance toward his enemies.
Thucydides: History of the Peloponnesian Wars, i, 404 B.C.

He is best secure from dangers who is on his guard even when he seems safe.
(Caret periculo, qui etiam tutus cavet.)
Publilius Syrus: Sententiae, c. 50 B.C.

Skepticism is the mother of security. Even though fools trust their enemies, prudent persons do not. The general is the principal sentinel of his army. He should always be careful of its preservation and see that it is never exposed to misfortune.

Frederick The Great: Instructions for His Generals, viii, 1747

To increase caution at the expense of the final goal is no military art.
Clausewitz: Principles of War, 1812

A general should direct his whole attention to the tranquillity of his cantonments, in order that the soldier may be relieved from all anxiety, and repose in security from his fatigues.
Attributed to Frederick The Great, 1712–1786

If I can deceive my own friends, I can make certain of deceiving the enemy.
Stonewall Jackson: Remark in 1862

He that is too secure is not safe.
English Proverb

The safety of the people must be the supreme law. (Salus populi suprema lex esto.)
Roman Proverb

Security Risks

Among the official class there are worthy men who have been deprived of office; others who have committed errors and have been punished. There are sycophants and minions who are covetous of wealth. There are those who wrongly remain in low office; those who have not obtained responsible positions, and those whose sole desire is to take advantage of times of trouble to extend the scope of their own abilities. There are those who are two-faced, changeable, and deceitful, and who are always sitting on the fence. As far as all such are concerned, you can secretly inquire about their welfare, reward them liberally with gold and silk, and so tie them to you. Then you may rely on them to seek out the real facts of the situation in their country.
Tu Mu, 803–852

Sedan (1 September 1870)

O, les braves gens! (Oh, what brave people!)
Wilhelm I of Prussia: As the French cavalry charged the Prussian lines, 1 September 1870

Seniority

I said an elder soldier, not a better . . .
Shakespeare: Julius Caesar, iv, 3, 1599

It is a military convention that infallibility is the privilege of seniority.
B.H. Liddell Hart: Thoughts on War, v, 1944

Armies are temples of ancestor worship.
B.H. Liddell Hart: Thoughts on War, vi, 1944

(*See also* Rank.)

Sentinel

The general is the principal sentinel of his army.
Frederick The Great: Instructions for His Generals, viii, 1747

It is the enemy who keeps the sentinel wakeful.
Madame Swetchine, 1782–1857

Faithless the watch that I kept: now I have none to keep.
I was slain because I slept: now I am slain I sleep.
Let no man reproach me again, whatever watch is unkept—
I sleep because I am slain. They slew me because I slept.
Rudyard Kipling: Epitaphs of the War (The Sleepy Sentinel) 1919

(*See also* Guard Duty.)

Sergeant

DUNCAN—What Bloody Man is That?
MALCOLM—It is the Sergeant.
Shakespeare: Macbeth, i, 2, 1605

Any officer can get by on his sergeants. To be a sergeant you have to know your stuff. I'd rather be an outstanding sergeant than just another officer.
Gunnery Sergeant Daniel Daly, USMC, 1873–1937

(*See also* First Sergeant.)

Service Reputation

Seeking the bubble reputation
Even in the cannon's mouth.

Shakespeare: As You Like It, ii, 1599

When a soldier was the theme, my name
Was not far off.

Shakespeare: Cymbeline, iii, 3, 1609

The only way to compel men to speak good of us is to do it.

Voltaire, 1694–1788, History of Charles XII

The real character of sea officers cannot be masked from each other; and I wish to be judged by that test.

Lord St. Vincent: Letter, 1789

A wounded reputation is seldom cured.

Henry George Bohn, 1796–1884, Handbook of Proverbs

The Army has its common law as well as its statute law; each officer is weighed in the balance by his fellows, and these rarely err. In the barrack, in the mess, on the scout, and especially in battle, a man cannot—successfully—enact the part of a hypocrite or flatterer, and his fellows will measure him pretty fairly for what he is.

W.T. Sherman, 1820–1891

I think very highly of this officer, and I am not at all impressed by the prejudices against him in certain quarters. Such prejudices attach frequently to persons of strong personality and original view . . . We cannot afford to confine Army appointments to persons who have excited no hostile comment in their career . . . This is a time to try men of force and vision and not be exclusively confined to those who are judged thoroughly safe by conventional standards.

Winston Churchill: Note to Chief of Imperial General Staff, 19 October 1940

Sharpsburg (Antietam) 17 September 1862

Sharpsburg was Artillery Hell.

Lieutenant General Stephen D. Lee, CSA, 1833–1908

(*See also* Antietam.)

Shelter

If you knows of a better 'ole, go to it.

Bruce Bairnsfather: Fragments from France, 1915 (caption on a cartoon showing two British soldiers pinned down under heavy shelling)

Sheridan, Philip H. (1831-1888)

The terrible grumble, and rumble, and roar,
Telling the battle was on once more,
And Sheridan twenty miles away.

Thomas Buchanan Read, 1822–1872, Sheridan's Ride

Sherman, William Tecumseh (1820–1891)

You have accomplished the most gigantic undertaking given to any general in this war, and with a skill and ability that will be acknowledged in history as unsurpassed if not unequalled.

U.S. Grant: Letter to Sherman after the capture of Atlanta, September 1864

It would seem as if in him all the attributes of man were merged in the enormities of the demon, as if Heaven intended in him to manifest depths of depravity yet untouched by a fallen race . . . Unsated still in his demoniac vengeance he sweeps over the country like a simoom of destruction.

Editorial, the Macon (Georgia) Telegraph, 5 December 1864

I don't think any of us claim to be great generals, in the strict sense of that term, or to have initiated anything new, but merely to have met an emergency forced on us, and to have ceased war the very moment it could be done.

Letter by Sherman to Colonel Edward Hanky, 10 May 1867

We drink to twenty years ago,
When Sherman led our banner;
His mistresses were fortresses,
His Christmas gift—Savannah!

George B. Corkhill: Toast at a dinner to Sherman in Washington 8 February, 1883 (Sherman captured Savannah on 22 December 1864 and signalled Lincoln, "I beg to present you as a Christman-gift the city of Savannah . . .")

Ship

Your ships are the wooden walls.

Themistocles: Interpreting to the people of Athens a saying by the Pythian Oracle, 480 B.C.

Of seas, ships are the grace.
Ancient Grecian hymn, 5th century B.C.

A ship and a woman are never sufficiently adorned, or too much.
Plautus: Poenulus, i, c. 200 B.C.

Ships are but boards, sailors but men.
Shakespeare: The Merchant of Venice, i, 3, 1596

What is a ship but a prison?
Robert Burton: Anatomy of Melancholy, 1621

Being in a ship is being in jail, with the chance of being drowned.
Samuel Johnson: To James Boswell, 16 March 1759

I wish to have no Connection with any Ship that does not sail *fast*, for I intend *to go in harm's way.*
John Paul Jones: Letter to le Ray de Chaumont, November 1778

A ship is like a lady's watch, always out of repair.
R.H. Dana, Jr.: Two Years Before the Mast, iii, 1840

Sink me the ship, Master Gunner—sink her, split her in twain!
Fall into the hands of God, not into the hands of Spain!
Tennyson: The Revenge, 1878

The Liner, she's a lady, and if a war should come,
The Man-o'-War's 'er 'usband, and 'e'd bid 'er stay at 'ome;
But, oh, the little cargo-boats that fill with every tide!
'E'd 'ave to up an' fight for them for they are England's pride.
Rudyard Kipling: The Liner She's a Lady, 1895

There is a port of no return, where ships
May ride at anchor for a little space
And then, some starless night, the cable slips,
Leaving an eddy at the mooring place . . .
Gulls, veer no longer. Sailor, rest your oar.
No tangled wreckage will be washed ashore.
Leslie Nelson Jennings, 1892– , Lost Harbor

The strength of the ship is the Service,
And the strength of the Service, the ship.
Ronald A. Hopwood, 1868–1949, The Laws of the Navy

Our ships are our natural bulwarks.
Woodrow Wilson: To Congress, 8 December 1914

There seems to be something wrong with our bloody ships today, Chatfield.
Sir David Beatty: At Jutland, after seeing HMS Queen Mary *blow up under German gunfire, 31 May 1916*

A ship is always referred to as "she" because it costs so much to keep her in paint and powder.
C.W. Nimitz: Remarks to the Society of Sponsors, U.S. Navy, 13 February 1940

There is no quiet Arlington for ships; their bones rust in unknown lands beneath the sea.
Lieutenant Commander Arnold S. Lott, USN: A Long Line of Ships, 1954

(*See also* Warship.)

Shipmate

. . . a loyal shipmate who belonged "to everybody's mess and nobody's watch."
Commander Frederick L. Sawyer, USN: Sons of Gunboats, 1946

Shipping

All was ruled by that harsh and despotic factor, shipping.
Winston Churchill: The Grand Alliance, 1950

Shock Troops

In the Han, the "Gallants from the Three Rivers" were "Sword Friends" of unusual talent. In Wu, the shock troops were called "Dissolvers of Difficulty"; in Ch'i, "Fate Deciders"; in the T'ang, "Leapers and Agitators." These were various names applied to shock troops; nothing is more

important in the tactics of winning battles than to employ them.
Ho Yen-Hsi, fl. 1000

(*See also* Elite Troops.)

Shore Bombardment

A ship's a fool to fight a fort.
Attributed to Horatio Nelson, 1758–1805

You must not always expect that ships, however well commanded, or however gallant their seamen may be, are capable of commonly engaging successfully with stone walls.
Wellington: To the House of Lords, 4 February 1841

The greatest single factor in the American success is naval gunfire.
Lieutenant General Yoshitsuga Saito: To a member of his staff, Saipan, June 1944

However firm and stout pillboxes you may build at the beach, they will be destroyed by bombardment of main armament of the battleships. Power of the American warships . . . makes every landing possible to whatever beachhead they like.
Lieutenant General Tadamichi Kuribayashi: Report to Japanese Imperial Headquarters from Iwo Jima, February 1945

The fire of your battleships was a main factor in hampering our counter-stroke. This was a big surprise, both in its range and in its effect. Army officers, who interrogated me after the war, did not seem to have appreciated this.
Gerd von Rundstedt: Conversation, 1947, about Normandy

Shot

Never say die, as long as there's a shot in the locker.
Seamen's saying, 18th century

Chance shots pick out the best men.
Frederick Marryat: Peter Simple, 1834

Here once the embattled farmers stood,
And fired the shot heard round the world.
R. W. Emerson: Concord Hymn 4 July 1836

Sick Call

There's a salve for every sore.
George Gascoigne, 1525–1577, Supposes, ii

Come and get your quinine, come and get your pills,
Oh come and get your quinine, come and get your pills.
Traditional words of Sick Call

Lay aft, the sick, lame, and lazy!
Traditional boatswain's mates' "word" for Sick Call

Siegecraft

So the people shouted when the priests blew with the trumpets: and it came to pass, when the people heard the sound of the trumpet, and the people shouted with a great shout, that the wall fell down flat, so that the people went up into the city, every man straight before him, and they took the city.
Joshua, VI, 20

He shall set engines of war against thy walls, and with his axes he shall break down they towers.
Ezekiel, XXVI, 9

Turn thou the mouth of thy artillery
As we will ours against these saucy walls.
Shakespeare: King John, ii, 2, 1596

Work, work your thoughts and therein see a siege;
Behold the ordinance on their carriages,
With fatal mouths gaping on girded Harfleur.
Shakespeare: King Henry V, iii, 1598

Fortified towns are hard nuts to crack, and your teeth are not accustomed to it. Taking strong places is a particular trade, which you have taken up without serving an apprenticeship to it. Armies and veterans need skillful engineers to direct them in their attack. Have you any? But some seem to think forts are as easy taken as snuff.
Benjamin Franklin: Letter to his brother in Boston before the siege of Louisburg, 1745

The art of conducting sieges has become a calling like those of the carpenter and shoemaker.

Frederick The Great: Instructions for His Generals, xiv, 1747

The strongest fortress and sternest virtue have weak points and require unremitting vigilance to guard them: let warrior and dame take warning.

Washington Irving: Chronicles of the Conquest of Granada, 1829

Signal

I have only one eye—I have a right to be blind sometimes: . . . I really do not see the signal!

Nelson: At Copenhagen, on having Sir Hyde Parker's signal to break off action pointed out to him, and having then put his telescope to his blind eye, 2 April 1801

. . .if the British march
By land or sea from the town tonight,
Hang a lantern aloft in the belfry arch
Of the North Church tower as a signal light,—
One if by land, two if by sea,
And I on the opposite shore will be,
Ready to ride and spread the alarm . . .

Henry Wadsworth Longfellow, 1807–1882, Paul Revere's Ride

I want none of this Nelson business in my squadron about not seeing signals.

David G. Farragut: Reprimanding Lieutenant W. S. Schley, USN, for failing to obey a signal promptly during the attack on Port Hudson, March 1863

Signals travel alone and, like telegrams, have to be carefully worded to be readily understood. Unlike telegrams, however, they are impersonal and public. They carry the authority of a ship or squadron. They are paid for with reputations, sometimes even with human lives.

Captain Jack Broome, RN: Make a Signal, 1955

Situation

In war, situations are the products of mutually exclusive and incompatible wills. Thus, they are practically always fluid.

Brigadier General S.B. Griffith, II, USMC: The Battle for Guadalcanal, x, 1963

It all depends on the terrain and the situation.

Tactical instructors' axiom

Situation normal, all fouled up. ("SNAFU")

Soldiers' saying, World War II (the origin of the slang noun, "snafu")

Slaughter

He smote them hip and thigh with a great slaughter.

Judges, XV

Go ye after him through the city, and smite: let not your eye spare, neither have ye pity; slay utterly old and young, both maids and little children.

Ezekiel, IX, 5–6

Millions of soules sit on the banks of Styx,
Waiting the black return of Charon's boat,
Hell and Elysian swarme with ghosts of men,
That I have sent from sundry foughten fields.

Christopher Marlowe: Tamburlaine the Great, 1590

I have seen the cannon
When it has blown his ranks into the air.

Shakespeare: Hamlet, iii, 4, 1600

The pursuit of victory without slaughter is likely to lead to slaughter without victory.

Duke of Marlborough, 1650–1722

But when it comes to slaughter,
You will do your work on water,
An' you'll lick the bloomin' boots of 'im that's got it.

Rudyard Kipling: Gunga Din, 1890

Sleep

No human being knows how sweet sleep is but a soldier.

John S. Mosby: War Reminiscences, iv, 1887

A man is no sailor if he cannot sleep when he turns-in, and turn out when he's called.

R.H. Dana, Jr.: Two Years Before the Mast, xxxiv, 1840

After battle sleep is best.
Roden Berkeley Wriothesley Noel, 1834–1894, The Old

Here the General slept before the battle of Tannenberg; here also the General slept after the battle; and, between you and me during the battle also.
Attributed to General Max Hoffman, c. 1915 (Of von Hindenburg at Tannenberg, 1914, at which Hoffman was the General Staff operations officer)

Smartness

Can that man be reckoned a good soldier who through negligence suffers his arms to deteriorate by dirt and rust?
Vegetius: De Re Militarii, ii, 378

Smartness is the cement, but not the bricks.
B.H. Liddell Hart: Thoughts on War, v, 1944

The old Army saying, "Who ever saw a dirty soldier with a medal?" is largely true.
George S. Patton, Jr.: War As I Knew It, 1947

We found it a great mistake to belittle the importance of smartness in turn-out, alertness of carriage, cleanliness of person, saluting, or precision of movement, and to dismiss them as naive, unintelligent, parade-ground stuff. I do not believe that troops can have unshakable battle discipline without showing those outward and formal signs which mark the pride men take in themselves and their units and the mutual confidence and respect that exists between them and their officers.
Sir William Slim: Defeat into Victory, 1956

Soldier

The sex is ever to a soldier kind.
Homer: Odyssey, xiv, c. 1000 B.C.

. . . mighty men of valor.
Joshua, I, 9

Readiness, obedience, and a sense of humor are the virtues of a soldier.
Brasidas of Sparta: Speech to the Lacadaemonian Army, battle of Amphipolis, 422 B.C.

He was fond of adventure, ready to lead an attack on the enemy by day or night, and, when he was in an awkward position, he kept his head.
Xenophon: Anabasis, ii, 390 B.C. (of Clearchus, the Spartan general)

It is a high thing, a bright honor, for a man to do battle with the enemy for the sake of his children, and for his land and his true wife; and death is a thing that will come when the spinning Destinies make it come.
Grecian lyric, author unknown, 4th century B.C.

War, as the saying goes, is full of false alarms, a fact which professional soldiers have had the best chance to learn; thus they appear brave because of other men's ignorance of the true situation.
Aristotle: The Nicomachean Ethic, iii, c. 340 B.C.

Soldiers fight and die to advance the wealth and luxury of the great, and they are called masters of the world without having a sod to call their own.
Tiberius Sempronius Gracchus: Speech in Rome, 133 B.C.

He who makes war his profession cannot be otherwise then vicious. War makes thieves, and peace brings them to the gallows.
Niccolo Machiavelli, 1469–1527

For who ought to be more faithful than a man that is entrusted with the safety of his country, and has sworn to defend it to the last drop of his blood? Who ought to be fonder of peace than those that suffer by nothing but war? Who are under greater obligations to worship God than Soldiers, who are daily exposed to innumerable dangers, and have most occasion for his protection?
Niccolo Machiavelli: The Art of War, preface, 1520

It is not gold, but good soldiers, that insure success in war . . . for it is impossible that good soldiers should not be able to procure gold, as it is impossible for gold to procure good soldiers.
Niccolo Machiavelli: Discorsi, ii, 1531

Soldiers in peace are like chimneys in summer.
William Cecil (Lord Burghley): Advice to His Son, c. 1555

The hardiest soldiers be either slain or maimed [or], if they escape all hazards, and return home again, if they be without relief of their friends they will surely desperately rob and steal, and either shortly be hanged or miserably die in prison.

Thomas Harman: A Caveat or Warning for Common Cursetors, i, 1566

There were three lusty soldiers
Went through a town of late;
The one loved Bess, the other Sis,
The third loved bouncing Kate.

Choice of Inventions, author unknown, c. 1575

As well the soldier dieth who standeth still, as he that gives the bravest onset.

Sir Philip Sidney, 1554–1586

They are soldiers,
Witty, courteous, liberal, full of spirit.

Shakespeare: III Henry VI, i, 2, 1590

I am a soldier unapt to weep
Or to exclaim on fortune's fickleness.

Shakespeare: I King Henry VI, v, 1591

And with a martial scorn,
With one hand beats cold death aside.

Shakespeare: Romeo and Juliet, iii, 1, 1594

. . . As soldiers will
That nothing do but meditate on blood,—
To swearing and stern looks.

Shakespeare: King Henry V, v, 2, 1598

. . . A soldier,
Full of strange oaths, and bearded like the pard;
Jealous in honor, sudden and quick in quarrel,
Seeking the bubble reputation
Even in the cannon's mouth.

Shakespeare: As You Like It, ii, 7, 1599

This is your devoted friend, sir, the manifold linguist, and the omnipotent soldier.

Shakespeare: All's Well That Ends Well, iv, 3

You may relish him more in the soldier than in the scholar.

Shakespeare: Othello, ii, 1, 1604

Rude am I in my speech,
And little bless'd with the soft phrase of peace.

Shakespeare: Othello, i, 3, 1604

A soldier's but a man;
A life's but a span;
Why, then, let a soldier drink.

Shakespeare: Othello, ii, 1604

He is a soldier fit to stand by Caesar.

Shakespeare: Othello, ii, 1604

Fie, my lord, fie! a soldier and afear'd?

Shakespeare: Macbeth, v, 1605

You are a soldier, therefore seldom rich.

Shakespeare: Timon of Athens, 1607

I love them [soldiers] for their virtues' sake and for their greatness of mind . . . If we may have peace, they have purchased it; and if we must have war they must manage it.

Robert Devereux (Earl of Essex): Change, c. 1600

He that gives a soldier the lie looks to receive the stab.

Thomas Dekker: The Seven Deadly Sins of London, ii, 1606

To take a soldier without ambition is to pull off his spurs.

Francis Bacon: Essays, xxxvi, 1625

. . . In the vaunting style of a soldier.

Francis Bacon, 1561–1626, Considerations Touching a War with Spain

Our God and the soldier we alike adore
Ev'n at the brink of danger, not before:
After deliverance, both alike requited,
God is forgotten, and the soldier slighted.

Francis Quarles: Emblems, 1635

Water, fire and soldiers quickly make room.

George Herbert: Outlandish Proverbs, 1640

On becoming soldiers we have not ceased to be citizens.

Humble Representation, addressed to Parliament by Cromwell's soldiers, 1647

The two parts of a soldier are Valour and Suffering.

Duke of Albemarle, 1608–1670

The Bettman Archive

NICCOLÒ MACHIAVELLI

1469–1527

"When princes think more of luxury than of arms, they lose their state."

Whose house doth burn, must soldier turn.
17th century adage, Thirty Years War

Soldiers and travelers may lie by authority.
James Howell: Proverbs, 1659

A soldier's life is a life of honor, but a dog would not lead it.
Prince Rupert of the Rhine, 1619–1682

Ay me! what perils do environ
The man that meddles with cold iron!
Samuel Butler: Hudibras, i, 1663

Such as have followed the wars are despised of every man until a very pinch of need doth come.
Unknown English writer, 17th century

You stink of brandy and tobacco, most soldier-like.
William Congreve: The Old Bachelor, iii, 1693

He shall turn soldier, and rather depend upon the outside of his head than the lining.
William Congreve: Love for Love, i, 1695

The military pedant always talks in a camp, and is storming towns, making lodgments and fighting battles from one end of the year to the other. Everything he speaks smells of gunpowder; if you take away his artillery from him, he has not a word to say for himself.
Joseph Addison: The Spectator, 30 June 1711

Of boasting more than bomb afraid,
A soldier should be modest as a maid.
Edward Young: Love of Fame, iv, 1724

The wise old soldier is never in haste to strike a blow.
Pietro Metastasio: Adriano, ii, 1735

The first man to be king was a fortunate soldier. Whoever serves his country well has no need of ancestors.
Voltaire, 1694–1778, Merope, i, 3

If my soldiers were to begin to think, not one would remain in the ranks.
Ascribed to Frederick The Great, 1712–1786

For the officer, honor is reserved, for the common man, obedience and loyalty . . . From honor flows intrepidity and equanimity in danger, zeal to win ability and experience, respect for superiors, modesty towards one's equals, condescension toward inferiors . . . Nothing therefore must incite the officer but honor, which carries its own recompense; but the enlisted man is driven and restrained and educated by reward and fear . . . The worst soldier is an officer without honor, an enlisted man without discipline
Saxon-Polish Field Service Regulations, 1752

A soldier worthy of the name he bears,
As brave and senseless as the sword he bears.
Mary Wortley Montagu: To James Steuart, 19 July, 1759

Military men belong to a profession which *may* be useful, but is *often* dangerous.
Henry Middleton: Remark while President of the General Continental Congress, 1774

Every man thinks meanly of himself for not having been a soldier.
Samuel Johnson: To James Boswell, 10 April 1778

. . . a rapacious and licentious soldiery.
Edmund Burke: To the House of Commons on Fox's East India Bill, 1783

For a soldier I 'listed, to grow great in fame,
And be shot at for sixpence a day.
Charles Dibdin, 1745–1814, Charity

This, this, my lad's a soldier's life:
He marches to the sprightly fife,
And in each town to some new wife
Swears he'll be ever true . . .
And follows the loud tattoo.
Charles Dibdin, 1745–1814, The Soldier's Life

Que du pauvre soldat, deplorable est la chance!
Quand la guerre finit, son malheur recommence.
(As for the poor soldier, his lot is deplorable! When the war's over, his ill fortune resumes.)
Fin des Travaux de Mars, 18th century French song

A soldier is a man whose business it is to kill those who never offended him, and who are the innocent martyrs of other men's iniquities. Whatever may become of the abstract question of the justifiableness of war, it seems impossible that the soldier should not be a depraved and unnatural thing.

William Godwin: The Enquirer, v, 1797

Soldier rest! thy warfare o'er.
Dream of fighting fields no more.
Sleep the sleep that knows not breaking,
Morn of toil, nor night of waking.

Walter Scott: The Lady of the Lake, i, 1810

Fell as he was in act and mind,
He left no bolder heart behind:
Then, give him, for a soldier meet,
A soldier's cloak for winding sheet.

Walter Scott: Rokeby, vi, 1813

In order to have good soldiers, a nation must always be at war.

Napoleon I: To Barry E. O'Meara, St. Helena, 26 October 1816

Soldiers are made on purpose to be killed.

Napoleon I: To General Gaspard Gourgaud, St. Helena, 1818

The first qualification of a soldier is fortitude under fatigue and privation. Courage is only the second; hardship, poverty, and want are the best school for a soldier.

Napoleon I: Maxims of War, 1831

Chacun disait: Voyez donc comme
Il est grand, comme il est beau!
Le bel habit! Le beau chapeau!
Morbleu, qu'un soldat est bel homme!
(Each one said, look how big he is, how handsome! The smart uniform, the handsome headgear! Goodness, but a soldier is a fine-looking man!)

Louis Benjamin Francoeur, 1773–1849

A modern general has said that the best troops would be as follows: an Irshman half drunk, a Scotchman half starved, and an Englishman with his belly full.

C.C. Colton: Lacon, 1820

Wrath is quickly changed to friendship in the hearts that throb beneath a soldier's coat.

Alessandro Manzoni, 1785–1873

Ah, what delight to be a soldier!

Augustin Eugene Scribe, 1791–1863, Dame Blanche

What the soldier said is not evidence.

Charles Dickens: Pickwick Papers, 1836

Ben Battle was a soldier bold,
And used to war's alarms;
But a cannon-ball took off his legs,
So he laid down his arms.

Thomas Hood: Faithless Nellie Gray, 1840

Soger (soldier) is the worst term of reproach that can be applied to a sailor. It signifies a *skulk*, a *shirk*,—one who is always trying to get clear of work, and is out of the way, or hanging back, when duty is to be done.

R.H. Dana Jr.: Two Years Before the Mast, xvii, 1840

"You're wounded!" "Nay," the soldier's pride
Touched to the quick, he said:
"I'm killed, Sire!" And his chief beside,
Smiling the boy fell dead.

Robert Browning: Incident in the French Camp, 1846

He fell on the field:
His country mourned him,
And his father was resigned.

Bulwer-Lytton: The Caxtons, xviii, 1849

Tell me what find we to admire
In epaulets and scarlet coats—
In men, because they load and fire,
And know the art of cutting throats?

William Makepeace Thackeray, 1811–1863

Were not here the real priests and martyrs of that loud-babbling, rotten generation?

Thomas Carlyle: Letter to R.W. Emerson, 25 June 1852 (on the Prussian soldiers of Frederick the Great's era)

Only the defeated and the deserters go to the war.

H.D. Thoreau: Walden, 1854

Theirs not to make reply,
Theirs not to reason why,
Theirs but to do and die.

Tennyson: The Charge of the Light Brigade, 1854

A soldier has a hard life, and but little consideration.

R.E. Lee: Letter to his wife, 5 November, 1855

The man-at-arms is the only man.

Henrik Ibsen: Lady Inger of Ostraat, 1855

The soldier as an abstract idea is a hero . . . but as a social fact he is a pariah.

Article in Blackwood's Edinburgh Magazine, 1859

Policemen are soldiers who act alone; soldiers are policemen who act in unison.

Herbert Spencer, 1820–1903, Social Statics

The soldier's trade is not slaying, but being slain. This, without well knowing its own meaning, the world honors it for.

John Ruskin: Unto This Last, i, 1862

I know that there must be soldiers; but as to every separate soldier I regret that he should be one of them.

Anthony Trollope: North America, i, 1862

It is enough for the world to know that I am a soldier.

W.T. Sherman, 1820–1891

The soldier—that is, the great soldier—of today is not a romantic animal, dashing at forlorn hopes, animated by frantic sentiment, full of fancies as to a love-lady or a sovereign; but a quiet, grave man, busied in charts, exact in sums, master of the art of tactics, occupied in trivial detail; thinking, as the Duke of Wellington was once said to do, most of the shoes of his soldiers; despising all manner of éclat and eloquence; perhaps, like Count Moltke, "silent in seven languages."

Walter Bagehot: The English Constitution, vii 1867

It were better to be a soldier's widow than a coward's wife.

Thomas Bailey Aldrich, 1836–1907, Mercedes, ii, 2

Hair is the glory of a woman but the shame of a soldier.

Sir Garnet Wolseley: The Soldier's Pocket Book, 1869

[The soldier] must be taught to believe that his duties are the noblest which fall to a man's lot. He must be taught to despise all those of civil life. Soldiers, like missionaries, must be fanatics.

Sir Garnet Wolseley: The Soldier's Pocket Book, 1809

It is nothing. For this, we are soldiers.

Captain Guy V. Henry, USA: After being shot through the face during the 3d Cavalry's action on the Rosebud River, 17 June 1876

I have one sentiment for soldiers living and dead: cheers for the living; tears for the dead.

Robert G. Ingersoll: Speech at Indianapolis, 21 September 1876

Soldiers should make it their function to exert themselves to the utmost of their loyalty and patriotism. They should strictly observe decorum. They should prize courage and bravery. They should treasure faith and confidence. They should practice frugality.

Emperor Meiji: Imperial Rescript, 4 January 1883

Love ain't enough for a soldier.

Rudyard Kipling: The Young British Soldier, 1890

Oh, it's Tommy this, and 'Tommy that,
an' "Tommy, wait outside"; But it's
"Special train for Atkins" when the
trooper's on the tide.

Rudyard Kipling: Tommy, 1890

Single men in barricks don't grow into
plaster saints.

Rudyard Kipling: Tommy, 1890

Yes, the large Birds o' Prey
They will carry us away,
An' you'll never see your soldiers any
more!

Rudyard Kipling: Birds of Prey, 1895

Servants of the staff and chain,
Mine and fuse and grapnel—
Some before the face of Kings,
Stand before the face of Kings;
Bearing divers gifts to Kings—
Gifts of case and shrapnel.

Rudyard Kipling: A School Song, 1899

A good soldier is always a bit of an old maid.

Rudyard Kipling: The New Army, 1915

Far and near and low and louder
On the roads of earth go by,
Dear to friends and food for powder,
Soldiers marching, all to die.
A.E. Housman, A Shropshire Lad, 1896

Living or dead, drunk or dry, Soldier, I wish you well.
A.E. Housman, 1859–1936

A soldier is an anachronism of which we must get rid.
George Bernard Shaw: The Devil's Disciple, iii, 1897

I never expect a soldier to think.
George Bernard Shaw: The Devil's Disciple, iii, 1897

When the military man approaches, the world locks up its spoons and packs off its womankind.
George Bernard Shaw: Man and Superman, iii, 1903

Oh, I did not raise my boy to be a soldier,
I brought him up to be my pride and joy . . .
Popular song, 1916

In this country of ours the man who has not raised himself to be a soldier, and the woman who has not raised her boy to be a soldier for the right—neither one of them is entitled to citizenship in the Republic.
Theodore Roosevelt: Speech at Camp Upton, 18 November 1917

You smug-faced crowds with kindling eyes
Who cheer when soldier lads march by,
Sneak home and pray you'll never know
The hell where youth and laughter go.
Siegfried Sassoon: Pray God, 1918

Soldiers are citizens of death's grey land.
Siegfried Sassoon, 1886–, Dreamers

Old soldiers never die:
They simply fade away.
British soldiers' song popular in World War I

A soldier's life is a hard one, interspersed with some real dangers.
André Maurois: Les Silences du Colonel Bramble, 1921

Boys are soldiers in their hearts already—and where is the harm? Soldiers are not pugnacious. Paul bade Timothy be a good soldier. Christ commended the centurion. Milton urged teachers to fit their pupils for all the offices of war. The very thought of danger and self sacrifice are inspirations.
Sir Ian Hamilton: The Soul and Body of an Army, xiii, 1921

A *soldier*!—a common *soldier!*—nothing but a body that makes movements when it hears a shout!
D.H. Lawrence, 1885–1930, Sons and Lovers

To ride boldly at what is in front of you, be it fence or enemy; to pray, not for comfort, but for combat; to remember that duty is not to be proved in the evil day, but then to be obeyed unquestioning; to love glory more than the temptations of wallowing ease.
Oliver Wendell Holmes, Jr., 1841–1935, speech

Three-quarters of a soldier's life is spent in aimlessly waiting about.
Eugen Rosenstock-Huessy, 1888–

A soldier is a slave—he does what he is told to do—everything is provided for him—his head is a superfluity. He is only a stick used by men to strike other men.
Elbert Hubbard: Roycroft Dictionary, 1923

Soldiers put one thing straight, but leave a dozen others crooked.
E.M. Forster: A Passage to India, xx, 1924

By your courage in tribulation, by your cheerfulness before the dirty devices of this world, you have won the love of those who have watched you.
Guy Chapman: A Passionate Prodigality, 1933

What are the qualities of the good soldier, by the development of which we make the man war-worthy—fit for any war? . . .
The following four—in whatever order you place them—pretty well cover the field: discipline, physical fitness, technical skill in the use of his weapons, battle-craft.
Sir A.P. Wavell: Lecture at the Royal United Service Institution, 15 February 1933

There's something about a soldier that is fine, fine, fine.

Noel Gay (Pseud. for Reginald Mixon Armitage): Popular song, 1933, in musical comedy, Soldiers of the King

The essential quality of the warrior is bravery; that of the soldier, discipline.
Robert Leurquin: In The Army Quarterly, April 1938

I needed hard riders and hard livers; men proud of themselves and without family.
T.E. Lawrence, 1888–1935

The functions of a citizen and a soldier are inseparable.
Benito Mussolini: Decree, 18 September 1934

It was commonly believed that any young man who joined the Army did so because he was too lazy to work, or else he had got a girl in the family way. Hardly anyone had a good word for the soldier, and mothers taught their daughters to beware of them.
Private Frank Richards: Old Soldier Sahib, 1936

Nothing has been made safe until the soldier has made safe the field where the building shall be built, and the soldier is the scaffolding until it has been built, and the soldier gets no reward but honor.
Eric Linklater, 1899–, Crisis in Heaven

Professional soldiers are sentimental men, for all the harsh realities of their calling. In their wallets and in their memories they carry bits of philosophy, fragments of poetry, quotations from the Scriptures, which, in times of stress and danger speak to them with great meaning.
General Matthew B. Ridgway, USA: My Battles in War and Peace, January 1956

How often have I seen them, unconscious ambassadors, showing their identity discs or photos of their wives and families, asking questions by signs, swapping cigarettes, buttons, and I am afraid, at times cap-badges.
Sir William Slim: Unofficial History, ix, 1959

Where else would you get a nonlineal computer weighing only 160 pounds, having a million precision elements, that can be mass produced by unskilled labor?
Attributed to Scott Crossfield, 1962

Plenty of foul-mouthed, blasphemous, hard-drinking, cynical and cruel men have made good soldiers.
Jon Manchip White: Marshal of France, 1962

A soldier is not a foreign minister. He cannot enter into negotiations, and he has to carry out his orders.
Nikita Khruschev: During an interview with a group of American visitors, the Kremlin, November 1963

A good soldier is never forgot,
Whether he die by musket or by the pot.
Epitaph at Winchester, England

The best soldiers are not warlike.
Chinese Proverb

Good iron is not used for nails; good men do not become soldiers.
Chinese Proverb

It is better to have no son than one who is a soldier.
Chinese Proverb

A soldier ought to fear nothing but God and dishonor.
English Proverb

Soldiers fight, and kings are heroes.
Hebrew Proverb

Among flowers, the best is the cherry blossom; among men, the best is the soldier.
Japanese Proverb

(*See also* Profession of Arms, Regular.)

Smokeless Powder

Smokeless powder has changed the picture and made the unknown both *complete and lasting*.
Ferdinand Foch: Precepts, 1919

(*See also* Ammunition, Gunpowder.)

Spain

The Spaniard, like the Turk whom he so strongly resembles in religion, does not leave his country to make war on other nations, but as soon as anyone sets foot

in his country, everyone is the enemy of the invader.
Stendhal (Henri Beyle): A Life of Napoleon, xli, 1818

Spanish-American War (1898)

You furnish the pictures and I'll furnish the war.
William Randolph Hearst: Cable to Frederic Remington, then in Cuba for the Hearst newspapers, March 1898

It has been a splendid little war; begun with the highest motives, carried on with magnificent intelligence and spirit, favored by that fortune which loves the brave.
John Hay: Letter to Theodore Roosevelt, 1898

The Spanish-American War was not a great war. A large number of our troops took the hazard of watermelons in Georgia and Florida, and fought the malaria and mosquitoes, but very few Spanish . . . The Spanish-American War yielded comparatively little in heroics [but] paid the most marvelous dividends in politics and in magazine articles of any war in the history of the country.
James L. Slayden: To the House of Representatives, 1906

An hour or two at Manila, an hour or two at Santiago, and the maps of the world were changed.
Rear Admiral A.S. Barker, USN, 1843–1916

(*See also* Dewey, Maine, Manila Bay.)

Spanish Armada

Our ships doth show themselves like gallants here. I assure you it will do a man's heart good to behold them; and would to God the Prince of Parma were upon the seas with all his forces, and we in the view of them; then I doubt not but that you should hear that we would make his enterprise very unpleasant to him.
Sir William Wynter: To the Principal Officers, 28 February 1588, during preparations against the Armada

I know I have the body of a weak and feeble woman, but I have the heart and stomach of a king, and of a king of England, too; and think foul scorn that Parma or Spain, or any prince of Europe should dare to invade the borders of my realm.
Elizabeth I: Speech to the troops at Tilbury during the approach of the Spanish Armada, 1588

He made the wynds and waters rise
To scatter all myne enemies . . .
Elizabeth I: Songe of Thanksgiving, 1588, composed after defeat of the Armada

Play out the game: there's time for that and to beat the Spanish after.
Attributed to Sir Francis Drake: when playing at long bowls on Plymouth Hoe, 19 July 1588, when the Spanish fleet was sighted

Their force is wonderful, great and strong, yet we pluck their feathers by little and little.
Lord Howard of Effingham: Of the Armada, 1588

We have them before us and mind with the Grace of God to wrestle a fall with them. There was never anything that pleased me better than seeing the enemy flying to northwards. God grant you have a good eye to the Duke of Parma, for with the Grace of God, if we live, I doubt not so to handle the matter with the Duke of Sidonia as he shall wish himself among his orange-trees.
Sir Francis Drake: To Walsingham, 30 July 1588

Spanish Revolution (1808)

. . . the frightful epoch when priests, women, and children throughout Spain plotted the murder of isolated soldiers.
Jomini: Précis de l'Art de la Guerre, 1838

Speed

In war we must be speedy.
Silius Italicus: Punica, i, c. 75 A.D.

By making your battle short, you will deprive it of time, so to speak, to rob you of men. The soldier who is led by you in this manner will gain confidence in you and expose himself gladly to all danger.
Attributed to Frederick The Great, 1712–1786

Speed is one of the characteristics of Strategic Marches . . . in this one quality lie all the advantages that a fortunate initiative may have procured . . . By rapidity of movement we can, like the Romans, make war feed war.

Dennis Hart Mahan: Out Post, 1847, (Professor Mahan was the father of Alfred Thayer Mahan.)

Speed is the essential requisite for a first-class ship of war—but essential only to go into action, not out of it.

John A. Dahlgren, 1809–1870

The greatest kindness in war is to bring it to a speedy conclusion.

Helmuth von Moltke ("The Elder"): Letter to Professor J.K. Bluntschli, 11 December 1880

The true speed of war is not headlong precipitancy, but the unremitting energy which wastes no time.

Mahan: Lessons of the War with Spain, 1899

It is of no use to get there first unless, when the enemy arrives, you have also the most men.

Mahan: Lessons of the War with Spain, 1899

In war, to strike quickly is the first step toward striking hard.

Gabriel Darrieus: War on the Sea, 1908

Swiftness in war comes from slow preparations.

Sir Ian Hamilton: Gallipoli Diary, I, 1920

In small operations, as in large, speed is the essential element of success. If the difference between two possible flanks is so small that it requires thought, the time wasted in thought is not well used.

George S. Patton, Jr.: War As I Knew It, 1947

(*See also* Expedition, Rapidity.)

Spit and Polish

Make ye my buckler's sheen outshine the radiant sun . . .

Plautus: The Braggart Captain, 3d century B.C.

Spy

. . . sent to spy out the land.

Numbers, XIII, 16

A sovereign should always regard an ambassador as a spy.

The Hitopadesa, iii, c. 500

As living spies we must recruit men who are intelligent but appear to be stupid; who seem to be dull but are strong in heart; men who are agile, vigorous, hardy, and brave; well-versed in lowly matters and able to endure hunger, cold, filth, and humiliation.

Sun Tzu, 400–320 B.C., The Art of War, xiii

An army without secret agents is exactly like a man without eyes or ears.

Chia Lin, fl. 700 A.D.

The life of spies is to know, not to be known.

George Herbert: Outlandish Proverbs, 1640

In general it is necessary to pay spies well and not be miserly in that respect. A man who risks being hanged in your service merits being well paid.

Frederick The Great: Instructions for His Generals, ix, 1747

I only regret that I have but one life to lose for my country.

Nathan Hale: Before being hanged by the British as a spy, Long Island, 22 September 1776

The fear of spies seems to be endemic in every crisis in every military campaign.

Alan Moorhead: Gallipoli, 1956

(*See also* Espionage, Intelligence.)

Stable Call

Oh, go to the stable,
All ye who are able,
And give your poor horses some hay and some corn.
For if you don't do it,
The Captain will know it,
And you'll catch the devil as sure as you're born.

Traditional words to Stable Call, c. 1870

(*See also* Cavalry.)

Staff

The futile employment ycleped *Staff* should be totally done away, and all the frippery of the Army sent to the devil.

Lord St. Vincent: Letter to Benjamin Tucker on naval reorganization, 1818

A good staff has the advantage of being more lasting than the genius of a single man.

Jomini: Précis de l'Art de la Guerre, 1838

Great captains have no need for counsel. They study the questions which arise, and decide them, and their *entourage* has only to execute their decisions. But such generals are stars of the first magnitude who scarcely appear once in a century. In the great majority of cases, the leader of the army cannot do without advice. This advice may be the outcome of the deliberations of a small number of qualified men. But within this small number, one and only one opinion must prevail. The organization of the military hierarchy must ensure subordination even in thought and give the right and duty of presenting a single opinion for the examination of the general-in-chief to one man and only one.

Helmuth von Moltke ("The Elder"): Letter, 1862

A bulky staff implies a division of responsibility, slowness of action and indecision; whereas a small staff implies activity and concentration of purpose.

W.T. Sherman, 1820–1891

Two branches of a staff can get more hostile to each other than to the enemy.

Captain Peter Wright: At the Supreme War Council, 1921

Without a staff, an army could not peel a potato.

Lieutenant General Hunter Liggett, USA, 1857–1935

The staff knew so much more of war than I did that they refused to learn from me of the strange conditions in which Arab irregulars had to act; and I could not be bothered to set up a kindergarten of the imagination for their benefit.

T.E. Lawrence: Seven Pillars of Wisdom, 1926

The military staff must be adequately composed: it must contain the best brains in the fields of land, air, and sea warfare, propaganda war, technology, economics, politics and also those who know the people's life.

Erich Ludendorff: Der totale Krieg, 1935

My war experience led me to believe that the staff must be the servants of the troops, and that a good staff officer must serve his commander and the troops but himself be anonymous.

Montgomery of Alamein: Memoirs, ii, 1958

The function of the staff is to serve the Line.

Military Maxim

(*See also* Administration, Official Correspondence, Plans, Planner, Staff Officer, Staff Secretary, Staff Work)

Staff Officer

. . . Firm and strict in discharging the duties of trust reposed in him. Be he too pliant in his disposition, he will most assuredly be imposed upon, and the efficient strength and condition of the Army will not be known to the Commander-in-Chief.

Attributed to George Washington, 1732–1799

Whenever the colonel or commanding officer is on the parade, you should always seem in a hurry, and the oftener you run or gallop from right to left, the more assiduous will you appear.

Francis Grose: Advice to the Officers of the British Army, 1782

I do not want to make an appointment on my staff except of such as are early risers.

Stonewall Jackson: Letter to his wife, 1862

A yes man on a staff is a menace to a commander. One with the courage to express his convictions is an asset.

MacKenzie Hill (Pseud., Major General Orlando Ward, USA), c. 1934

There is always a danger that anything contrary to Service prejudices will be obstructed and delayed by officers of the second grade in the machine. The way to

deal with this is to make signal examples of one or two. When this becomes known you get a better service afterwards.
Winston Churchill: Note for Secretary of State for War, 8 September 1940

(*See also* Planner, Staff, Staff Secretary.)

Staff Secretary

You should have a clever secretary to write your dispatches, in case you should not be so well qualified yourself. This gentleman may often serve to get you out of a scrape.
Francis Grose: Advice to the Officers of the British Army, 1782

And so, while the great ones depart to their dinner,
The Secretariat works—and gets thinner and thinner.
British military jingle, mid-20th century

Staffwork

No military or naval force, in war, can accomplish anything worthwhile unless there is back of it the work of an efficient, loyal, and devoted staff.
Lieutenant General Hunter Liggett, USA, 1857–1935

The final test of completed staffwork is this: If you yourself were the commander, would you be willing to sign the paper you have prepared? Would you stake your professional reputation on its being right? If your answer would be "No," take the paper back and rework it, because it is not yet completed staffwork.
Completed Staffwork, monograph by unknown U.S. Army officer, c. 1930

(*See also* Administration, Plans, Programs.)

Stalemate

Next to unilateral disarmament, stalemate is the most misleading and misguided military theme yet conceived.
General Thomas D. White, USAF: In Newsweek, 24 February 1964

I don't believe in stalemates. I don't think there is such a thing.
General Curtis E. LeMay, USAF: Remark, 1965

Stalingrad (August 1942 – February 1943)

The duty of the men at Stalingrad is to be dead.
Adolf Hitler: During a luncheon conference, January 1943

Standing Army

To keep watchdogs, who from want of discipline or hunger, or from some evil habit or other, would turn upon the sheep and worry them, would be a foul and monstrous thing in a shepherd . . . and therefore every care must be taken that our auxiliaries, being stronger than our citizens, may not grow too much for them and become savage beasts.
Plato, 427–347 B.C., The Republic

To erect a standing authority of military men might even overthrow the civil power.
John Winthrop, 1588–1649

The raising or keeping of a standing army within the kingdom in time of peace, unless it be with consent of Parliament, is against law.
The English Bill of Rights, vi, December 1689

Mouths without hands; maintained at vast expense,
In peace a charge, in war a weak defense.
John Dryden: Cymon and Iphigenia, 1699

My life's amusements have been just the same
Before and after standing armies came.
Horace, 65–8 B.C., Satires, ii, 2

History affords abundant instances of established armies making themselves the masters of those countries which they were designed to protect.
Samuel Adams, 1722–1803

Altho' a *large* standing Army in time of Peace hath ever been considered dangerous to the liberties of a Country, yet a few Troops, under certain circumstances, are not only safe, but indispensably necessary.
George Washington: Sentiments on a Peace Establishment, 1783

Great peace establishments will, if we do not take care, prove our ruin.
Lord North: Letter to Lord Sandwich, 5 September 1772

A standing army may be likened to a standing member—an excellent assurance of domestic tranquility, but a dangerous temptation to foreign adventure.
Attributed to Elbridge Gerry: During the Constitutional Convention, 1787

A standing army is one of the greatest mischiefs that can happen.
James Madison: During debate on adoption of the Constitution, 1789

The distribution of our little army to distant garrisons . . . is the most eligible arrangement of that perhaps necessary evil that can be contrived. But I never want to see the face of one in our cities and intermingling with the people.
Albert Gallatin: Letter, c. 1801

I believe there was not a member in the Federal Convention who did not feel indignation at such an institution.
James Madison, 1751–1836 (Madison's reference is to a standing army.)

Standing armies can never consist of resolute robust men; they may be well disciplined machines, but they will seldom contain men under the influence of strong passions, or with very vigorous faculties.
Mary Wollstonecraft, d. 1797

The Greeks by their laws, and the Romans by the spirit of their people, took care to put into the hands of the rulers no such engine of oppression as a standing army.
Thomas Jefferson: Letter to Thomas Cooper, 1814

It is against sound policy for a free people to keep up large military establishments and standing armies in time of peace, both from the enormous expenses with which they are attended, and the facile means which they afford to ambitious and unprincipled rulers to subvert the government, or trample upon the rights of the people.
Joseph Story: Commentaries on the Constitution of the United States, 1833

A standing army can never be turned into a moral institution.
William Eward Gladstone, 1809–1898

The terrible power of a standing army may usually be exercised by whomever can control its leaders, as a mighty engine is set in motion by turning of a handle.
Winston Churchill: The River War, iii, 1899

(*See also* Army, Militarism, Pacifism.)

Standing Operating Procedure

Serve God daily, love one another, preserve your victuals, beware of fire, and keep good company.
Sir John Hawkins, 1532–1595, standing orders to his ships

One should, once and for all, establish standard combat procedures known to the troops, as well as to the general who leads them.
Maurice de Saxe: Mes Rêvèries, xxxi, 1732

Thus there arises a certain methodism in warfare to take the place of art, wherever the latter is absent.
Clausewitz: Principles of War, 1812

Fundamental princples of action against different arms must be laid down so definitely that complicated orders in each particular case will not be required.
Friederich von Bernhardi, 1849–1930

Steadfastness

The race is not to the swift, nor the battle to the strong.
Ecclesiastes, IX, 11

The boy stood on the burning deck
Whence all but he had fled;
The flame that lit the battle's wreck,
Shone round him o'er the dead.
Felicia Hemans: Casabianca, 1826 (Casabianca, the boy, 10 years old, was the son of the Captain of the French ship, Orient, killed at the Battle of the Nile, 1 August 1798.)

(*See also* Resolution, Tenacity.)

Stonewall Brigade

I never found anything impossible with this Brigade.

Stonewall Jackson: To Brigadier General R.B. Garnett, CSA, 1862

Straggler

The straggler is generally a thief and always a coward, lost to all sense of shame; he can only be kept in ranks by a strict and sanguinary discipline.

D.H. Hill: Action report after Antietam, November 1862

Stratagem

Victory always falls to the side having the last crown . . . Would it not be much better to save money by buying up the enemy's army whenever the occasion arose?

John Law: Letter, 1718 (Law was then Finance Minister of France.)

The art of war is divided between force and strategem. What cannot be done by force must be done by strategem.

Frederick The Great: Instructions for His Generals, xii, 1747

(*See also* Ruse de Guerre.)

Strategy

Supreme excellence consists of breaking the enemy's resistance without fighting.

Sun Tzu, 400–320 B.C., The Art of War

Those skilled in war bring the enemy to the field of battle and are not brought there by him.

Sun Tzu, 400–320 B.C., The Art of War

The most complete and happy victory is this: to compel one's enemy to give up his purpose, while suffering no harm oneself.

Belisarius, 505–565

In planning, never a useless move; in strategy, no step taken in vain.

Ch'ên Hao, fl. 8th century A.D.

Success in war is obtained by anticipating the plans of the enemy, and by diverting his attention from our own designs.

Francesco Guicciacardini, 1483–1540

Few sieges and many combats.

Marshal Turenne, 1611–1675 (Turenne's favorite maxim)

It is better to lose a province than split the forces with which one seeks victory.

Frederick The Great, 1712–1786

Move upon your enemy in one mass on one line so that when brought to battle you shall outnumber him, and from such a direction that you compromise him.

Napoleon I, 1769–1821

In war, as in love, we must achieve contact ere we triumph.

Napoleon I: Political Aphorisms, 1848

[Strategy] means the combination of individual engagements to attain the goal of the campaign.

Clausewitz: Principles of War, 1812

The theory of warfare tries to discover how we may gain a preponderance of physical forces and material advantages at the decisive point. As this is not always possible, theory also teaches us to calculate moral factors.

Clausewitz: Principles of War, 1812

. . . the art of the employment of battles as a means to gain the object of the war.

Clausewitz: On War, 1832

Strategy is . . . the art of making war upon the map, and comprehends the whole of the theater of operations.

Jomini: Précis de l'Art de la Guerre, 1838

Always mystify, mislead, and surprise the enemy.

Stonewall Jackson, 1824–1863

. . . the practical adaptation of the means placed at a general's disposal to the attainment of the object in view.

Helmuth von Moltke ("The Elder"), 1800–1891

He who writes on strategy and tactics should force himself to teach an exclusive national strategy and tactics—which are the only ones liable to benefit the nation for whom he is writing.

Colmar von der Goltz: The Nation in Arms, 1883

What is grand strategy? Common sense applied to the art of war.

W.T. Sherman: Speech at Portland, Maine, 3 July 1890

As in a building, which, however fair and beautiful the superstructure, is radically marred and imperfect if the foundation be insecure—so, if the strategy be wrong, the skill of the general on the battlefield, the valor of the soldier, the brilliancy of victory, however otherwise decisive, fail of their effect.

Mahan: Naval Administration and Warfare, 1908

A victory on the battlefield is of little account if it has not resulted either in breakthrough or encirclement. Though pushed back, the enemy will appear again on different ground to renew the resistance he momentarily gave up. The campaign will go on.

Alfred von Schlieffen, 1833–1913

Make the right wing strong!

Alfred von Schlieffen: Last words, 1913 (referring to the 1905 Plan of the German General Staff for encirclement of the French armies)

The soundest strategy is to postpone operations until the moral disintegration of the enemy renders the delivery of the mortal blow both possible and easy.

V.I. Lenin, 1870–1924

General strategy is the art of controlling, in time of war or peace, the sum total of the powers of a nation. It welds them together and makes them a homogeneous unit in the service of a single will. It transcends and disciplines all the special strategies—political, land, sea, air, economic, colonial, moral, etc.

Raoul Castex, 1937

The management of operations so as to determine the times, areas, and results of campaigns in order to win the war is called grand strategy.

Quincy Wright: A Study of War, 1942

The highest type of strategy—sometimes called grand strategy—is that which so integrates the policies and armaments of the nation that resort to war is either rendered unnecessary or is undertaken with the maximum chance of victory.

Edward Meade Earle: Makers of Modern Strategy, 1944

Grand strategy must always remember that peace follows war.

B.H. Liddell Hart: Thoughts on War, ii, 1944

In Napoleon, the Power Age found its prophet . . . Its Koran was written by Karl von Clausewitz . . . It has become the war creed of all nations.

J.F.C. Fuller: Armament and History, 1945

. . . the art of distributing and applying the military means to fulfill the ends of policy.

B.H. Liddell Hart: Strategy, 1954

The true aim is not so much to seek battle as to seek a strategic situation so advantageous that if it does not of itself produce the decision, its continuation by a battle is sure to achieve this.

B.H. Liddell Hart: Strategy, 1954

The theory of war and strategy is the core of all things.

Mao Tse-tung: Problems of War and Strategy, 1954

The principles of strategy are simple. Their application is immensely difficult. A strategic doctrine, necessary as it may be . . . can never be applied to all situations.

Robert Strausz-Hupé: To the National Military-Industrial Conference, Chicago, February 1958

The notion of strategy implies an organized authority capable of sustained action along lines of policy.

Paul H. Nitze, Address to the Army War College, 27 August 1958

Against a brave and well-led enemy all war hinges on the ability to destroy the enemy's main armed forces in pitched battle, or, of course, to subvert his will to fight at all.

John Masters: The Road Past Mandalay, 1961

(*See also* Generalship.)

Strength

Even the bravest cannot fight beyond his strength.

Homer: The Iliad, xiii, c. 1000 B.C.

In war, numbers alone confer no advantage. Do not advance relying on sheer military power.

Sun Tzu, 400–320 B.C., The Art of War, ii

"I wish," quoth my uncle Toby, "you had seen what prodigious armies we had in Flanders."

Laurence Sterne: Tristram Shandy, 1760

I am not strong enough even to get beaten.

Marquis De La Fayette: To Washington, Virginia, 1781

The first rule is to enter the field with an army as strong as possible.

Clausewitz: On War, 1832

The best strategy is always to be strong.

Clausewitz: On War, 1832

I have the best possible reason for knowing the strength of the Confederate Army to be one million of men, for whenever one of our generals engages a rebel army he reports that he has encountered a force twice his strength. Now I know we have half a million soldiers, so I am bound to believe that the rebels have twice that number.

Abraham Lincoln: To a deputation from New England 1862

The cult of numbers is the supreme fallacy of modern warfare.

B.H. Liddell Hart: Thoughts on War, 1944

As important as having strength is being known to have it.

McGeorge Bundy: In Foreign Affairs, 1964

Take care that you be strong. (Cura ut valeas.)

Roman Maxim

Stupidity

Stupidity in an officer is a permanent and total disability. It's the unforgivable fault.

Admiral R. K. Turner, USN: Letter to Colonel Warren T. Clear, 1961

Submarine Warfare

He goes a great voyage that goes to the bottom of the sea.

George Herbert: Jacula Prudentum, 1651

. . . a mode of warfare which they who commanded the seas did not want, and which if successful would deprive them of it.

Lord St. Vincent: Comment on William Pitt's negotiations with Robert Fulton for construction of a submarine, October 1805

The submarine is always in a fog.

Sir Charles Beresford, 1846–1919

Swift flame—then shipwrecks only
Beach in the ruined light;
Above them reach up lonely
The headlands of the night.

Frederic Ridgely Torrence, 1875–1950

Stumbling we grope and stifle here below
In the gross garb of this too cumbering flesh,
And draw such hard-won breaths as may be drawn,
Until, perchance, with pearls, we rise and go
To doff our diver's mail and taste the fresh,
The generous winds of the eternal dawn.

Robert Haven Schauffler, 1879–1945

Like the destroyer, the submarine has created its own type of officer and man—with language and traditions apart from the rest of the Service, and yet at heart unchangingly of the Service.

Rudyard Kipling: The Fringes of the Fleet, 1915

Sunk without trace. (Spurlos versenckt.)

Count Luxburg: Recommendation, as German chargé d'affaires, Buenos Aires, with regard to Argentine shipping carrying Allied cargo, 1917

To the end that prohibition of the use of submarines as commerce destroyers shall be accepted universally as part of the law of nations, the signatory powers herewith accept that prohibition as binding between themselves, and invite all other nations to adhere thereto.

Washington Naval Treaty, 6 February 1922

It was like a dark stain spreading all over the huge sea: the area of safety dimished, the poisoned water, in which no ship could count on safety from hour to hour, seemed swiftly to infect a wider and wider circle.

The Bettman Archive

HELMUTH VON MOLTKE

1800–1891

"War forms part of the order of things instituted by God."

Nicholas Monsarrat, 1910–, The Cruel Sea

The atomic bomb was the funeral pyre of an enemy who had been drowned.
Theodore Roscoe: Submarine Operations in World War II, 1949

The submarine is the weapon of the future . . . The submarine alone can assure command of the sea.
Admiral Georges Cabanier: In Revue Maritime, October 1961

(*See also* Antisubmarine Warfare.)

Subsistence

Few victories are won on an empty belly.
Sir John Hawkwood, 1320–1394

If you want the soldiers to perform great fatigue-duties and do not furnish anything to sustain them, it will come about that, human bodies not being made out of iron, they will leave you on the road, or if you come to battle they will be so weak that they can serve you only very little. But if you carry refreshments with you and accompany them with remonstrances, you will not only make them march but run if you desire.
Blaise Montluc: Commentaires, 1592

The stomach carries the feet.
Cervantes, 1547–1616

Understand that the foundation of an army is the belly. It is necessary to procure nourishment for the soldier wherever you assemble him and wherever you wish to conduct him. This is the primary duty of a general.
Frederick The Great: Instructions for His Generals, iv, 1747

As it is the business of a good non-commission officer to be active in taking up all deserters, when on the march, or at any other time, you observe any ducks, geese, or fowls that have escaped the bounds of their confinement, immediately apprehend them, and take them along with you, that they may be tried for their offense at a proper season. This will prevent the soldiers from marauding.
Francis Grose: Advice to the Officers of the British Army, 1782

Articles of provision are not to be trifled with, or left to chance; and there is nothing more clear than that the subsistence of the troops must be certain upon the proposed service, or the service must be relinquished.
Wellington: Despatch, 18 February 1801

An army crawls on its stomach.
Count Johann Radetsky, 1766–1858

An army travels on its stomach.
Attributed to Napoleon I, 1769–1821

Generals should mess with the common soldiers. The Spartan system was a good one.
Napoleon I: To General Gaspard Gourgaud, St. Helena, 1818

Biscuits make war possible.
Attributed to Napoleon I, 1769–1821

Where a million of people find subsistence, my army won't starve.
W.T. Sherman: Letter to U.S. Grant from Atlanta, 20 September 1864

The soup makes the soldier.
French Proverb

(*See also* Cook, Mess Sergeant, Rations.)

Success

Whosoever desires constant success must change his conduct with the times.
Niccolo Machiavelli: Discorsi, ix, 3, 1531

'Tis not in mortals to command success,
But we'll do more, Sempronius, we'll
deserve it.
Joseph Addison: Cato, i, 2, 1713

Although I cannot insure success, I will endeavor to deserve it.
John Paul Jones: Letter, 1780

I am sure you will deserve success. To mortals is not given the power of commanding it.
Lord St. Vincent: Letter to Nelson before sailing against Tenerife, 14 July 1797

Success in war, like charity in religion, covers a multitude of sins.
Sir William Napier, 1785–1860

The Service cannot afford to keep a man who does not succeed.
Stonewall Jackson: Letter, 1861

I shall expect nothing short of success.
P.H. Sheridan: Order to his division commanders prior to moving against Stuart, 8 March 1864

Success, like charity, covers a multitude of sins.
Mahan: Naval Strategy, 1911 (cf. Napier, above)

Success has a soporiferous influence on generalship.
J.F.C. Fuller: Decisive Battles, xx, 1939

Sudanese

So 'ere's to you, Fuzzy-Wuzzy, at your 'ome in the Soudan;
You're a pore benighted 'eathen, but a first-class fightin' man.
Rudyard Kipling: Fuzzy-Wuzzy, 1890

Superannuation

. . . the scandal of old men at war and old men in love—but at what age a general ceases to be a danger to the enemy and a Don Juan, is not easy to determine.
Sir Archibald Wavell, 1883–1950

Superiority

One who has few must prepare against the enemy; one who has many makes the enemy prepare against him.
Sun Tzu, 400–320 B.C., The Art of War, vi

And so it is certain that a small country cannot contend with a great, that few cannot contend with many; that the weak cannot contend with the strong.
Mencius, 370–290 B.C.

God is ordinarily for the big battalions against the little ones.
Roger Bussy-Rabutin: Letter, 18 October 1677

When men are equally inured and disciplined in war, 'tis, without a miracle, number that gains the victory.
Admiral Sir Cloudesley Shovell, 1650–1707

It is not big armies that win battles; it is the good ones.
Maurice de Saxe: Mes Rêveries, iv, 1732

God is always on the side of the heaviest battalions.
Voltaire: Letter to M. de Riche, 6 February 1770 (cf. Bussy-Rabutin, above, 1677)

The most unjust war, if supported by the greatest force, always succeeds; hence the most just ones, when supported only by their justice, as often fail.
St. John de Crèvecoeur: Letters from an American Farmer, lx, 1782

We must get the upper hand, and if we once have that, we shall keep it with ease, and shall certainly succeed.
Wellington: Despatch, 17 August 1803

Numbers only can annihilate.
Nelson: Letter, 1804

The fundamental object in all military combinations is to gain local superiority by concentration.
Mahan: Naval Strategy, 1911

Supply

Without supplies neither a general nor a soldier is good for anything.
Clearchus of Sparta: Speech to the Greek army in Asia Minor, 401 B.C.

Without supplies no army is brave.
Frederick The Great: Instructions for His Generals, ii, 1747

What I want to avoid is that my supplies should command me.
Comte de Guibert: Essai General de la Tacticque, 1770

For want of a nail, the shoe was lost—for want of a shoe, the horse was lost—for want of a horse the rider was lost—for want of a rider the battle was lost.
Benjamin Franklin, 1706–1790, Poor Richard's Almanac

Seeing what had been thrown away, I wondered how the battle had been fought.
Remark by a Union officer after Cold Harbor, June 1864

Whatever isn't nailed down is mine.
Whatever I can pry loose isn't nailed down.
Ascribed to Collis P. Huntington, 1821–1900

Supply cannot be achieved without command.
Winston Churchill: Memorandum, 6 June 1935

Mobility is the true test of a supply system.
B.H. Liddell Hart: Thoughts on War, iv, 1944

The onus of supply rests equally on the giver and the taker.
George S. Patton, Jr.: War As I Knew It, 1947

(*See also* Logistics.)

Support

Every unit that is not supported is a defeated unit.
Maurice de Saxe: Mes Rêveries, xiii, 1732

Surgeon

War is the only proper school of the surgeon.
Hippocrates: Wounds of the Head, c. 415 B.C.

Honest, sober, and of good counsel, skillful in the science, able to heal all kinds of sores, wounds, and griefs, to take a pellet out of the flesh and bone, and to slake the fire of the same.
Description of a 16th century military surgeon by unknown contemporary

As for his performance on arms and legs, he does it after a way, 'tis true; but, betwixt you and me, the slaughter-house on Tower Hill would scarce grant him their journeyman's wages. The poorest patients are sure to fare best where he is, because he leaves them to nature, the less dangerous doctor of the two.
Ned Ward 1667–1731, The Wooden World (on the typical naval surgeon, 18th century)

Whenever you are ignorant of a soldier's complaint, you should first take a little blood from him, and then give him an emetic and a cathartic—to which you may add a blister. This will serve, at least, to diminish the number of your patients.
Francis Grose: Advice to the Officers of the British Army, 1782

You medical people will have more lives to answer for in the other world than even we generals.
Napoleon I: Letter to Barry E. O'Meara, St. Helena, 29 September 1817

Doctors is all swabs.
R.L. Stevenson: Treasure Island, 1883 (The speaker is Billy Bones.)

(*See also* Field Hospital, Health, Sanitation, Sick Call.)

Surprise

The execution of a military surprise is always dangerous, and the general who is never taken off his guard himself, and never loses an opportunity of striking at an unguarded foe, will be most likely to succeed in war.
Thucydides: History of the Peloponnesian War, iii, 404 B.C.

The enemy must not know where I intend to give battle. For if he does not know where I intend to give battle, he must prepare in a great many places . . . If he prepares to the front his rear will be weak, and if to the rear, his front will be fragile. If he prepares to the left, his right will be vulnerable and if to the right, there will be few on his left. And when he prepares everywhere he will be weak everywhere.
Sun Tzu, 400–320 B.C., The Art of War, vi

He who knows best how to manage an army is sudden in his movements; his plans are very deep-laid, and no one knows whence he may attack.
Hsun Tzu, 320–235 B.C., A Debate on Military Affairs

Go into emptiness, strike voids, bypass what he defends, hit him where he does not expect you.
Ts'ao Ts'ao, 155–220 A.D.

Everything which the enemy least expects will succeed the best.
Frederick The Great: Instructions for His Generals, xi, 1747

Any officer or non-commission officer who shall suffer himself to be surprised . . . must not expect to be forgiven.
James Wolfe: General Orders, Quebec Expedition, 1759

. . . nothing so pregnant with dangerous consequences, or so disgraceful to an Officer in arms, as a surprise.
Lord St. Vincent: Memorandum to the Fleet, off Cadiz, 1798

To be defeated is pardonable; to be surprised—never!
Napoleon I: Maxims of War, 1831

War is composed of nothing but surprises. While a general should adhere to basic principles, he should never miss an opportunity to profit by such surprises. It is the essence of genius. In war there is but one favorable moment; genius grasps it.
Napoleon I: Maxims of War, 1831

Always mystify, mislead, and surprise the enemy if possible.
Stonewall Jackson, 1824–1863

The officers and men who permit themselves to be surprised deserve to die, and the commanding general will spare no efforts to secure them their deserts.
D.H. Hill: Address to troops on assumption of command in North Carolina, February 1863

Inaction leads to surprise, and surprise to defeat, which is after all only a form of surprise.
Ferdinand Foch: Precepts, 1919

Surprise—the pith and marrow of war!
Sir John Fisher: Memories, 1919

The impossible can only be overborne by the unprecedented.
Sir Ian Hamilton: Gallipoli Diary, I, 1920

Movement generates surprise, and surprise gives impetus to movement.
B.H. Liddell Hart: Article, "Strategy," Encyclopaedia Britannica, 1929 edition

War is the realm of the unexpected.
B.H. Liddell Hart: Defense of the West, 1950

It is extremely important to keep the enemy in the dark about where and when our forces will attack.
Mao Tse-tung, 1893–, On Protracted War

Human experience largely consists of
surprises
Superseding surmises.
Ogden Nash, 1902– , It Would Have Been Quicker to Walk

Hit 'em where they ain't.
"Wee Willie" Keeler, 1872–1923

We have inflicted a complete surprise on the enemy. All our columns are inserted in the enemy's guts.
Orde Wingate: Order of the Day to all ranks, 3d Indian Division, 11 March 1944

Surrender

The Guard dies; it does not surrender. (La Garde meurt et ne se rend pas.)
General Pierre de Cambronne: Of the Imperial Guard at Waterloo, 18 June 1815 (It is also widely asserted that his reply was simply, "Merde!," from which this expression is known in French society as "le mot de Cambronne.")

There is but one honorable mode of becoming prisoner of war. That is, by being taken separately; by which is meant, by being cut off, entirely, and when we can no longer make use of our weapons. In this case there can be no conditions, for honor can impose none. We yield to irresistible necessity.
Napoleon I: Maxims of War, 1831

To avoid one's own danger by making the situation more dangerous for others is clearly cowardice.
Napoleon I: Maxims of War, 1831

Tell him to go to hell.
Zachary Taylor: In reply to a demand for surrender from Santa Ana, at Buena Vista, 22 February 1847

Never surrender.
Sir Henry Lawrence: Death bed order to the garrison of Lucknow during the Indian Mutiny, July 1857 (Lawrence was mortally wounded during the defense.)

No terms except an unconditional and immediate surrender can be accepted. I propose to move immediately upon your works.

U.S. Grant: To Major General S.B. Buckner, CSA, at Fort Donelson, 16 February 1862

Then there is nothing left for me but to go and see General Grant, and I would rather die a thousand deaths.

R.E. Lee: At Appomattox Court House, after deciding to surrender the Army of Northern Virginia, 9 April 1865

We shall never surrender.

Winston Churchill: To the House of Commons, 4 June 1940

Nuts!

Major General Anthony C. Mc Auliffe USA: In reply to a German demand that he surrender his beleaguered airborne force, Bastogne, 23 December 1944.

Sutler

. . . I shall sutler be
Unto the camp, and profits shall accrue.

Shakespeare: King Henry V, ii, 1, 1598

It is hard enough to get necessities to the troops without giving space to their candies, pies, soft drinks and gee-gaws.

Major General M.C. Meigs, USA: Letter to Secretary of War Stanton, 1864 (of the sutler problem in the Union Army)

(*See also* Quartermaster.)

Sword

The sword itself often incites a man to fight.

Homer, Odyssey, xvi, c. 1000 B.C.

I will cause war to come. By the sword shalt thou live.

Genesis, XXVII, 40

I came not to send peace, but a sword.

Matthew, X, 34

All they that take to the sword shall perish with the sword.

Matthew, XXVI, 52

He that hath no sword, let him sell his garment and buy one.

Luke, XVI, 36

Our swords shall play the orators for us.

Christopher Marlowe: Tamburlaine the Great, i, 2, 1587

The swords of soldiers are his teeth, his fangs . . .

Shakespeare: King John, ii, 2, 1596

The air was cut by purple sabres,
Into solid ruby the blades were turned.

Mogul war song, India, 16th century, A.D.

One sword keeps another in its sheath.

George Herbert: Outlandish Proverbs 1640

They say they can obtain land and people for the King with the pen; but I say it can only be done with the sword.

Frederick William I of Prussia, 1688–1740

There are but two powers in the world, the sword and the mind. In the long run, the sword is always beaten by the mind.

Attributed to Napoleon I, 1769–1821

The sword does not plow deep.

Francis Lieber, 1800–1872

Beneath the rule of men entirely great
The pen is mightier than the sword.

Bulwer-Lytton, 1803–1873, Richelieu, ii

I let them take whate'er they would, but kept my father's sword.

Caroline Elizabeth Norton, 1808–1877, The Soldier of the Rhine

He knew me and named me
The War-Thing, the Comrade,
Father of honor, and giver of kingship,
The fame-smith, the song-master,
Bringer of women.

William Ernest Henley, 1849–1903, The Song of the Sword

Step by step, in the past, man has ascended by means of the sword.

Mahan: The Peace Conference, 1899

Until the world comes to an end the ultimate decision will rest with the sword.

Kaiser Wilhelm II: Speech in Berlin, 1913

The world continues to offer glittering prizes to those who have stout hearts and sharp swords.

Lord Birkenhead: Address, 7 November 1923

There is no refutation in God's truth and man's duty for the flash of a clean sword.

Frederick Brown Harris: In the Washington Star, 26 December 1955

The ore, the furnace, and the hammer are all that is needed for a sword.

Indian Proverb

(*See also* Weapons.)

Tactics

Know the enemy and know yourself; in a hundred battles you will never be in peril. When you are ignorant of the enemy but know yourself, your chances of winning or losing are equal. If ignorant both of your enemy and of yourself, you are certain in every battle to be in peril.

Sun Tzu, 400–320 B.C., The Art of War, iii

Now an army may be likened to water, for just as flowing water avoids the heights and hastens to the lowlands, so an army avoids strength and strikes weakness. And as water shapes its flow in accordance with the ground, so an army manages its victory in accordance with the situation of the enemy. And as water has no constant form, there are in war no constant conditions.

Sun Tzu, 400–320 B.C., The Art of War, vi.

Uproar in the East; strike in the West.

Sun Tzu, 400–320 B.C., The Art of War

Taking advantage of spots where [the enemy] is unprepared, make repeated sorties . . . When he comes to aid the right, attack his left; when he goes to succor the left, attack the right; exhaust him by causing him continually to run about.

Tu Mu, 803–852, Wai Liao Tzu

The nature of water is that it avoids heights and hastens to the lowlands . . . Now the shape of an army resembles water. Take advantage of the enemy's unpreparedness; attack him when he does not expect it; avoid his strength and strike emptiness, and, like water, none can oppose you.

Chang Yu, c. 1000

Within a single square mile a hundred different orders of battle can be formed. The clever general perceives the advantages of the terrain instantly; he gains advantage from the slightest hillock, from a tiny marsh; he advances or withdraws a wing to gain superiority; he strengthens either his right or left, moves ahead or to the rear, and profits from the merest bagatelles.

Frederick The Great: Instructions for His Generals, vi, 1747

Let no one imagine that it is sufficient merely to move an army about, to make the enemy regulate himself according to your movements. A general who has too presumptuous confidence in his skill runs the risk of being grossly duped. War is not an affair of chance.

Frederick The Great: Instructions for His Generals, xi, 1747

It is an invariable axiom of war to secure your own flanks and rear and endeavor to turn those of your enemy.

Frederick The Great: Instructions for His Generals, 1747

A general should show boldness, strike a decided blow, and maneuver upon the flank of his enemy. The victory is in his hands.

Napoleon I: Maxims of War, 1831

When you determine to risk a battle, reserve to yourself every possible choice of success, more particularly if you have to deal with an adversary of superior talent; for if you are beaten, even in the midst of your magazines and communications, woe to the vanquished!

Napoleon I: Maxims of War, 1831

It is an accepted maxim of war, never to do what the enemy wishes you to do, for this reason alone, that he desires it; avoid a battlefield he has reconnoitered and studied, and, with even more reason, ground that he has fortified and where he is entrenched.

Napoleon I: Maxims of War, 1831

A general-in-chief should ask himself frequently in the day, "What would I do if the enemy's army appeared now in my front, or on my right, or my left?" If he has any difficulty in answering these questions, his position is bad, and he should seek to remedy it.

Napoleon I: Maxims of War, 1831

Grand tactics is the art of posting troops upon the battlefield according to the characteristics of the ground, of bringing them into action, and of fighting them upon the ground.

Jomini: Précis de l'Art de la Guerre, 1838

To move swiftly, strike vigorously, and secure all the fruits of victory, is the secret of successful war.

Stonewall Jackson: Letter, 1863

Hit 'em where they ain't.

"Wee Willie" Keeler, 1872–1923

(*See also* Surprise.)

To remain separated as long as possible while operating and to be concentrated in good time for the decisive battle, that is the task of the leader of large masses of troops.

Helmuth von Moltke ("The Elder"),. Instructions for Superior Commanders of Troops, 1869

The problem is to grasp, in innumerable special cases, the actual situation which is covered by the mist of uncertainty, to appraise the facts correctly and to guess the unknown elements, to reach a decision quickly and then to carry it out forcefully and relentlessly.

Helmuth von Moltke ("The Elder"), 1800–1891

Tactics is an art based on the knowledge of how to make men fight with maximum energy against fear, a maximum which organization alone can give.

Ardant du Picq, 1821–1870, Battle Studies

Changes in tactics have not only taken place after changes in weapons, which necessarily is the case, but the interval between such changes has been unduly long. An improvement of weapons is due to the energy of one or two men, while changes in tactics have to overcome the inertia of a conservative class.

Mahan, 1840–1914

The unresting progress of mankind causes continual change in the weapons; and with that must come a continual change in the manner of fighting.

Mahan, 1840–1914

Unless very urgent reasons to the contrary exist, strike at one end rather than at the middle, because both ends can come up to help the middle against you quicker than one end can get to help the other; and, as between the two ends, strike at the one upon which the enemy most depends for reinforcements and supplies to maintain his strength.

Mahan: Sea Power in its Relations with the War of 1812, I, 1905

. . . The art of making good combinations preliminary to battles as well as during their progress.

An unattributed quotation by Alfred Thayer Mahan in the introduction to Influence of Sea Power upon History

The enemy's front is not the objective. The essential thing is to crush the enemy's flanks . . . and complete the extermination by attack upon his rear.

Alfred von Schlieffen: Cannae, 1913

Nine-tenths of tactics are certain, and taught in books: but the irrational tenth is like the kingfisher flashing across the pool and that is the test of generals. It can only be ensured by instinct, sharpened by thought practicing the stroke so often that at the crisis it is as natural as a reflex.

T.E. Lawrence, 1888–1935, The Science of Guerrilla Warfare

I rate the skilful tactician above the skilful strategist, especially him who plays the bad cards well.

Sir A.P. Wavell: Soldiers and Soldiering, 1939

In war the power to use two fists is an inestimable asset. To feint with one fist and strike with the other yields an advantage, but a still greater advantage lies in being able to interchange them—to convert the feint into the real blow if the opponent uncovers himself.

B.H. Liddell Hart: Thoughts on War, i, 1944

To pin an opponent is the vital prelude to a decisive maneuver; this dual act gives a double meaning to the old maxim—"divide and conquer."

B.H. Liddell Hart, 1895–

There is only one tactical principle which is not subject to change. It is: to use the means at hand to inflict the maximum amount of wounds, death and destruction on the enemy in the minimum of time.

George S. Patton, Jr.: War As I Knew It, 1947

There is no approved solution to any tactical situation.

George S. Patton, Jr.: War As I Knew It, 1947

The commander must decide how he will fight the battle *before it begins.* He must then decide how he will use the military effort at his disposal to force the battle to swing the way he wishes it to go; he must make the enemy dance to his

tune from the beginning, and never vice versa.
Montgomery of Alamein: Memoirs, vi, 1958

Any competent field officer, given the full intelligence possessed by military historians, could evolve maneuvers to defeat all the great commanders of history.
Jac Weller: Wellington in the Peninsula, xiv, 1961

March divided and fight concentrated.
Military maxim

Find, fix, fight, follow, finish.
Military maxim

It all depends on the terrain and the situation.
Traditional instructors' cliche, U.S. service schools

(*See also* Generalship.)

Tamerlane (1336–1405)

We'll lead you to the stately tent of war,
Where you shall hear the Scythian Tamburlaine
Threatening the world with high astounding terms
And scourging kingdoms with his conquering sword.
Christopher Marlowe: Tamburlaine the Great, prologue, 1587

The God of War resigns his roume to me,
Meaning to make me Generall of the world;
Jove, viewing me in armes, looks pale and wan,
Fearing my power should pull him from his throne.
Christopher Marlowe: Tamburlaine the Great, 1587

Tank

. . . A pretty mechanical toy.
Lord Kitchener: After observing British tank tests, 1915

The tank marks as great a revolution in land warfare as an armored steamship would have marked had it appeared amongst the toilsome triremes of Actium.
Sir Ian Hamilton: The Soul and Body of an Army, xii, 1921

The Tank was the beginning of the bullet-proof army.
Winston Churchill: The World Crisis, II, 1923

Where tanks are, is the front . . . Wherever in future wars the battle is fought, tank troops will play the decisive role.
Heinz Guderian: Achtung! Panzer! 1937

With the development of tank forces the old linear warfare is replaced by circular warfare.
B.H. Liddell Hart: Thoughts on War, i, 1944

(*See also* Mechanized Warfare.)

Taps

High over all the lonely bugle grieves.
Henry A. Beers, 1847–1926, The Singer of One Song

The last post is the *Nunc Dimittis* of the dead soldier. It is the last bugle call . . . but it gives promise of reveille . . . the greatest reveille which ultimately the Archangel Gabriel will blow.
Stephen Graham: A Private in the Guards, 1919

Target

What mark is so fair as the breast of a foe?
Byron: Childe Harold's Pilgrimmage, ii, 1816

Taylor, Zachary (1784–1850)

I think I hear his cheerful voice,
"On column! Steady! Steady!"
So handy and so prompt was he,
We called him Rough and Ready.
Soldiers' ditty, Seminole Wars, c. 1837 ("Old Rough and Ready" became General Taylor's Army nickname.)

Tenacity

We fight, get beat, rise, and fight again.
Nathanael Greene: Letter regarding the campaign in the Carolinas, 22 June 1781

Brown Brothers

ERWIN ROMMEL

1891–1944

"The best form of 'welfare' for the troops is first-class training."

If the military leader is filled with high ambition and if he pursues his aims with audacity and strength of will, he will reach them in spite of all obstacles.

Clausewitz: Principles of War, 1812

J'y suis, j'y reste. (Here I am, here I stay.)

Marshal MacMahon: Reply before the Malakoff, Sebastopol, when warned of the possibility of the explosion of a Russian mine underfoot, September 1855

Tenacity of purpose and untiring energy in execution can repair a first mistake and baffle deeply laid plans.

Alfred Thayer Mahan, 1840–1914

Victory will come to the side that outlasts the other.

Ferdinand Foch: Order during the battle of the Marne, 7 September 1914

I knew my ground, my material and my allies. If I met fifty checks I could yet see a fifty-first way to my object.

T.E. Lawrence, 1888–1935

Oh yesterday our little troop was ridden through and through,
Our swaying, tattered pennons fled, a broken, beaten few,
And all a summer afternoon they hunted us and slew;
But tomorrow,
By the living God, we'll try the game again

John Masefield, 1878– , Tomorrow

I shall return. (On departure from Corregidor, 11 March 1942)

* * *

I have returned. (On landing at Leyte, 20 October 1944)

Douglas MacArthur, 1880–1964

Terrain

Those who do not know the conditions of mountains and forests, hazardous defiles, marshes and swamps, cannot conduct the march of an army. Those who do not use native guides are unable to obtain the advantages of the ground.

Sun Tzu, 400–320 B.C., The Art of War

The nature of the ground is the fundamental factor in aiding the army to set up its victory.

Mei Yao-Ch'en, 1002–1060

A general should possess a perfect knowledge of the localities where he is carrying on a war.

Niccolo Machiavelli: Discorsi, xl, 1531

In peace, soldiers must learn the nature of the land, how steep the mountains are, how the valleys debouch, where the plains lie, and understand the nature of rivers and swamps—then by means of the knowledge and experience gained in one locality, one can easily understand any other that it may be necessary to observe.

Niccolo Machiavelli, 1469–1527

The British may harass us and distress us, but the Carolinas alone can subdue us.

Nathanael Green: Letter 18 March 1781

An army can pass wherever a man can set foot.

Napoleon I, 1769–1821

There is in every battlefield a decisive point the possession of which, more than any other, helps to secure victory by enabling its holder to make a proper application of the principles of war.

Jomini: Précis de l'Art de la Guerre, 1838

With brave infantry and bold commanders mountain ranges can usually be forced.

Jomini: Précis de l'Art de la Guerre, 1838

Young officers of all Services must learn terrain or learn Russian.

Major General Alden K. Sibley, USA: Lecture at the Naval War College, 9 February 1962

Index the terrain with mental signposts from which you can maneuver or call for fire.

Major General Orlando Ward, USA: Letter to a Marine officer, May 1965

Theater of Operations

. . . the whole chess-table of war.

Jomini: Précis de l'Art de la Guerre, 1838

Thermopylae

Go tell the Spartans, thou that passeth by,
That here, obedient to the laws, we lie.
Simonides of Ceos, 556–458 B.C., epitaph for the Spartan soldiers who fell at Thermopylae

Every great crisis of human history is a pass of Thermopylae, and there is always a Leonidas and his three hundred to die in it, if they cannot conquer.
George William Curtis, 1824–1892, The Call of Freedom

Thin Red Line

The English, silent and impassive, with grounded arms, loomed like a long red wall.
Marshal Bugeaud: Of the British line at Talavera, where Bugeaud served as a junior officer, 28 July 1809.

The Russians dashed on toward that thin red-line streak tipped with a line of steel.
William Howard Russell: Dispatch to the London Times, describing the British infantry at Balaklava, 25 October 1854

Soon the men of the column began to see that though the scarlet line was slender, it was very rigid and exact.
Alexander Kinglake: Invasion of the Crimea, ii, 1863

But it's "Thin red line of 'eroes" when
the drums begin to roll . . .
Rudyard Kipling: Tommy, 1892

(*See also* Army, British.)

Time

Take all the swift advantage of the hours.
Shakespeare, 1564–1616

Forbear waste of Time, precious Time.
Oliver Cromwell, 1599–1658

In military operations, time is everything.
Wellington: Despatch, 30 June 1800

Time is everything; five minutes makes the difference between victory and defeat.
Nelson, 1758–1805

Lose not an hour.
Nelson: To the council of war off Copenhagen, 23 March 1801

Go, sir, gallop, and don't forget that the world was made in six days. You can ask me for anything you like, except time.
Napoleon I: To a staff officer, 1803

He who gains time gains everything.
Benjamin Disraeli, 1804–1881, Tancred, iv, 3

. . . too late? Ah! two fatal words of this war! Too late in moving here. Too late in arriving there. Too late in coming to this decision. Too late in starting with enterprises. Too late in preparing. In this war the footsteps of the Allied Forces have been dogged by the mocking specter of "Too Late"!
Lloyd George: To the House of Commons, 20 December 1915

Time is the essence in war, and while a defeat may be balanced by a battle won, days and hours—even minutes—frittered away, can never be regained.
Brigadier General S.B. Griffith, II, USMC: The Battle for Guadalcanal, ii, 1963

Time and Space Factors

Now those skilled in war must know where and when a battle will be fought. They measure the roads and fix the date. They divide the army and march in separate columns. Those who are distant start first, those who are nearby, later. Thus the meeting of troops from distances of a thousand *li* takes place at the same time. It is like people coming to a city market.
Tu Yü, 735–812 A.D.

The advantage of time and place in all martial actions is half a victory, which being lost is irrecoverable.
Sir Francis Drake: Letter to Elizabeth I, 1588

Our cards were speed and time, not hitting power, and these gave us strategical rather than tactical strength. Range is more to strategy than force.
T.E. Lawrence: "Guerrilla Warfare," Encyclopaedia Britannica, 1929

Timidity

Of all the dangers in employing troops, timidity is the greatest and all calamities which overtake an army arise from hesitation.

Wu Ch'i, 430–381 B.C.

. . . misplaced caution, more ruinous than the most daring venture.

Mahan: Life of Nelson, x, 1897

All men are timid on entering any fight whether it is the first fight or the last fight. All of us are timid. Cowards are those who let their timidity get the better of their manhood.

Geroge S. Patton, Jr.: Letter to Cadet George S. Patton, III, USMA, 6 June 1944

(*See also* Cowardice, Fear.)

Torpedoes

Damn the torpedoes! Captain Drayton, go ahead! Jouett, full speed!

David G. Farragut: During the battle of Mobile Bay, as the U.S. ships entered a minefield, 5 August 1864

Hit, and hard hit! The blow went home,
The muffled, knocking stroke—
The steam that overruns the foam—
The foam that thins to smoke—
The smoke that clokes the deep aboil—
The deep that chokes her throes
Till, streaked with ash and sleeked with oil,
The lukewarm whirlpools close!

Rudyard Kipling: The Destroyers, 1898

It may well be a hundred to one against a hit with a heavy torpedo upon a ship, but the chance is always there, and the disproportion is grievous. Like a hero being stung by a malarious mosquito!

Winston Churchill: Memorandum for the First Sea Lord, 21 October 1939

Total War

The whole art of war is being transformed into mere prudence, with the primary aim of preventing the uncertain balance from shifting suddenly to our disadvantage and half-war from developing into total war.

Clausewitz: On War, viii, 1832 (This appears to be the earliest appearance of the phrase, "total war.")

We are not only fighting hostile armies, but a hostile people, and must make young and old, rich and poor, feel the hard hand of war, as well as the organized armies.

W.T. Sherman, 1820–1891

The immediate object of fighting is to kill and to go on killing, until there is nothing left to kill.

Statement by unidentified French officer, 1914

The prevailing forms of social organization have given war a character of national totality—that is, the entire population and all the resources of a nation are sucked into the maw of war.

Giulio Douhet: Command of the Air, 1921

. . . Brutish mutual extermination.

Winston Churchill: My Early Life, 1930

Total war is not a succession of mere episodes in a day or week. It is a long drawn out and intricately planned business, and the longer it continues the heavier are the demands on the character of the men engaged in it.

George C. Marshall: Address at Trinity College, Hartford, Connecticut, 15 June 1941

Before we're through with them, the Japanese language will be spoken only in hell.

W.F. Halsey: Remark in 1943, during the Pacific War

Absolute war is a war in which the conductor allows the fighting instinct to usurp control of his reason.

B.H. Liddell Hart: Thoughts on War, ii, 1944

There is no such thing as total peace, but there is such a thing as total war and total annihilation of rights.

Max Ascoli: The Power of Freedom, 1949

The day of total war has passed . . . From now on limited military operations are the only ones which could conceivably serve any coherent purpose.

George F. Kennan: The Realities of American Foreign Policy, 1954

To try to win a war, to set victory as an aim, is pure madness, since total war with nuclear weapons will be fatal to both sides.
B.H. Liddell Hart: Defense or Deterrence, 1962

. . . the high end of the spectrum of conflict.
Robert S. McNamara: Speech in New York, 18 November 1963

(*See also* Atrocities, Horrors of War, Nuclear War, Scorched Earth, War.

Traditions

Remember the past: let the elder men among us emulate their own earlier deeds, and the younger who are the sons of those valiant fathers do their best not to tarnish the virtues of their race.
Pagondas of Thebes: Speech to the Boeotian army before the battle of Delium, 424 B.C.

Hold the traditions which ye have been taught.
II Thessalonians, II, 15

To give reputation to the army of any state, it is necessary to revive the discipline of the ancients, cherish and honor it, and give it life, so that in return it may give reputation to the state.
Niccolo Machiavelli: Discorsi, xviii, 1531

A nation's traditions are its wealth.
Stendhal (Henri Beyle): A Life of Napoleon, xxiii, 1818

Every trifle, every tag or ribbon that tradition may have associated with the former glories of a regiment should be retained, so long as its retention does not interfere with efficiency.
Colonel Clifford Walton, History of the British Standing Army, 1660–1700, 1894

The value of tradition to the social body is immense. The veneration for practices, or for authority, consecrated by long acceptance, has a reserve of strength which cannot be obtained by any novel device.
Mahan, 1840–1914, The Military Rule of Obedience

May we not who are partakers of their brotherhood claim that in a small way at least we are partakers of their glory? Certainly it is our duty to keep these traditions alive and in our memory, and to pass them on untarnished to those who come after us.
Rear Admiral Albert Gleaves, USN, 1859–1937

Urged by a bitterer shout within,
Men of the trumpets and the drums
Seek, with appropriate discipline,
That Glory, past the pit or wall
Which contradicts and stops the breath,
And with immortalizing gall
Builds the most stubborn things on death.
Oliver St. John Gogarty, 1878– , Marcus Curtius

It is characteristic of good soldiers to cherish any little peculiarity of uniform or equipment which differentiates their own Regiment or Corps from others.
Colonel Cyril Field, RMLI: Britain's Sea Soldiers II, 1924

Ships, men, and weapons change, but tradition, which can neither be bought nor sold, nor created, is a solid rock amidst shifting sands.
Sir Bruce Fraser, 1888–

It takes the Navy three years to build a ship. It would take three hundred to rebuild a tradition.
Sir Andrew Browne Cunningham: To his staff, disapproving the recommendation that the Royal Navy save its ships by retiring from Crete and abandon the Marines and soldiers ashore, May 1941

The spirit of discipline, as distinct from its outward and visible guises, is the result of association with martial traditions and their living embodiment.
B.H. Liddell Hart: Thoughts on War, v, 1944

Fortune is rightly malignant to those who break with the traditions and customs of the past.
Winston Churchill: Note to the Foreign Secretary, 23 April 1945

This modern tendency to scorn and ignore tradition and to sacrifice it to administrative convenience is one that wise men will resist in all branches of life, but more especially in our military life.
Sir Archibald Wavell: Address to the officers of the Canadian Black Watch, Montreal, 1949

However praiseworthy it may be to uphold tradition in the field of soldierly ethics, it is to be resisted in the field of military command.

Erwin Rommel: The Rommel Papers, ix, 1953

It should be the first duty and pride of a midshipman to learn and to conform to the customs and traditions of the Naval Service.

Regulations, U.S. Naval Academy

Trafalgar (21 October 1805)

The signal has been made that the Enemy's combined fleet are coming out of port . . . May God Almighty give us success over these fellows and enable us to get a Peace.

Nelson: Last letter to Lady Hamilton, unfinished, 19 October 1805 (delivered after Nelson's death; at the bottom of the sheet, Emma Hamilton wrote: "Oh miserable wretched Emma, Oh glorious and happy Nelson.")

Trafalgar was not only the greatest naval victory, it was the greatest and most momentous victory won either by land or by sea during the whole of the Revolutionary War. No victory, and no series of victories, of Napoleon produced the same effect upon Europe . . . Nelson's last triumph left England in such a position that no means remained to injure her.

Charles Alan Fyffe: History of Modern Europe, 1883

(*See also* Nelson.)

Training

To lead an untrained people to war is to throw them away.

Confucius: Analacts, xiii, c. 500 B.C.

We must remember that one man is much the same as another, and that he is best who is trained in the severest school.

Thucydides: History of the Peloponnesian Wars, i, c. 404 B.C.

For they had learned that true safety was to be found in long previous training, and not in eloquent exhortations uttered when they were going into action.

Thucydides: History of the Peloponnesian Wars, v, c. 404 B.C. (of the Lacadaemonian army at the battle of Mantinea, 418 B.C.)

No speech of admonition can be so fine that it will at once make those who hear it good men if they are not good already; it would surely not make archers good if they had not had previous practice in shooting; neither could it make lancers good, nor horsemen; it cannot even make men able to endure bodily labor, unless they have been trained to it before.

Cyrus The Younger of Persia, d. 401 B.C.

The Romans are sure of victory . . . for their exercises are battles without bloodshed, and their battles bloody exercises.

Josephus, 37–100 A.D.

If officers are unaccustomed to rigorous drilling they will be worried and hesitant in battle. If generals are not thoroughly trained they will inwardly quail when they face the enemy.

Wang Ling, c. 200 A.D.

The main end and design of all the care and pains that are bestowed in keeping up good order and discipline is to fit and prepare an army to engage an enemy in a proper manner.

Niccolo Machiavelli: Arte della Guerra, 1520

The finest edge is made with the blunt whetstone.

John Lyly: Euphues, 1579

Let such teach others who themselves excel.

Alexander Pope: Essay on Criticism, i, 1711

The troops should be exercised frequently, cavalry as well as infantry, and the general should often be present to praise some, to criticize others, and to see with his own eyes that the orders . . . are observed exactly.

Frederick The Great: Instructions for His Generals, v, 1747

To bring Men to a proper degree of Subordination, is not the work of a day, a month, or even a year.

George Washington: Letter to the President of Congress, 24 September 1776

A government is the murderer of its

citizens which sends them to the field uninformed and untaught, where they are to meet men of the same age and strength, mechanized by education and discipline for battle.

Henry ("Light Horse Harry") Lee, 1756–1818

Hardship, poverty, and want are the best school for a soldier.

Napoleon I: Maxims of War, 1831

A good general, a well organized system, good instructions, and severe discipline, aided by effective establishments, will always make good troops, independently of the cause for which they fight. At the same time, a love of country, spirit of enthusiasm, a sense of national honor, and fanaticism will operate upon young soldiers with advantage.

Napoleon I: Maxims of War, 1831

The battle of Waterloo was won on the playing fields of Eton.

Attributed to Wellington, c. 1825, but almost certainly not authentic (Another version is that he said, while watching a cricket match at Eton, "The battle of Waterloo was won here.")

Naval education and training lie at the foundation of naval success; and the power that neglects this essential element of strength will, when the battle is fought, find that its ships, however formidable, are but built for a more thoroughly trained and educated enemy.

Stephen R. Mallory: Annual Report (as Confederate Secretary of the Navy) to Jefferson Davis, 1864

It cannot be too often repeated that in modern war, and especially in modern naval war, the chief factor in achieving triumph is what has been done in the way of thorough preparation and training before the beginning of war.

Theodore Roosevelt: Graduation address, U.S. Naval Academy, June 1902

There must be developed in the men that handle [destroyers] that mixture of skill and daring which can only be attained if the boats are habitually used under circumstances that imply the risk of accident. The business of a naval officer is one which above all others, needs daring and decision.

William S. Sims, 1858–1936, endorsement (pre-World War I) on correspondence regarding destroyers

All a soldier needs to know is how to shoot and salute.

Attributed to John J. Pershing, 1860–1948

Untutored courage is useless in the face of educated bullets.

George S. Patton, Jr.: in Cavalry Journal, April 1922

If the exercise is subsequently discussed in the officers' mess, it is probably worth while; if there is argument over it in the sergeants' mess, it is a good exercise; while if it should be mentioned in the corporals' room, it is an undoubted success.

Sir A.P. Wavell: In Journal of the Royal United Service Institution, May 1933

In no other profession are the penalties for employing untrained personnel so appaling or so irrevocable as in the military.

Douglas MacArthur: Annual Report Chief of Staff, U.S. Army, 1933

Train in difficult, trackless, wooded terrain. War makes extremely heavy demands on the soldier's strength and nerves. For this reason make heavy demands on your men in peacetime.

Erwin Rommel: Infantry Attacks, i, 1937

The best form of "welfare" for the troops is first-class training.

Erwin Rommel: Rommel Papers, ix, 1953

Long training tends to make a man more expert in execution, but such expertness is apt to be gained at the expense of fertility of ideas, originality, and elasticity.

B.H. Liddell Hart: Defense of the West, 1950

War is highly competitive; we are trying to train people to endure the hardships and strain of war and we would be doing ourselves and our country a disservice to adopt measures which would soften the fibre of men in uniform.

Admiral Robert B. Carney, USN, 1895– , remarks while Chief of Naval Operations to Navy officers stationed in Washington.

The more you sweat in peace, the less you bleed in war.
Chinese Proverb

Flog two to death, and train one.
Russian Proverb

(*See also* Drill Instructor, Recruit Training.)

Transportation

Build no more fortresses, build railways.
Helmuth von Moltke ("The Elder"), 1801–1891

Victory is the beautiful, bright-colored flower. Transport is the stem without which it could never have blossomed.
Winston Churchill: The River War, viii, 1899

You can't have any more of anything than you can haul.
Colonel J. Monroe Johnson, USA, d. 1964

(*See also* Logistics, Shipping.)

Treason

Treason doth never prosper: what's the reason?
For if it prosper, none dare call it treason.
John Harington, 1561–1612, Epigrams

Nothing can excuse a general who takes advantage of the knowledge acquired in the service of his country, to deliver up her frontier and her towns to foreigners. This is a crime reprobated by every principle of religion, morality, and honor.
Napoleon I: Maxims of War, 1831

Trench Warfare

. . . the misery of the soaking trenches.
John Masefield, 1878– , August 1914

I knew a simple soldier boy
Who grinned at life in empty joy,
Slept soundly through the lonesome dark,
And whistled early with the lark.

In winter trenches, cowed and glum,
With crumps and lice and lack of rum,
He put a bullet through his brain.
No one spoke of him again.
Siegfried Sassoon: Pray God, 1918

The front was a Moloch that consumed bodies, but souls were often tempered in its fire.
B.H. Liddell Hart: Lawrence, 1934

You cannot see the enemy, but you know he is there and that it is wiser to keep your head down.
Lawrence Durell: Balthazar, i, 3, 1958

(*See also* Western Front, World War I)

Trenton (26 December 1776)

Necessity, dire necessity, will, nay must, justify my attack.
George Washington: During final preparations for the attack, 25 December 1776

All our hopes were blasted by the unhappy affair at Trenton.
Lord George Germain: To the House of Lords, January 1777

Troubridge, Thomas (1758–1807)

Look at Troubridge there! He tacks his ship to battle as if the eyes of all England were upon him.
Sir John Jervis (Lord St. Vincent): At the opening of the battle of Cape St. Vincent, 14 February 1797

Truce

When, without a previous understanding, the enemy asks for a truce, he is plotting.
Sun Tzu, 400–320 B.C., The Art of War, ix

(*See also* Negotiations.)

Trumpet

If ye go to war . . . ye shall blow an alarm with the trumpet.
Numbers, X, 9

And it shall come to pass, that when they make a long blast with the ram's horn, and when ye hear the sound of the trumpet,

all the people shall shout with a great shout and the wall of the city shall fall down flat.
Joshua, VI, 5

Blow ye the cornet in Gibeah, and the trumpet in Ramah.
Hosea, V, 8

If the trumpet give an uncertain sound, who shall prepare himself to the battle?
I Corinthians, XIV, 8

With harsh-resounding trumpets' dreadful bray . . .
Shakespeare: King Richard II, i, 3, 1595

Sound trumpets, and set forward combatants.
Shakespeare: King Richard II, i, 3, 1595

Trumpet, blow loud,
Send thy brass voice through all these lazy tents.
Shakespeare: Troilus and Cressida, i, 3, 1601

Make all our trumpets speak; give them all breath,
Those clamorous harbingers of blood and death.
Shakespeare: Macbeth, v, 6, 1605

So he passed over, and all the trumpets sounded for him on the other side.
John Bunyan: The Pilgrim's Progress, 1678

The silver, snarling trumpets 'gan to chide.
John Keats: The Eve of St. Agnes, 1820

He has sounded forth his trumpet that shall never call retreat . . .
Julia Ward Howe: Battle Hymn of the Republic, 1862

Tsu-Shima (27 May 1905)

The rise or fall of the Empire depends on today's battle. Let every man do his utmost.
Admiral Heihachiro Togo, IJN: Togo's famous "Z-Signal" to the Japanese Fleet

Unconditional Surrender

No terms except an unconditional and immediate surrender can be accepted. I propose to move immediately upon your works.

U.S. Grant: To Major General S.B. Buckner, CSA, Fort Donelson, 16 February 1862

We, the United Nations, demand from the Nazi, Fascist and Japanese tyrannies unconditional surrender. By this we mean that their will power to resist must be completely broken, and that they must yield themselves absolutely to our justice and mercy.

Winston Churchill: Speech in the Guildhall, London, 30 June 1943 (The policy of unconditional surrender had been agreed to by Churchill and Roosevelt in conference on 20 January 1943.)

. . . These two words—a putrefying albatross around the necks of America and Britain.

J.F.C. Fuller, 1878–1966

(*See also* Surrender.)

Underdeveloped Countries

Nations are never so grateful as their benefactors expect.

Wellington: Letter to Canning, 15 December 1814

Becky Sharp's acute remark that it is not difficult to be virtuous on ten thousand a year has its application to nations; and it is futile to expect a hungry and squalid population to be anything but violent and gross.

Thomas H. Huxley: Joseph Priestley, 1874

In planning a war against an uncivilized nation . . . your first objective should be the capture of whatever they prize most, and the destruction and deprivation of which will probably bring the war most rapidly to a conclusion.

Sir Garnet Wolseley, 1833–1913

Take up the White Man's burden—
The savage wars of peace—
Fill full the mouth of Famine
And bid the sickness cease;
And when your goal is nearest
The end for others sought,
Watch Sloth and heathen Folly
Bring all your hope to nought.

Rudyard Kipling: The White Man's Burden, 1899

Up to a point the underdeveloped races can copy; certainly up to the point where we stand now. We have to carry on into regions where they cannot follow us without themselves becoming so civilized that their armies will no longer be a menace.

Sir Ian Hamilton: The Soul and Body of an Army, viii, 1921

Is it progress if a cannibal uses a knife and fork?

Stanislaus J. Lec, 1909–

Independence may be a joy to the overfed politician in the capital but to the man in the bush it means murder, pillage, and rape. If he is lucky it means none of these things, but merely poverty and loneliness.

Major Richard Lawson: Strange Soldiering, 1963

Underdevelopment carries its own kind of invulnerability.

Bernard Fall: In Current History, December 1964 (on the inefficacy of strategic air warfare against North Vietnam)

(*See also* Counterinsurgency.)

Unforeseen Factors

In war we must always leave room for strokes of fortune, and accidents that cannot be foreseen.

Polybius: Histories, ii, c. 125 B.C.

When I am without orders, and unexpected occurrences arise, I shall always act as I think the honor and glory of my King and Country demand.

Nelson: Letter to Hugh Elliott, November 1804

(*See also* Chance, Luck.)

Unification

Here is such a controversy between the Sailors and the Gentlemen, and such stomaching between the Gentlemen and Sailors, that it doth even make me mad to hear it. But, my Masters, I must have it left, For I must have the Gentleman to haul and draw with the mariner, and the Mariner with the Gentleman.

Sir Francis Drake, 1540–1596 (The "Gentlemen" in this passage were embarked "Gentlemen-at-Arms," i.e., soldiers.)

The remedy is worse than the disease.

Francis Bacon, 1561–1626, essays (Of Seditions and Troubles)

That their cause is but one, and they both can unite,
Needs no other example than this to be seen;
Who is bolder in danger, experter in fight,
Than that Maritime Soldier, the Honest MARINE?
He pulls and he hauls,
He fights till he falls,
And from fore-tack and musket he never will waver;
But, when the fray's o'er,
With his Dolly on-shore,
Drinks the Navy and Army of Britain forever.

Song, The Navy and Army of Britain, 1795

I should as soon expect a musket to swim, as expect a good understanding between seamen and soldiers.

Letter by unknown officer, Walcheren Expedition, 1809

We are intended to seek and fight the enemy's fleet, and I shall not be diverted from my efforts by any sinister attempt to render us subordinate to, or an appendage of, the Army.

Commodore Isaac Chauncey, USN: Letter to Major General Jacob Brown, USA, on Lake Ontario, 1813

One of the tragedies of unification is that there are not, at the top, men who really know enough about each of the Services to evaluate all of those Services.

Hanson W. Baldwin: in the New York Times, 1949

Our military forces are one team—in the game to win regardless of who carries the ball. This is no time for "Fancy Dans" who won't hit the line with all they have on every play, unless they can call the signals. Each player on this team—whether he shines in the spotlight of the backfield or eats dirt in the line—must be an all-American.

General Omar N. Bradley, USA: Statement to the House Armed Services Committee, 19 October 1949

Uniform

I have an insuperable bias in favor of an elegant uniform and a soldierly appearance, so much so that I would rather risk my life and reputation at the head of the same men in an attack, clothed and appointed as I could wish, merely with bayonets and a single charge of ammunition, than to take them as they appear in common with sixty rounds of cartridges.

Anthony Wayne: To George Washington, 1776

A well dressed soldier has more respect for himself. He also appears more redoubtable to the enemy and dominates him; for a good appearance is itself a force.

Joseph Joubert, 1754–1824

I think it indifferent how a soldier is clothed, provided it is in a uniform manner; and that he is forced to keep himself clean and smart, as a soldier ought to be.

Wellington: Letter to the War Office from Portugal, 1811

A soldier must learn to love his profession, must look to it to satisfy all his tastes and his sense of honor. That is why handsome uniforms are useful.

Napoleon I: To General Gaspard Gourgaud, St. Helena, 1815

A good uniform must work its way with the women, sooner or later.

Charles Dickens: Pickwick Papers, xxxvii, 1837

The better you dress a soldier, the more highly he will be thought of by women.

Sir Garnet Wolseley: Soldier's Pocket-Book, 1869

The secret of uniform was to make a crowd solid, dignified, impersonal: to give it the singleness and tautness of an upstanding man.

T.E. Lawrence : Revolt in the Desert, xxxv, 1927

This death's livery which walled its bearers from ordinary life, was sign that they had sold their wills and bodies to the State.

T.E. Lawrence, 1888–1935

It is proverbial that well dressed soldiers are usually well behaved soldiers.
John A. Lejeune: Reminiscences of a Marine, 1930

Soldiers, in the main, are not impressive clad in the drabness, the futilities of civilian dress.
Joseph Hergesheimer: Sheridan, 1931

Uniform Code of Military Justice

The first thing we do, let's kill all the lawyers.
Shakespeare: II King Henry VI, iv, 2, 1591

(*See also* Court Martial, Law, Law Specialist, Sea Lawyer.)

Unity of Command

An army should have but one chief; a greater number is detrimental.
Niccolo Machiavelli: Discorsi, xv, 1531

Nothing is so important in war as an undivided command.
Napoleon I: Maxims of War, 1831

It is not a question of one general being better than another, but of one general being better than two.
Lloyd George, 1863–1945

Little is done where many command.
Dutch Proverb

One bad general is better than two good ones.
French Proverb

Universal Military Training

Shall your brethren go to war, and ye sit here?
Exodus, XXXII, 6

Cease to hire your armies. Go yourselves, every man of you, and stand in the ranks.
Demosthenes: Third Philippic, 341 B.C.

Mouths without hands; maintained at vast expense,
In peace a charge, in war a weak defense.
John Dryden: Cymon and Iphigenia, 1699

Every citizen [should] be a soldier. This was the case with the Greeks and Romans, and must be that of every free state.
Thomas Jefferson: Letter to James Monroe, 1813

In the countries where everyone is a soldier, everyone is a bad soldier.
Louis Adolphe Thiers, 1797–1877

. . . An untried Swiss system.
Sir Ian Hamilton: The Soul and Body of an Army, iv, 1921

Military training forms generations which obey, not because they are ordered, and that fight because it is their desire.
Benito Mussolini, 1883–1945

The backbone of our military force should be the trained citizen who is first and foremost a civilian, and who becomes a soldier or a sailor only in time of danger—and only when the Congress considers it necessary . . . In such a system, however, the citizen reserve must be a trained reserve. We can meet the need for a trained reserve in only one way—by universal training.
Harry S. Truman: Message to Congress, 23 October 1945

(*See also* Conscript, Conscription, Manpower, Militia, Reservists.)

Valley Forge (Winter, 1777–1778)

You might have tracked the army from White Marsh to Valley Forge by the blood of their feet.
George Washington, 1732–1799

Valor

The Valiant profit more
Their country than the finest, cleverest speakers.
Valor once known will soon find eloquence
To trumpet forth her praise.
Plautus, 254–184 B.C.

In valor there is hope.
Tacitus, 55–117 A.D., Annals, ii

Valor is the contempt of death and pain.
Tacitus, 55–117 A.D.

Valor is superior to numbers.
Vegetius: De re Militari, iii, 378

The better part of valor is discretion.
Shakespeare: I King Henry IV, v, 4, 1597

Cowards die many times before their deaths;
The valiant never taste of death but once.
Shakespeare: Julius Caesar, ii, 2, 1599

'Tis held that valor is the chiefest virtue, and most dignifies the haver.
Shakespeare: Coriolanus, ii, 2, 1607

No thought of flight,
None of retreat, no unbecoming deed
That argued fear; each on himself relied,
As only in his arm the moment lay
Of Victory.
Milton: Paradise Lost, vi, 1667

He that is valiant and dare fight,
Though drubbed, can lose no honor by't.
Samuel Butler: Hudibras, i, 1663

It is not always from valor that men are brave, nor from virtue that women are chaste.
François de la Rochefoucauld: Maxims, 1665

Valor, among private soldiers, is a dangerous trade which they follow in order to earn their living
Francçis de la Rochefoucauld: Maxims, 1665

No man can answer for his own valor or courage until he has been in danger.
François de la Rochefoucauld: Maxims, 1665

If valor can make amends for the want of numbers, we shall probably succeed.
James Wolfe: Letter to Pitt the Elder from Halifax, 1759

For Valor.
Inscription on the Victoria Cross, instituted 29 January 1856

Among the men who fought on Iwo Jima, uncommon valor was a common virtue.
C.W. Nimitz: Pacific Fleet communique, March 1945

By valor, not by trickery. (Animo, non astutia.)
Latin Proverb

(*See also* Audacity, Bravery, Courage, Daring, Heroism.)

Van Tromp, Cornelius (1629–1691)

Van Tromp was an Admiral bold,
The Dutchman's pride was he,
And he cried, "I'll reign on the rolling main,
As I do on the Zuyder Zee!"
Naval song, 19th century

Vegetius (Flavius Vegetius Renatus) fl. 378

A god, said Vegetius, inspired the legion, but for myself, I find that a god inspired Vegetius.
Prince de Ligne: Letter, 1770

Verdun (1916)

They shall not pass. (On ne passe pas.)
Henri Philippe Petain: To General de Castelnau, at Verdun, 26 February 1916 (This remark became the watch-word of the defenders of Verdun.)

Verdun has become a battle of madmen inside a volcano.
Statement by a French officer, 1916

Veteran

One who in the past had suffered much in the wars and from the waves; now he slept at peace forgetful of what he had suffered.
Homer: The Odyssey, xiii, c. 1000 B.C.

He also made other laws, one of which provides that those who are maimed in war shall be maintained at the public charge.
Plutarch, 46–120 A.D., Lives (Solon) (of Peisistratus)

He that outlives this day, and comes safe home,
Will stand a tip-toe when this day is named.
Shakespeare: King Henry V, iv, 3, 1598

Hacked, hewn with constant service, thrown aside,
To rust in peace, and rot in hospitals.
Thomas Southerne: The Loyal Brother, 1682

As long as there are a few veterans, you can do what you want with the rest.
Maurice de Saxe: Mes Rêveries, vii, 1732

The broken soldier, kindly bid to stay,
Sat by his fire, and talk'd the night away;
Wept o'er his wounds, or, tales of sorrow done,
Shoulder'd his crutch, and show'd how fields were won.
Oliver Goldsmith: The Deserted Village, i, 1770

That's the rum Old Commodore:
That's the tough Old Commodore,
The fighting Old Commodore, he!
But the bullets and the gout
Have so knocked his hull about,
That he'll never more be fit for sea!
The Old Commodore, naval song, 19th century

There are no greater patriots than those good men who have been maimed in the service of their country.
Napoleon I: Political Aphorisms, 1848

And, 'mid the dead and dying, were some grown old in war . . .
Caroline Elizabeth Norton, 1808–1877, The Soldier of the Rhine

Thus the war terminated, and with it all remembrance of the veteran's services.
Sir William Napier: History of the War in the Peninsula, bk xxiv, Ch 5, 1850

Let us strive on to finish the work we are in; to bind up the nation's wounds; to care for him who shall have borne the battle, and for his widow, and for his orphan . . .
Abraham Lincoln: Second Inaugural Address, 4 March 1865

Uncover your head and hold your breath:
This boon not every lifetime hath—
To look on men who have walked with death,
And have not been afraid.
Elizabeth Akers Allen: The Return of the Regiment, 1865 (of the New York 7th Regiment)

I have considered the pension list of the republic a roll of honor.
Grover Cleveland: Veto message on the Dependent Pension Bill, 5 July 1888

A man who is good enough to shed his blood for his country is good enough to be given a square deal afterwards.
Theodore Roosevelt, 1858–1919, Life of Thomas Hart Benton

The Federal Government should treat with the utmost consideration every disabled soldier, sailor or Marine of the World War, whether his disability be due to wounds received in line of action or to health impaired in service; and for the dependents of the brave men who died in line of duty the Government's tenderest concern and richest bounty should be their requital.
Democratic National Platform, 1920

It takes very little yeast to leaven a lump of dough . . . It takes a very few veterans to leaven a division of doughboys.
George S. Patton, Jr.: War As I Knew It, 1947

Veterans of foreign wars do not live long in the gratitude of the republic.
American political aphorism

(*See also* Old Soldier.)

Victory

Victory often changes her side.
Homer: The Iliad, vi, c. 1000 B.C.

A multitude slain!—And their death
Is a matter of grief and for tears;
The victory after a conflict
Is a theme for a funeral rite.
Lao Tze: The Way of Life, 6th Century B.C.

His enemies shall lick the dust.
Psalm LXXII

Whoever wants to keep alive must aim at victory. It is the winners who do the killing and the losers who get killed.
Xenophon: Speech to the Greek army in Persia after the defeat of Cyrus at Cunaxa, 401 B.C.

Know the enemy, know yourself; your victory will never be endangered. Know the ground, know the weather; your victory will then be total.
Sun Tzu, 400–320 B.C., The Art of War, x

A skilled commander seeks victory from the situation, and does not demand it from his subordinates.
Sun Tzu, 400–320 B.C., The Art of War

One more such victory and we are undone.
Pyrrhus Of Epirus: After his victory over the Romans at Asculum, 297 B.C. (the origin of the phrase, "Pyrrhic victory")

A good general not only sees the way to victory; he also knows when victory is impossible.
Polybius: Histories, i, c. 125 B.C.

Victory is by nature insolent and haughty.
Cicero: Pro Marcello, 46 B.C.

Victory in war does not depend entirely upon numbers or mere courage; only skill and discipline will insure it.
Vegetius: De Re Militari, i, 378 A.D.

A victory gained before the situation has crystalized is one the common man does not comprehend. Thus its author gains no reputation for sagacity. Before he has bloodied his blade the enemy state has submitted.
Tu Mu, 803–852, Wei Liao Tzu

The greatest happiness is to vanquish your enemies, to chase them before you, to rob them of their wealth, to see those dear to them bathed in tears, to clasp to your bosom their wives and daughters.
Genghis Khan, d. 1227

What satisfaction in this world, what pleasure can equal that of vanquishing and triumphing over one's enemy? None without doubt.
Cervantes: Don Quixote, 1604

Victory, with advantage, is rather robbed than purchased.
Sir Philip Sidney, 1554–1586

The more hard the fight is, the more haughty is the conquest; and the more doubtful the battle, the more doughty the victory.
George Pettie: Petite Palace of Pettie His Pleasure, 1576

Now are our brows bound with victorious wreaths,
Our bruised arms hung up for monuments.
Shakespeare: King Richard III, i, 1, 1592

Nothing can seem foul to those that win.
Shakespeare: I King Henry IV, v, 1597

A victory is twice itself when the achiever brings home full numbers.
Shakespeare: Much Ado About Nothing, i, 1598

All victories breed hate.
Baltasar Gracian: The Art of Wordly Wisdom, vii, 1647

Peace hath her victories
No less renown'd than war.
John Milton: To the Lord Generall Cromwell, 1652

The enemy came. He was beaten. I am tired. Good night.
Turenne: After the battle of Tünen, 14 June 1658

How beautiful is victory, but how costly!
Chevalier de Boufflers, 1738–1815

The battle, sir, is not to the strong alone; it is to the vigilant, the active, the brave.
Patrick Henry: To the Virginia Convention, 23 March 1775

A victory is very essential to England at the moment.
Sir John Jervis (Lord St. Vincent):

Before engaging the Spanish fleet off Cape St. Vincent, 14 February 1797

"But what good came of it at last?"
Quoth little Peterkin.
"Why, that I cannot tell," said he;
"But 'twas a famous victory."
Robert Southey: The Battle of Blenheim, 1800

Madam, there is nothing so dreadful as a great victory—excepting a great defeat.
Attributed to the Duke of Wellington, 1769–1852 (possibly derived from Wellington's reply to a lady who talked of the glories of a victory: "The greatest tragedy in the world, Madam, except a defeat.")

We have met the enemy, and they are ours—two ships, two brigs, one schooner and one sloop.
Oliver Hazard Perry: Despatch to William Henry Harrison announcing his victory over the British on Lake Erie, 10 September 1813

I do not deserve more than half the credit for the battles I have won. Soldiers generally win battles; generals get credit for them.
Napoleon I: To Gaspard Gourgaud, St. Helena, 1818

As victory is silent, so is defeat.
Thomas Carlyle, 1795–1881, The French Revolution, I, ii

Man does not enter battle to fight, but for victory. He does everything he can to avoid the first and obtain the second.
Ardant du Picq, 1821–1870, Battle Studies

The laurels of victory are at the point of the enemy bayonets. They must be plucked *there*; they must be carried by a hand-to-hand fight if one really means to conquer.
Ferdinand Foch: Precepts 1919

A battle won is a battle in which one will not confess himself beaten.
Ferdinand Foch: Precepts, 1919

The will to conquer is victory's first condition, and therefore every soldier's first duty.
Ferdinand Foch: Principles of War, 1920

Faith in victory determines victory.
Marshal Lyautey: Address on admission to the Academie Française, 8 July 1920

Victory smiles upon those who anticipate the changes in the character of war, not upon those who wait to adapt themselves after they occur.
Giulio Douhet: Command of the Air, 1921

Victory at all costs, victory in spite of all terror, victory however long and hard the road may be; for without victory there is no survival.
Winston Churchill: To the House of Commons, 13 May 1940

The problems of victory are more agreeable than those of defeat, but are no less pressing.
Winston Churchill: To the House of Commons, 1942

I repeat that the United States can accept no result save victory, final and complete.
Franklin D. Roosevelt: Broadcast to the American people, 9 December 1941

Victory is a moral, rather than a material effect.
B.H. Liddell Hart: Thoughts on War, i, 1944

Gaining military victory is not in itself equivalent to gaining the object of war.
B.H. Liddell Hart: Thoughts on War, ii, 1944

In war there is no second prize for the runner-up.
General Omar N. Bradley, USA: In Military Review, February 1950

In war there can be no substitute for victory.
Douglas MacArthur: Letter to Representative Joseph Martin, Jr., 6 April, 1951

Victory has a hundred fathers and defeat is an orphan.
John F. Kennedy: After the Bay of Pigs, April 1961 (Said by him to be "an old saying," but cannot be found or attributed eleswhere)

La victoire, c'est la volonté. (Victory is a thing of the will.)
French saying (said to have been a favorite of Marshal Foch)

On the day of victory no one is tired.
Arab Proverb

Victory or death. (Aut vincere aut mori.)
Latin Proverb

Virginia Military Institute

The Virginia Military Institute will be heard from today.
Stonewall Jackson: At First Bull Run, 21 July 1861

Volunteers

One volunteer is worth ten pressed men.
18th century saying

Our voluntary service regulars are the last descendants of those rulers of the ancient world, the Roman legionaries.
Sir Ian Hamilton: Gallipoli Diary, I, 1920

Von Seeckt, Hans (1866–1936)

. . . a cold, aloof inscrutable man, who knew exactly what he wanted, von Seeckt combined the penetrating practical intelligence of an excellent staff officer with the fundamental romanticism so often found in good regimental soldiers.
D.J. Goodspeed: The Conspirators, 1962

War

Men grow tired of sleep, love, singing and dancing, sooner than of war.
Homer, c. 1000 B.C.

War spares not the brave but the cowardly.
Anacreon, 563–478 B.C.

War is father of all things.
Heraclitus, 540–480 B.C.

War is sweet to those who have never experienced it.
Pindar, 518–439 B.C.

War will go on its way whithersoever chance may lead, and will not restrict itself to the limits which he who meddles with it would fain prescribe.
Speech by the Lacadaemonian Ambassadors to the Athenians, 425 B.C.

Nobody is driven into war by ignorance, and no one who thinks that he will gain anything from it is deterred by fear.
Hermocrates of Syracuse: Speech to the Sicilian envoys at Gela, 424 B.C.

Men will always judge any war in which they are actually fighting to be the greatest at the time.
Thucydides: History of the Peloponnesian Wars, i, c. 404 B.C.

The art of war is, in the last result, the art of keeping one's freedom of action.
Xenophon 430–c. 355 B.C.

Only the dead have seen the end of war.
Plato, 428–347 B.C.

War is a matter of vital importance to the state; the province of life or death; the road to survival or ruin.
Sun Tzu, 400–320 B.C., The Art of War, i

There has never been a protracted war from which a country has benefited.
Sun Tzu, 400–320 B.C., The Art of War, ii

Wars are to be undertaken in order that we may live in peace without suffering wrong.
Cicero, 106–43 B.C.

In war trivial causes produce momentous events.
Julius Caesar: De Bello Gallico, i, 51 B.C.

Wars, horrid wars! (Bella, horrida bella.)
Virgil, 70–19 B.C.

Wars are the dread of mothers. (Bella detestata matribus.)
Horace: Carmina, i, c. 20 B.C.

The war that is necessary is just, and hallowed are the arms where no hope exists but in them.
Livy: History of Rome, c. 10 A.D.

The outcome corresponds less to expectations in war than in any other case whatever.
Livy: History of Rome, xxx, c. 10 A.D.

The fear of war is worse than war itself.
Seneca: Hercules Furens, c. 50

The fortunes of war are always doubtful.
Seneca: Phoenissae, c. 60

Ye shall hear of wars and rumors of wars.
Matthew, XXIV, 6

War ought neither to be dreaded nor provoked. (Bellum nec timendum nec provocandum.)
Pliny The Younger, 61–112

War is delightful to the unexperienced. (Dulce bellum inexpertis.)
Inscription on the tomb of Henry III in Westminster Abbey

War is sweet to those who don't know it.
Erasmus: Adagia, 1508 (cf. inscription above, tomb of Henry III)

A necessary war is a just war.
Niccolo Machiavelli: The Prince, xxvi, 1513

Accurs'd be he that first invented war!
Christopher Marlowe: Tamburlaine the Great, ii, 4 1587

O war! Thou son of Hell!
Shakespeare: II King Henry VI, v, 1590

Grim-visaged War hath smoothed his wrinkled front.
Shakespeare: King Richard III, i, 1, 1592

. . . To reap the harvest of perpetual peace
By this one bloody trial of sharp war.
Shakespeare: King Richard III, v, 1592

Brown Brothers

XENOPHON

fl. c. 430 B. C.

"'Willing obedience always beats forced obedience."

Oh! now doth death line his dead chaps with steel;
The swords of soldiers are his teeth, his fangs;
And now he feasts, mousing the flesh of men,
In undetermin'd differences of kings.
Shakespeare: King John, ii, 1596

. . . thou hast talk't
Of sallies and retires, of trenches, tents,
Of palisadoes, frontiers, parapets,
Of basilisks, of cannon, culverin, . . .
Shakespeare: I King Henry IV, ii, 3, 1597

Farewell the plumed troop, and the big wars,
That make ambition virtue! O, farewell!
Farewell the neighing steed, and the shrill trump,
The spirit-stirring drum, the ear-piercing fife,
The royal banner and all quality,
Pride, pomp and circumstance of glorious war!
Shakespeare: Othello, iii, 1604

Religious canons, civil laws are cruel;
Then what should war be?
Shakespeare: Timon of Athens, iv, 1607

Contumelious, beastly, mad-brained war.
Shakespeare: Timon of Athens, v, 2 1607

O great corrector of enormous times,
Shaker of o'er-rank states, thou grand decider
Of dusty and old titles, that healest with blood,
The earth when it is sick, and curest the world
O' the pleurisy of people.
Shakespeare and John Fletcher: The Two Noble Kinsmen, v, 1613

A civil war is like the heat of fever; but a foreign war is like the heat of exercise, and serveth to keep the body in health.
Francis Bacon: Essays, xxix, 1625

. . . the perfect type of Hell.
Fulke Greville (Lord Brooke): A Treatie of Warre, c. 1625

In war, hunting, and love, men for one pleasure a thousand griefs prove.
George Herbert: Outlandish Proverbs, 1640

He that makes a good war makes a good peace.
George Herbert: Outlandish Proverbs, 1640

War makes thieves, and peace hangs them.
George Herbert: Outlandish Proverbs, 1640

War is death's feast.
George Herbert: Outlandish Proverbs, 1640

When war begins, Hell openeth.
George Herbert: Jacula Prudentum, 1651

Few wage honorable war.
Baltasar Gracian: The Art of Worldly Wisdom, xxiv 1647

Every man is bound by nature, as much as in him lieth, to protect in war the authority by which he is himself protected in time of peace.
Thomas Hobbes: Leviathan (Conclusion), 1651

Force and fraud are in war the two cardinal virtues.
Thomas Hobbes: Leviathan, i, 1651 (also attributed to Machiavelli)

The beginning of all war may be discerned not only by the first act of hostility, but by the counsels and preparations foregoing.
Milton: Eikonoklastes, x, 1649

. . . the brazen throat of war.
Milton: Paradise Lost, viii, 1667

For what can war but endless war still breed?
Milton, 1608–1674, Sonnet on the Lord General Fairfax

Can anything be more ridiculous than that a man has a right to kill me because he dwells on the other side of the water, and because his prince has a quarrel with mine, although I have none with him?
Blaise Pascal: Pensées, iv, 1670

When war begins, the Devil makes Hell bigger.
John Ray: English Proverbs, 1670

War seldom enters but where wealth allures.

John Dryden: The Hind and the Panther, ii, 1687

One minute gives invention to destroy
What to rebuild will a whole age employ.

William Congreve: The Double Dealer, i, 1694

War is the trade of Kings.

John Dryden: King Arthur, ii, 1691

War is the best academy in the world, where men study by necessity and practice by force, and both to some purpose, with duty in the action, and a reward in the end.

Daniel Defoe: Essay upon Projects, 1692

War, he sung, is toil and trouble;
Honor but an empty bubble.

John Dryden: Alexander's Feast, 1697

La guerre est ma patrie,
Mon harnais, ma maison,
Et en toute saison,
Combattre, c'est ma vie.
(War is my fatherland, my harness, my home, and in every season combat is my life.)

Le Soldat François, 18th century song

War! that mad game the world so loves to play.

Jonathan Swift: 1667–1745

War is a trade for the ignorant and a science for the expert.

Chevalier Folard, 1669–1752

I have loved war too well.

Louis XIV of France: On his deathbed, 1715

One to destroy is murder by the law,
And gibbets keep the lifted hand in awe;
To murder thousands takes a specious name,
War's glorious art, and gives immortal fame.

Edward Young: Love of Fame, vii, 1728

War is a science replete with shadows in whose obscurity one cannot move with assured step. Routine and prejudice, the natural result of ignorance, are its foundation and support. All sciences have principles and rules. War has none. The great captains who have written of it give us none. Extreme cleverness is required merely to understand them.

Maurice de Saxe: Mes Rêveries, 1732

The circumstances of war are sensed rather than explained.

Maurice de Saxe: 1696–1750, Letters and Memoirs, iv

It's the fashion now to make war and presumably it will last a good long while.

Frederick The Great: Letter to Voltaire, 1742

The true strength of a prince does not consist so much in his ability to conquer his neighbors, as in the difficulty they find in attacking him.

Montesquieu: The Spirit of the Laws, 1748

Even war is pusillanimously carried on in this degenerate age; quarter is given; towns are taken, and the people spared; even in a storm, a woman can hardly hope for the benefit of a rape.

Lord Chesterfield: Letter to his son, 12 January 1757

The king who makes war on his enemies tenderly distresses his subjects most cruelly.

Samuel Johnson (attributed), 1709–1784

There never was a good war or a bad peace.

Benjamin Franklin: Letter to Josiah Quincy, 11 September 1773

War begets poverty, poverty peace.

Attributed to J. Horne Tooke, 1736–1812

The art of war, as it is certainly the noblest of arts, so in the progress of improvement it necessarily becomes one of the most complicated among them.

Adam Smith: An Inquiry into the Nature and Causes of the Wealth of Nations, Bk V, Ch 1, 1776

It is not a field of a few acres of ground, but a cause, that we are defending, and whether we defeat the enemy in one battle, or by degrees, the consequences will be the same.

Thomas Paine: The American Crisis, No. iv, 12 September 1777

It is the object only of war that makes it honorable.

Thomas Paine: The American Crisis, No. v, 21 March 1778

The only way to save our empires from the enroachment of the people is to engage in war, and thus substitute national passions for social aspirations.

Catherine of Russia, 1729–1796

War is a game, but unfortunately, the cards, counters, and fishes suffer by an ill run more than the gamesters.

Horace Walpole: Letter to Viscount Beauchamp, 13 July 1788 ("Fishes" was an 18th century term for pieces in a game.)

War is not the most favorable moment for divesting the monarchy of power. On the contrary, it is the moment when the energy of a single hand shows itself in the most seducing form.

Thomas Jefferson: Letter to Sl. John de Crêvecoeur, 1788

War is as much a punishment to the punisher as to the sufferer.

Thomas Jefferson: Letter to Tench Coxe, May 1794

War never leaves, where it found a nation.

Edmund Burke, 1724–1797, Letters on a Regicide Peace, i

It would puzzle a keen casuist to prove the reasonableness of the greater number of wars that have dubbed heroes.

Mary Wollstonecraft: A Vindication of the Rights of Woman, ix, 1792

And is not war a youthful king,
A stately hero clad in mail?
Beneath his footsteps laurels spring;
Him earth's majestic monarchs hail
Their friend, their playmate! and his bold bright eye
Compels the maiden's love-confessing sigh.

S.T. Coleridge: A Christmas Carol, 1799

War is a rough, violent trade.

Schiller: The Piccolomini, i, 1799

With men the state of nature is not a state of peace, but of war; if not an open war, then at least ready to break out.

Immanuel Kant: Perpetual Peace, ii, 1795

War requires no particular motive; it appears ingrafted on human nature; it passes even for an act of greatness, to which the love of glory alone, without any other motive impels.

Immanuel Kant: Perpetual Peace, ii, 1795

Now tell us all about the war,
And what they fought each other for.

Robert Southey: The Battle of Blenheim, 1800

The surest way to prevent war is not to fear it.

John Randolph: To the House of Representatives, 5 March 1808

The art of war is the most difficult of all arts; therefore military glory is universally considered the highest, and the services of warriors are rewarded by a sensible government in a splendid manner and above all other services.

Napoleon I, 1769–1821

War,—the trade of barbarians, and the art of bringing the greatest physical force to bear on a single point.

Attributed to Napoleon I, 1769–1821

War to the knife!

Jose de Palafox: Reply to a French demand that he surrender the city of Saragossa, 1808

How easy it is to shed human blood; how easy it is to persuade ourselves that it is our duty to do so—and that the decision has cost us a severe struggle; how much in all ages have wounds and shrieks and tears been the cheap and vulgar resources of the rulers of mankind.

Sydney Smith: Peter Plymley's Letters, 1808

There is more of misery inflicted upon mankind by one year of war than by all the civil peculations and oppressions in a century. Yet it is a state into which the mass of mankind rush with a greatest avidity, hailing official murderers, in scarlet and gold, and cock's feathers, as the greatest and most glorious of human creatures.

Sydney Smith: In the Edinburgh Review, 1813

. . . the ruin, the disgrace, the woe of war.
Shelley: Queen Mab, iv, 1813

War is the statesman's game, the priest's delight.
The lawyer's jest, the hired assassin's trade.
Shelley: Queen Mab, iv, 1813

If there had never been a war, there could never have been tyranny in the world.
Shelley: A Philosophical View of Reform, 1819

Kubla heard from far
Ancestral voices prophesying war.
S.T. Coleridge: Kubla Khan, 1816

Wars are to the body politic what drams are to the individual. There are times when they may prevent a sudden death, but if frequently resorted to, or long persisted in, they heighten the energies, only to hasten the dissolution.
C.C. Colton: Lacon, 1820

War kills men, and men deplore the loss; but war also crushes bad principle and tyrants, and so saves societies.
C.C. Colton: Lacon, 1820

War ought never to be undertaken but under circumstances which render all intercourse of courtesy between the combatants impossible. It is a bad thing that men should hate each other; but it is far worse that they should contract the habit of cutting one another's throats without hatred. War is never lenient but where it is wanton; when men are compelled to fight in self-defense, they must hate and avenge: this may be bad; but it is human nature.
T.B. Macaulay: On Milford's History of Greece, 1824

To carry the spirit of peace into war is a weak and cruel policy. When an extreme case calls for that remedy which is in its own nature most violent, and which in such cases, is a remedy only because it is violent, it is idle to think of mitigating and diluting. Languid war can do nothing which negotiation or submission will not do better: and to act on any other principle is not to save blood and money, but to squander them.
T.B. Macaulay: Hallam, 1828

Strike—till the last armed foe expires;
Strike—for your altars and your fires;
Strike—for the green graves of your sires;
God—and your native land!
Fitz-Green Halleck: Marco Bozzaris, 1825

War is undertaken for the sake of peace, which is its only lawful end and purpose.
James Kent: Commentaries on American Law, i, 1826 (cf. Cicero, ante.)

The conduct of war resembles the workings of an intricate machine with tremendous friction, so that combinations which are easily planned on paper can be executed only with effort.
Clausewitz: Principles of War, 1812

War is an act of violence pushed to its utmost limits.
Clausewitz: On War, 1832

War is an act of violence whose object is to constrain the enemy to accomplish our will.
Clausewitz: On War, 1832

War admittedly has its own grammar, but not its own logic.
Clausewitz: On War, 1832

War is nothing but a duel on a large scale.
Clausewitz: On War, 1832

War is not only chameleon-like in character, because it changes its colors in some degree in each particular case, but it is also, as a whole, in relation to the predominant tendencies that are in it, a wonderful trinity, composed of the original violence of its elements, hatred and animosity, which may be looked upon as blind instinct; of the play of probabilities and chance, which make it a free activity of the soul; and of the subordinate nature of a political instrument, by which it belongs purely to reason.
Clausewitz: on War, 1832

War, like all other situations of danger and of change, calls forth the exertion of admirable intellectual qualities and great virtues.
William Cullen Bryant, 1794–1878, The Value and Uses of Poetry

The lower people everywhere desire war. Not so unwisely; there is then a demand for lower people—to be shot!
Thomas Carlyle: Sartor Resartus, iii, 1836

War in its ensemble is not a science, but an art.

Jomini: Précis de l'Art de la Guerre, 1838

The most just war is one which is founded upon undoubted rights and which, in addition, promises to the state advantages commensurate with the sacrifices required and the hazards incurred.

Jomini: Précis de l'Art de la Guerre, 1838

And ever since historian writ,
And ever since a bard could sing,
Doth each exalt with all his wit
The noble art of murdering.

W.M. Thackeray: The Chronicles of the Drum, 1840

War crushes with bloody heel all justice, all happiness, all that is God-like in man. In our age there can be no peace that is not honorable; there can be no war that is not dishonorable.

Charles Sumner: The True Grandeur of Nations (speech in Boston, 4 July 1845; cf. Franklin, ante, 1773.)

The Bible nowhere prohibits war. In the Old Testament we find war and even conquest positively commanded, and although war was raging in the world in the time of Christ and His Apostles, still they said not a word of its unlawfulness and immorality.

Major General H.W. Halleck, USA: Elements of Military Art and Science, 1846

Ez fer war, I call it murder—
Ther you hev it plain and flat;
I don't want to go no furder
Than my Testyment fer that.

James Russell Lowell: The Biglow Papers, i, 1848

I understand well the respect of mankind for war, because war breaks up the Chinese stagnation of society, and demonstrates the personal merits of all men.

R.W. Emerson: The Conservative, 1841

War disorganizes, but it is to reorganize.

R.W. Emerson, 1803–1882, The Man of Letters

Think ye I have made this ball
A field of havoc and war,
Where tyrants great and small
Might harry the weak and poor?

R.W. Emerson: Boston Hymn, 1863

War is not the greatest calamity.

R.W. Emerson: Remark during the Civil War

I shall always respect war hereafter. The cost of life, the dreary havoc of comfort and time, are overpaid by the vistas it opens of eternal life, eternal, law, reconstructing and uplifting society,—breaks up the old horizon, and we see through the rifts a wider vista.

R.W. Emerson: Letter to Thomas Carlyle, 26 September 1864

War is the science of destruction.

John S.C. Abbott, 1805–1877, History of Napoleon Bonaparte (preface)

War is assassination, war is theft.

Madame Girardin, 1804–1855

The difference of race is one of the reasons why I fear war may always exist; because race implies difference, difference implies superiority, and superiority leads to predominance.

Benjamin Disraeli: To the House of Commons, 1 February 1849

If there be greater calamity to human nature than famine, it is that of an exterminating war.

Benjamin Disraeli: Speech, Mansion House, 9 November 1877

It is magnificent, but it is not war.

Pierre Bosquet: On observing the charge of the Light Brigade, Balaklava, 25 October 1854

I believe that war is at present productive of good more than of evil.

John Ruskin: Modern Painters, iv, 1856

Both peace and war are noble or ignoble according to their kind and occasion.

John Ruskin: The Two Paths, v, 1859

War is the foundation of all the arts, because it is the foundation of all the high virtues and faculties of men.

John Ruskin, 1819–1900

Suppose you go to war, you cannot fight always; and when, after much loss on both sides, and no gain on either, you cease fighting, the identical old questions as to terms of intercourse are again upon you.
Abraham Lincoln: Inaugural address, 4 March 1861

It is painful enough to discover with what unconcern they speak of war and threaten it. I have seen enough of it to make me look upon it as the sum of all evils.
Stonewall Jackson: Letter, April 1861

War is an organized bore.
Oliver Wendell Holmes, Jr.: To a visitor, after being wounded, 1862

It is well that war is so terrible—we would grow too fond of it.
R.E. Lee: To James Longstreet, battle of Fredericksburg, 13 December 1862 (cf. Louis XIV, ante, 1715.)

What a cruel thing is war: to separate and destroy families and friends, and mar the purest joys and happiness God has granted us in this world; to fill our hearts with hatred instead of love for our neighbors, and to devastate the fair face of this beautiful world.
R.E. Lee: Letter to his wife, 25 December 1862

It would be superfluous in me to point out to your Lordship that this is war.
Charles Francis Adams: Note, as American Minister, to Lord Russell, the British Foreign Minister, 5 September 1863

War means fighting, and fighting means killing.
Attributed to Nathan Bedford Forrest, 1821–1877

A nation is not worthy to be saved if, in the hours of its fate, it will not gather up all its jewels of manhood and life, and go down into the conflict, however bloody and doubtful, resolved on measureless ruin or complete success.
James A. Garfield: To the House of Representatives 25 June 1864

All the people retire before us and desolation is behind. To realize what war is one should follow in our tracks.
W.T. Sherman: Letter, 26 June 1864

I begin to regard the death and mangling of a couple thousand men as a small affair, a kind of morning dash—and it may well be that we become so hardened.
W.T. Sherman: Letter to his wife, July 1864

If the people [of Georgia] raise a howl against my barbarity and cruelty, I will answer that war is war, and not popularity-seeking. If they want peace, they and their relatives must stop the war.
W.T. Sherman: Letter to Major General H.W. Halleck, 4 September 1864

You cannot qualify war in harsher terms than I will. War is cruelty, and you cannot refine it.
W.T. Sherman: Letter to the Mayor of Atlanta, 12 September 1864

You might as well appeal against the thunderstorm as against these terrible hardships of war.
W.T. Sherman: Letter to the Mayor of Atlanta, 12 September 1864

War is hell.
Attributed to W.T. Sherman (This phrase, not known to have been spoken or written exactly by Sherman, is probably derived from his speech at Columbus, Ohio, 12 August 1880: "There is many a boy here today who looks on war as all glory, but, boys, it is all hell."

War is simply power unrestrained by constitution or compact.
Attributed to W.T. Sherman, 1820–1891

When this cruel war is over . . .
C.C. Sawyer: Title of popular song, 1864

Know'st thou not, there is but one theme
for ever-enduring bards?
And that is the theme of war, the fortune
of battles,
The making of perfect soldiers?
Walt Whitman: As I Ponder'd in Silence, 1870

I hate this slaughter. I never desired the honors of war, and would gladly have left such glory to others. Nevertheless it is my hard fate to go from battlefield to battlefield, from one war to another before ascending the throne of my ancestors.
Prince Frederick of Prussia, 1870

War makes the victor stupid and the vanquished vengeful.
F.W. Nietzche: Human All-Too-Human, i, 1878

The man who has renounced war has renounced a grand life.
F.W. Nietzche: The Twilight of the Idols, 1889

The success of a war is gauged by the amount of damage it does.
Victor Hugo: Ninety-Three, 1879

War forms part of the order of things instituted by God.
Helmuth von Moltke ("The Elder"): Letter to J.K. Bluntschili, 11 December 1880

Eternal peace is a dream, and not even a beautiful one. War is part of God's world order. In it are developed the noblest virtues of man: courage and abnegation, dutifulness and self-sacrifice. Without war the world would sink into materialism.
Helmuth von Moltke ("The Elder"): Letter to J.K. Bluntschli, 11 December 1880

War talk by men who have been in a war is always interesting.
S.L. Clemens (Mark Twain): Life on The Mississippi, xlv, 1883

War is a perpetual struggle with embarrassments.
Colmar von der Goltz: Das Volk in Waffen, 1883

To abolish war we must remove its cause, which lies in the imperfection of human nature.
Colmar von der Goltz: The Nation in Arms, 1883

War loses a great deal of its romance after a soldier has seen his first battle.
John S. Mosby: War Reminiscences, xiv, 1887

War, with its many acknowledged sufferings, is above all harmful when it cuts a nation off from others and throws it back upon itself.
Mahan: The Influence of Sea Power upon History, 1890

Let there be dismissed at once, as preposterous, the hope that war can be carried on without some one or something being hurt.
Mahan: Some Neglected Aspects of War, 1907

Where evil is mighty and defiant, the obligation to use force—that is, war—arises.
Mahan: Naval Strategy, 1911

As long as war is regarded as wicked it will always have its fascinations. When it is looked upon as vulgar, it will cease to be popular.
Oscar Wilde: Intentions, 1891

. . . The coward's art of attacking mercilessly when you are strong, and keeping out of harm's way when you are weak. That is the whole secret of successful fighting. Get your enemy at a disadvantage; and never, on any account, fight him on equal terms.
George Bernard Shaw: Arms and the Man, ii, 1894

Oh, war! war! the dream of patriots and heroes! A fraud. A hollow sham, like love.
George Bernard Shaw: Arms and the Man, iii, 1894

War is kind.
Stephen Crane: Title of a poem, 1895

War, when you are at it, is horrible and dull. It is only when time has passed that you see that its message was divine.
Oliver Wendell Holmes, Jr.: Speech, "The Soldier's Faith," at Harvard, 30 May 1895

God will see to it that war shall always recur, as a drastic medicine for ailing humanity.
Heinrich von Treitschke: Politik, i, 1897

War, with its abominably casual, inaccurate methods of destroying good and bad together, but at last unquestionably able to hew a way out of intolerable situations, when through man's delusion or perversity every better way is blocked.
William James, 1842–1910

What we now need to discover in the social realm is the moral equivalent of war: something heroic that will speak to men as universally as war does, and yet will be as compatible with their spiritual selves as war has proved itself to be incompatible.

William James: The Varieties of Religious Experience, xiv, 1902

War and revolution never produce what is wanted, but only some mixture of the old evils with the new ones.

William Graham Sumner: War, 1903

War makes rattling good history; but peace is poor reading.

Thomas Hardy: The Dynasts, ii, 1906

Yes; quaint and curious war is!
You shoot a fellow down
You'd treat if met where any bar is,
Or help to half-a-crown.

Thomas Hardy, 1840–1928, The Man He Killed

War is the contention between two or more states through their armed forces for the purpose of overpowering each other and imposing such conditions of peace as the victor pleases.

L.F.L. Oppenheim: International Law, ii, 1906

War is not merely justifiable, but imperative, upon honorable men, upon an honorable nation, where peace can only be obtained by sacrifice of conscientious conviction or of national welfare.

Theodore Roosevelt: Message to Congress, 3 December 1906

War is a dreadful thing, and unjust war is a crime against humanity. But it is such a crime because it is unjust, not because it is war.

Theodore Roosevelt: Speech at the Sorbonne, 23 April 1910

War is a biological necessity of the first importance.

Friedrich von Bernhardi: Germany and the Next War, 1911

The essence of war is violence; moderation in war is imbecility.

Sir John Fisher: Letter to Lord Esher, 25 April 1912

Until the world comes to an end the ultimate decision will rest with the sword.

Wilhelm II of Germany: Speech in Berlin, 1913

War is only a sort of dramatic representation, a sort of dramatic symbol of a thousand forms of duty.

Woodrow Wilson: Speech at Brooklyn, 11 May 1914

War is a phase in the life-effort of the state towards completer self-realization, a phase of the eternal nisus, the perpetual omnipresent strife of all being towards self-fulfilment.

J.A. Cramb: The Origins and Destiny of Imperial Britain, ii, 1900

It is at this period [le Grand Siècle] that war did indeed become . . . a game of chess. When, following a series of complex maneuvers, one of the two adversaries had lost or won several pawns—towns or fortresses—there came the great battle. Standing on some hilltop from which he could survey the whole field, the entire chessboard, a Marshal skillfully caused his splendid regiments to advance or retreat. Check and Mate, the loser put his pawns away: the regiments marched off to winter quarters, and everybody went about his business pending the next game or campaign.

J. Boulenger: Le Grand Siècle, Paris, 1915

I have seen war, and faced modern artillery, and I know what an outrage it is against simple men.

T.M. Kettle: The Ways of War, 1915

There are poets and writers who see naught in war but carrion, filth, savagery, and horror . . . They refuse war the credit of being the only exercise in devotion on the large scale existing in this world. The superb moral victory over death leaves them cold. Each one to his taste. To me this is no valley of death—it is a valley brim full of life at its highest power.

Sir Ian Hamilton: Diary entry, Gallipoli, 30 May 1915

As I reflected upon the intensive application of man to war in cold, rain and mud; in rivers, canals, and lakes; underground and in the air, and under the sea; infected with vermin, covered with scabs, adding the stench of his own filthy body to that of his decomposing comrades; hairy, begrimed, bedraggled, yet with unflagging zeal striving eagerly to kill his fellows; and as I felt within myself the mystical urge of the sound of great cannon, I realized that war is a normal style of man.

G.W. Crile: A Mechanistic View of War and Peace, 1915

War means an ugly mob-madness, crucifying the truth-tellers, choking the artists, sidetracking reforms, revolutions, and the working of social forces.
John Reed: Whose War? April 1917

... and many young men in it, mostly in messrooms and wardrooms, used to say to each other—"It's a damned bad war, but it's better than no war at all."
Joseph Conrad: The Tale, 1917

War is getting at the vitals of the enemy, that is, to shoot him in the heart.
Colonel William Mitchell, USA: Letter from France, 1917

There mustn't be any more war. It disturbs too many people.
An old French peasant woman: to Aristide Briand, c. 1917

Sooner or later every war of trade becomes a war of blood.
Eugene V. Debs: Speech at Canton, Ohio, 16 June 1918

We cannot joke with war.
V.I. Lenin: Letter to the Bolshevik Central Committee, April 1918

Generals cannot be entrusted with anything—not even with war.
Georges Clemenceau, 1841–1929 (often given as "War is too important to be left to the generals")

Ah, God, sweet is war, with its songs, with its prolonged leisures.
Guillaume Appolinaire, 1880–1918

The first hundred years are the hardest.
Soldier saying, American Expeditionary Force, World War I, c. 1918

War is the highest expression of the racial will to life.
Erich Ludendorff: Meine Kriegserinnerungen, 1919

In war, only what is simple can succeed.
Paul von Hindenburg, 1847–1934

... A Rabelaisian game of chess where the board has a million squares and the pieces consist of a dozen Kings and Queens, a thousand Knights, and so many pawns that no one can exactly count them.
Sir Ian Hamilton: The Soul and Body of an Army, xi, 1921

Once in a generation, a mysterious wish for war passes through a people. Their instinct tells them there is no other way of progress and of escape from habits that no longer fit them. Whole generations of statesmen will fumble over reforms for a lifetime which are put into full-blooded execution within a week of a declaration of war. There is no other way. Only by intense sufferings can the nations grow, just as a snake once a year must with anguish slough off the once beautiful coat which has now become a strait jacket.
Sir Ian Hamilton: Gallipoli Diary, 1920

There is nothing certain about war except that one side won't win.
Sir Ian Hamilton: Gallipoli Diary, 1920

I'd prefer to eat a little dirt rather than have another war.
Attributed to Thomas R. Marshall, 1854–1925

War is a simple art: its essence lies in its accomplishment.
Ferdinand Foch, 1851–1929

The military mind always imagines that the next war will be on the same lines as the last. That has never been the case and never will be.
Ferdinand Foch, 1851–1929

There is no "science" of war, and there never will be any. There are many sciences war is concerned with. But war itself is not a science; war is practical art and skill.
Leon Trotszky: How the Revolution Developed its Military Power, 1924

To delight in war is a merit in the soldier, a dangerous quality in the captain, and a positive crime in the statesman.
George Santayana, 1863–1925

The power to wage war is the power to wage war successfully.
Charles Evans Hughes, C.J.: For the Supreme Court of the United States, 1927

War is a transfer of property from nation to nation.
Leon Samson: The New Humanism, 1930

Where armies take the field it is fate knocking at the door, it is nature deciding the life and death of nations.
Alfred Machin, 1888-

War alone brings up to its highest tension all human energy and puts the stamp of nobility upon the people who have the courage to face it.
Benito Mussolini, 1883–1945, article in The Italian Encyclopaedia

There is no better teacher of war than war.
Mao Tse-tung: On the Study of War, 1936

I have seen war. I have seen war on land and sea. I have seen blood running from the wounded. I have seen men coughing out their gassed lungs. I have seen the dead in the mud. I have seen cities destroyed. I have seen two hundred limping, exhausted men come out of the line—the survivors of a regiment of a thousand that went forward 48 hours before. I have seen children starving. I have seen the agony of mothers and wives. I hate war.
Franklin D. Roosevelt: Speech at Chatauqua, New York, 14 August 1936

War is a contagion.
Franklin D. Roosevelt: Speech at Chicago, 5 October 1937

I do not think that a philosophical view of the world would regard war as absurd.
Oliver Wendell Holmes, Jr.: Dissenting opinion in U.S. vs. Schwimmer, 1928

War is an art, proficiency in which depends more on experience than on study, and more on natural apitude and judgment than on either.
Lloyd George: War Memoirs, VI, 1937

Before a war military science seems a real science, like astronomy, but after a war it seems more like astrology.
Rebecca West, 1892–

War can protect; it cannot create.
Alfred North Whitehead, 1861–1947

In war, whichever side may call itself the victor, there are no winners, only losers.
Neville Chamberlain: Speech at Kettering, England, 2 July 1938

It is simply not true that war never settles anything.
Felix Frankfurter, 1882–1965

Although war is evil, it is occasionally the lesser of two evils.
McGeorge Bundy: Essay while at Yale College, 1940

War is force—force to the utmost—force to make the enemy yield to our own will—to yield because they see their comrades killed or wounded—to yield because their own will to fight is broken. War is men against men—mechanized war is still men against men. Machines are mere masses of inert metal without the men who man them.
Ernest J. King: Address to the graduating class, U.S. Naval Academy, 19 June 1942

War is very simple, direct and ruthless. It takes a simple, direct, and ruthless man to wage war.
George S. Patton, Jr.: Diary entry, 15 April 1943

War is a very simple thing, and the determining characteristics are self confidence, speed and audacity. None of these things can ever be perfect, but they can be good.
George S. Patton, Jr.: War As I Knew It, 1947

Older men declare war. But it is youth that must fight and die. And it is youth who must inherit the tribulation, the sorrow, and the triumphs that are the aftermath of war.
Herbert C. Hoover: Address to the Republican National Convention, Chicago, 27 June 1944

There has never been a war yet which, if the facts had been put calmly before the ordinary folk, could not have been prevented. The common man is the greatest protection against war.
Ernest Bevin: To the House of Commons, November 1945

Every new war embarked upon by the nations for the purpose of detaching themselves from one another, merely results in their being bound and mingled together in a more inextricable knot. The more we seek to thrust each other away, the more do we interpenetrate.

Teilhard de Chardin: The Planetization of Mankind, 1945

War is a wasteful, boring, and muddled affair.

Sir A.P. Wavell: Unpublished "Recollections," 1946

The main ethical objection to war for intelligent people is that it is so deplorably dull and usually so inefficiently run . . . Most people seeing the muddle of war forget the muddles of peace and the general inefficiency of the human race in ordering its affairs.

Sir A.P. Wavell: Unpublished "Recollections," 1947

War, which used to be cruel and magnificent, has now become cruel and squalid. It is all the fault of democracy and science. From the moment that either of these meddlers and muddlers was allowed to take part in actual fighting, the doom of War was sealed. Instead of a small number of well-trained professionals championing their country's cause with ancient weapons and a beautiful intricacy of archaic movement, we now have entire populations, including even women and children, pitted against each other in brutish mutual extermination, and only a set of blear-eyed clerks left to add up the butcher's bill. From the moment when Democracy was admitted to, or rather forced itself upon, the battlefield, War ceased to be a gentleman's pursuit.

Winston Churchill: My Early Life, 1930

Death and sorrow will be the companions of our journey; hardship our garment; constancy and valor our only shield. We must be united, we must be undaunted, we must be inflexible.

Winston Churchill: To the House of Commons, 8 October 1940

In mortal war, anger must be subordinated to defeating the main immediate enemy.

Winston Churchill: The Gathering Storm, 1948

There is no merit in putting off a war for a year if, when it comes, it is a far worse war or one much harder to win.

Winston Churchill: The Gathering Storm, 1948

Nothing in our age has been more significant than this new technique of aggression and of pacific conquest, which has taken the place of orthodox war.

Jan Christian Smuts: Speech at Cambridge University, 11 June 1948

Nothing is easy in war. Mistakes are always paid for in casualties and troops are quick to sense any blunder made by their commanders.

Dwight D. Eisenhower, 1890– , Crusade in Europe

War is a science which depends on art for its application.

B.H. Liddell Hart: Strategy (article in Encyclopaedia Britannica, 1929 edition)

Whereas the other arts are, at their height, individual, the art of war is essentially orchestrated.

B.H. Liddell Hart: Thoughts on War, iv, 1944

War is always a matter of doing evil in the hope that good may come of it.

B.H. Liddell Hart: Defense of the West, 1950

War is basically a conflict of wills.

Sir John Slessor: Strategy for the West, 1054

The essence of war as an extension of politics does not depend upon changes in technology or armaments.

In Voennaya Strategiya (Military Strategy), USSR military handbook, c. 1957

Unlike mathematics, war is an empirical matter. War is history; this means that its laws are deductions to be made only after the event.

Jean Dutourd: Taxis of the Marne, 1957

Thus war, the horseman, turned back to his crimson courts and dragged brave gallants by their belts, girls by their braids, and hung small children from his saddle-horns in clusters. Behind him the blind followed, stumbling with long staffs, and some way back the cripples, the armless, the half-wits, and mothers in long rows who walked alive toward Hades.

Nikos Kazantzakis: The Modern Odyssey, a Sequel, 1958

Brown Brothers

V. I. LENIN

1870–1924

"Insurrection is an art as much as war."

Nothing is worse than war?
Dishonor is worse than war.
Slavery is worse than war.
Winston Churchill, 1874–1965

The most crucial problem men face today is their unfortunate habit of killing one another.
Owen M. Weatherly: The Ten Commandments in Modern Perspective, vi, 1961

Mankind must put an end to war—or war will put an end to mankind.
John F. Kennedy: To the General Assembly of the United Nations, 25 September 1961

War is a shabby, really impractical thing, anyway, and it takes a genius to conduct it with any sort of economy and efficiency.
William Faulkner: At West Point, 20 April 1962

Any historian of warfare knows it is in good part a comedy of errors and a museum of incompetence.
Richard Hofstadter: "The Paranoid Style in Politics," Harper's, 1964

War is the great auditor of institutions.
Correlli Barnett: The Swordbearers, 1964

Modern warfare is an interlocking system of actions—political, economic, psychological, military—that aims at the overthrow of the established authority in a country and its replacement by another regime.
Colonel Roger Trinquier: Modern Warfare, 1964

A great war always creates more scoundrels than it kills.
Author unidentified

When after many battles past,
Both, tired with blows, make peace at last,
What is it, after all, the people get?
Why, taxes, widows, wooden legs, and debt.
Author unidentified

War is a business that ruins those who succeed in it.
Author unidentified

It's not much of a war, but it's the only one we've got.
Marine Corps saying (Cf. Conrad, ante, 1917)

Most wars are caused by either priests or women.
Czech Proverb

When elephants fight, it is the country that suffers.
East African Proverb

War to extremity. (Guerre à outrance.)
French Proverb

It's war. (C'est la guerre.)
French Saying

A great war leaves a country with three armies—an army of cripples, an army of mourners, and an army of thieves.
German Proverb

War is not sugar plums.
Hindi Proverb

The fear of war is worse than war.
Italian Proverb

All are not soldiers that go to war.
Spanish Proverb

War is blind.
West African Proverb

War Aims

The aim of war is to be able to live unhurt in peace.
Cicero: De Officiis, i, 78 B.C.

It is not the object of war to annihilate those who have given provocation for it, but to cause them to mend their ways; not to ruin the innocent and the guilty alike, but to save both.
Polybius, 200–118 B.C., Histories, v

The purpose of all war is peace.
St. Augustine: De Civitate Dei, xv, 427

To reap the harvest of perpetual peace,
By this one bloody trial of perpetual war.
Shakespeare: King Richard III, v, 1952

We shall never make war except for peace.
William McKinley: Speech at El Paso, Texas, 6 May 1901

The legitimate object of war is a more perfect peace.

W.T. Sherman: Epigram, 23 February 1882 (Cf. Cicero, ante, 78 B.C., and St. Augustine, ante, 427)

It must be a peace without victory.

Woodrow Wilson: To the Senate, 22 January 1917

War means fighting for definite results. I am not waging war for the sake of waging war. If I obtain through the armistice the conditions we wish to impose upon Germany, I am satisfied.

Ferdinand Foch: To the Supreme War Council, November 1918

The kind of war we plan to fight must fit the kind of peace we want.

Arthur W. Radford: To the House Armed Services Committee, 5 October 1949

(*See also* Peace.)

War and Peace

Peace is the dream of the wise; war is the history of man.

Sir Richard Burton: quoting an old proverb, c. 1865

(*See also* Peace, War.)

War Correspondents

. . . the low and grovelling correspondents of *The Times.*

General Sir James Simpson: In the Crimea, 1854

I will never again command an army in America if we must carry along paid spies. I will banish myself to some foreign country first.

W.T. Sherman: Letter to his wife, February 1863 (referring to correspondents who had been traducing him and his operations)

So you think the papers ought to say more about your husband! My brigade is not a brigade of newspaper correspondents.

Stonewall Jackson: To his wife after First Bull Run, July 1861

. . . Those newly invented curses to armies . . . that race of drones who are an encumbrance to an army; they eat the rations of the fighting men, and do not work at all.

Sir Garnet Wolseley: The Soldier's Pocket Book, 1869

You furnish the pictures and I'll furnish the war.

William Randolph Hearst: Instructions to Frederic Remington in Cuba just prior to the sinking of USS Maine, *March 1898*

A war expert . . . is a man ye niver heerd iv before. If ye can think iv annywan whose face is onfamiliar to ye an' ye don't raymimber his face, an' he's got a job on a pa-aper ye didn't know was published, he's a war expert.

Finley Peter Dunne ("Mr. Dooley"): Mr. Dooley's Philosophy, 1900

The best way to write about a war is to go along and make notes.

Lieutenant Commander Arnold S. Lott, USN: Most Dangerous Sea, 1959

Being a lady war correspondent is like being a lady wrestler—you can be one of them at a time, but not both simultaneously.

Dickey Chapelle: Remark at Danang, 2 November 1965 (Miss Chapelle was killed in action on the following day.)

(*See also* Public Information, Public Relations.)

War Crimes

The question for us rightly considered is not, what are their crimes? but, what is our interest? If I prove them ever so guilty, I will not on that account bid you put them to death, unless it is expedient.

Diodotus of Athens: Speech to the Athenians, 427 B.C.

War Damages

Damage by enemy action stands on a different footing from any other kind of damage because the nation undertakes the task of defending the lives and property of its subjects and taxpayers against assaults from outside.

Winston Churchill: To the House of Commons, 5 September 1940

War of 1812 (1812–1815)

I have indeed been fortunate but not more so than I am confident my brother officers will be if they fall in with the enemy. Where there is anything like equal force you will find [the British] are not invincible. They are not now fighting Frenchmen and Spaniards.

Isaac Hull: To a brother officer after taking HMS Guerrière, *September 1812*

(*See also* Chippewa, Constitution, Impressment, Lake Champlain, Lake Erie, Lawrence (James), Lundy's Lane, New Orleans.)

War Powers

In times of peace the people look most to their representatives; but in war, to the executive solely.

Thomas Jefferson: Letter to C.A. Rodney, 1810

Any order of the President, or under his authority, made at any time during the existence of the present rebellion, shall be a defense in all courts for any seizure, arrest or imprisonment made, done, or committed, or acts omitted to be done under and by virtue of such order.

Act of Congress, 3 March 1863

Warhorse

The horse is a vain thing for safety; nor shall he deliver anyone by his strength.

Psalm XXXIII

Hast thou given the horse strength? Hast thou clothed his neck with thunder? . . . He paweth the valley, and rejoiceth in his strength; he goeth on to meet the armed men . . . He saith among the trumpets, Ha, ha; and he smelleth the battle afar off, the thunder of the captains and the shouting.

Job, XXXIX, 19–25

Steed threatens steed, in high and boastful neighs,
Piercing the night's dull ear.

Shakespeare: King Henry V, iv, 1598

Mars's fiery steed . . .

Shakespeare: All's Well That Ends Well, ii, 3, 1602

By the brand on my withers, the finest of tunes
Is played by the Lancers, Hussars, and Dragoons,
And it's sweeter than "Stables" or "Water" to me,
The Cavalry Canter of "Bonnie Dundee!"

Then feed us and break us and handle and groom,
And give us good riders and plenty of room,
And launch us in column of squadron to see
The way of the War-Horse to "Bonnie Dundee!"

Rudyard Kipling: Parade-Song of the Camp Animals, 1913

In dreary, doubtful, waiting hours,
Before the brazen frenzy starts,
The horses show him nobler powers;
O patient eyes, courageous hearts!

Julian Grenfell, 1888–1915, Into Battle

(*See also* Cavalry.)

Warrior

Therefore arise, thou Son of Kunti! Brace
Thine arm for conflict; nerve thy heart to meet,
As things alike to thee, pleasure or pain,
Profit or ruin, victory or defeat.
So minded, gird thee to the fight, for so
Thou shalt not sin!

Bhagavad-Gita, Chant of Krishna to Prince Arunja before battle, c. 5th century B.C.

. . . this Hotspur, Mars
In swathling clothes . . .

Shakespeare: I King Henry IV, iii, 1, 1597

Who is the happy warrior? Who is he
That every man in arms should wish to be?

William Wordsworth: Character of the Happy Warrior, 1806

He lay like a warrior taking his rest.

Charles Wolfe: The Burial of Sir John Moore at Corunna, 1817

Great Warriors, like great earthquakes, are principally remembered for the mischief they have done.

Christian Nestell Bovée, 1820–1904

I see many soldiers; could I but see as many warriors!
F.W. Nietzche: Thus Spake Zarathrustra, i, 10, 1885

(*See also* Knight, Regular, Soldier.)

Warship

. . . Ships,
Fraught with the ministers and instruments
Of cruel war.
Shakespeare: Troilus and Cressida, prologue, 1601

A man-of-war is the best ambassador.
Oliver Cromwell, 1599–1658

Take it all in all, a ship of the line is the most honorable thing that man, as a gregarious animal, has ever produced.
John Ruskin: The Harbors of England, 1856

The backbone and real power of any navy are the vessels which, by due proportion of defensive and offensive powers, are capable of giving and taking hard knocks.
Mahan: The Interest of America in Sea Power, 1896

(*See also* Ships.)

Washington, George (1732–1799)

These are high times when a British general is to take counsel of a Virginia buckskin.
Major General Edward Braddock: In rejecting suggestions by George Washington regarding forthcoming operations against the French and Indians, 1755

He has been beaten whenever he has engaged; and that this is left to befall him again, is a problem which, I believe, most military men are utterly at a loss to solve.
In The Gentleman's Magazine, London, August 1778 (of George Washington)

We must have your name. There will be more efficacy in it than in many an army.
John Adams: Letter to George Washington at the Outset of the Naval War with France, 1798

. . . A citizen, first in war, first in peace, first in the hearts of his countrymen.
Henry ("Light Horse Harry") Lee: Resolution proposed to, and passed by, Congress on the death of Washington, 1799

. . . The Cincinnatus of the West,
Whom envy dared not hate.
Byron: Ode to Napoleon Bonaparte, ii, 1814 (Byron's reference here is to Washington.)

(*See also* American Revolution.)

Waterloo (18 June 1815)

I tell you that Wellington is a bad general, that the English are bad troops, and that this affair is only a *déjeuner.*
Attributed to Napoleon I: On the morning of Waterloo, 18 June 1815

Waterloo is cast in my teeth . . . I ought to have died in Moscow.
Napoleon I: Letter to Gaspard Gourgaud, St. Helena, 1816

. . . a damned nice thing—the nearest-run thing you ever saw in your life.
Wellington: To Mr. Thomas Creevey at Brussels, the day after Waterloo, 19 June 1815

You will have heard of our battle of the 18th. Never did I see such a pounding match . . . Napoleon did not maneuver at all. He just moved forward in the old style, and was driven off in the old style.
Wellington: Letter to Sir William Beresford, 2 July 1815

Meeting an acquaintance of another regiment, a very little fellow, I asked him what had happened to them yesterday. "I'll be hanged," says he, "if I know anything at all about the matter, for I was all day trodden in the mud and galloped over by every scoundrel who had a horse; and, in short, I only owe my existence to my insignificance!"
Captain John Kincaid, reminiscence of Waterloo in Adventures with the Rifle Brigade

Waterloo is a battle of the first rank won by a captain of the second.
Victor Hugo: Les Miserables, i, 16, 1862

Waterloo was the change of front of the universe.
Attributed to Victor Hugo, 1802–1885

Wavell, Archibald Percival (1883–1950)

The only one who showed a touch of genius was Wavell.
Erwin Rommel: The Rommel Papers, xxiii, 1953

Wayne, Anthony (1745–1796)

. . . Brave homme, mais tres ardent. (A good man, but very hot-headed.)
Comte de Rochambeau: Letter to LaFayette, July 1781

General Wayne had a constitutional attachment to the sword, and this cast of character had acquired strength from indulgence.
Henry ("Light Horse Harry") Lee, 1756–1818

Weapons

Wisdom is better than weapons of war.
Ecclesiastes, IX, 18

Weapons are tools of ill omen.
Li Ch'üan, c. 8th century, A.D.

Cannon and fire-arms are cruel and damnable machines; I believe them to have been the direct suggestion of the Devil.
Martin Luther: Table Talk, dccccxx, 1569

A man that will fight may find a cudgel in every hedge.
John Clarke, 1609–1676, Paroemiologia Anglo-Latina

There is no weapon too short for a brave man.
Richard Steele: The Guardian, 25 August 1713

The means of destruction are approaching perfection with frightful rapidity.
Jomini: Précis de l'Art de la Guerre, 1838

I'll forge bright steel for liberty!
Attributed to Paul O'Donaghue, Irish rebel blacksmith, 19th century

The instruments of battle are valuable only if one knows how to use them.
Ardant du Picq, d. 1870, Battle Studies

The unresting progress of mankind causes continual change in the weapons; and with that must come a continual change in the manner of fighting.
Mahan: The Influence of Sea Power Upon History, i, 1890

Every development or improvement in firearms favors the defensive.
Giulio Douhet: The Command of the Air, 1921

New weapons operating in an element hitherto unavailable to mankind will not necessarily change the ultimate character of war. The next war may well start in the air but in all probability it will wind up, as did the last war, in the mud.
Report of the President's Board to Study Development of Aircraft for the National Defense, 1925

A weapon is defensive or offensive depending on which end of it is pointing at you.
Attributed to Aristide Briand, c. 1930

Every improvement in weapon power has aimed at lessening the danger on one side by increasing it on the other. Therefore, every improvement in weapons has eventually been met by a counter-improvement which has rendered the improvement obsolete, the evolutionary pendulum of weapon power, slowly or rapidly, swinging from the offensive to the protective and back again in harmony with the pace of civil progress, with each swing in a measurable degree eliminating danger.
J.F.C. Fuller: Armaments and History, 1945

There are two universal and important weapons of the soldier which are often overlooked—the boot and the spade. Speed and length of marching has won many victories; the spade has saved many defeats and gained time for victory.
Sir A. P. Wavell: The Good Soldier, 1945

We cannot expect the enemy to oblige by planning his wars to suit our weapons; we must plan our weapons to fight war where, when, and how the enemy chooses.
Vice Admiral Charles Turner Joy, USN, 1895–1956

There is still one absolute weapon . . . That weapon is man himself.
General Matthew B. Ridgway, USA: Speech at Cleveland, 10 November 1953

Non-sophistication does not preclude firepower.
Attributed by Bernard Fall to a French officer: Street without Joy, xv, 1964

(*See also* Armament, Arms, Bayonet, Cannon, Flamethrower, Gun, Mines, Missiles, Musket, Needle Gun, Rockets, Sword, Torpedoes.

Weather

Extraordinary rains pretty generally fall after great battles.
Plutarch 46–120 A.D. Lives (Marcus Cato)

Who's there, besides foul weather?
Shakespeare: King Lear, 1605

It is always necessary to shape operation plans . . . on estimates of the weather, and, as this is always changing, one cannot imitate in one season what has turned out well in another.
Frederick The Great: Instructions for His Generals, iii, 1747

I cannot command winds and weather.
Nelson: To the British Minister, Genoa, April 1796

Man is a creature fit for any climate, and necessity and determination soon reconcile him to anything.
Ferdinand Wrangell: Narrative of an Expedition to the Polar Sea in the Years 1820–1823

The seamen said it blew great guns.
Charles Dickens: David Copperfield, lv, 1850

Russia has two generals whom she can trust—General Janvier and General Février.
Tsar Nicholas I Of Russia: As reported in Punch 10 March 1853

The Admiral cannot take up a position that only in ideal conditions of tide and moon can the operation be begun. It has got to be begun as soon as possible, as long as conditions are practicable, even though they are not the best. People have to fight in war on all sorts of days, and under all sorts of conditions.
Winston Churchill: Note regarding the Dakar operation to General Ismay, 19 August 1940

When freshly blow the nor'western gales,
Then on courses snug we fly;
Soon lesser breezes will fill the sails,
And royals proudly sweep the sky.
Seamen's rhyme

Welles, Gideon (1802–1878)

Retire, O Gideon, to an onion farm,
Ply any trade that's innocent and slow.
Do anything, where you can do no harm.
Go anywhere you fancy—only go.
Verses attacking Welles, while Secretary of the Navy, in Leslie's, 5 June 1862

Wellington, Sir Arthur Wellesley, Duke of (1769–1852)

. . . a fine fellow with the best nerves of anyone I ever met with.
Major General Sir Lowry Cole: Letter from Portugal, 1811 (Cole was one of Wellington's division commanders.)

I should pronounce him to be a man of little genius, without generosity, and without greatness of soul.
Napoleon I: Letter to Barry E. O'Meara, St. Helena, 20 September 1817

. . . that long-nosed Bugger that beats the French.
Remark by a British private in the Peninsula, 1811

West Point

A man could fight the battles of his country, and lead his country's armies, without being educated at West Point. Jackson never went to West Point School, nor Brown—no, nor Governor Carroll; nor Colonel Cannon . . . though he mentioned it not to boast of it. He and thousands of other poor men had never been to West Point.
Representative David ("Davy") Crockett, 1786–1836

It but rarely happens that a graduate from West Point is not a gentleman in his deportment, as well as a soldier in his education.

Colonel Archibald Henderson, USMC: Letter to the Secretary of the Navy, 23 November 1823

I give it as my fixed opinion, that but for our graduated cadets, the war between the United States and Mexico might, and probably would, have lasted some four or five years, with, in its first half, more defeats than victories falling to our share.

Winfield Scott: Speech at a dinner after the capture of Mexico City, 8 December 1847

The standards for the American Army will be those of West Point. The rigid attention, the upright bearing, attention to detail, uncomplaining obedience to instruction, required of the cadet, will be required of every officer and soldier of our armies in France.

John J. Pershing: General Order to the AEF, October 1917

Duty, Honor, Country.

Motto of the U.S. Military Academy

Western Front

My God, did we really send men to fight in that?

Lieutenant General Sir Launcelot Kiggell: On seeing the terrain and the mud after the battle of Passchendaele, 1917

War sank into the lowest depths of beastliness and degeneration . . . For years the armies had to eat, drink, sleep amidst their own putrefactions. Bit by bit the old campaigner's memories and young soldier's dreams were engulfed in machinery and mud.

Sir Ian Hamilton: The Soul and Body of an Army, vii, 1921

The army report confined itself to a single sentence: All quiet on the Western Front.

Erich Maria Remarque: In Westen Nichts Neues, 1929

Soldiers of the Western Front, your hour has come. The fight which begins today will determine Germany's destiny for a thousand years.

Adolph Hitler: Order of the Day, 10 May 1940

(*See also* Chemical Warfare, Trench Warfare, World War I.)

Whiz Kids

The age of chivalry is gone; and that of sophisters, economists, and calculators, has succeeded.

Edmund Burke: Reflections on the Revolution in France, iii, 1790

The civilian is too inclined to think that war is only like working out an arithmetical problem with given numbers. It is anything but that.

Erich Ludendorff: My War Memories, 1919

The Pentagon Whiz Kids are, I think, conscientious, patriotic people who are experts at calculating odds, figuring cost-effectiveness and squeezing the last cent out of contract negotiations. But they are heavy-handed butchers in dealing with that delicate, vital thing called "morale." This is the stuff that makes ships like the *Enterprise*, puts flags on top of Iwo Jima and wins wars. But I doubt if Mr. McNamara and his crew have any morale setting on their computers.

Rear Admiral Daniel V. Gallery: Eight Bells and All's Well, 1965

(*See also* Civil-Military Relations, Civilian Interference, Research and Development.)

Will to Fight

Blessed be the Lord my strength, who teacheth my hands to war, and my fingers to fight.

Psalm CXXXXIV

The Spartans do not ask how many the enemy number, but where they are.

Agis of Sparta, 415 B.C.

A despised enemy has often maintained a sanguinary contest, and renowned states have been conquered by a very slight effort.

Hannibal: Address to his troops on the eve of the battle of the Ticino, 218 B.C.

Walled towns, stored arsenals and armories, goodly races of horse, chariots of war, elephants, ordnance, artillery, and the like; all this is but a sheep in a lion's skin except the breed and disposition of the people be stout and warlike.

Francis Bacon: Essays, xxix, 1625

War consisteth not in battle only, or the act of fighting; but in a tract of time, wherein the will to contend by battle is sufficiently known.

Thomas Hobbes: Leviathan, i, 1651

When men are irritated and their passions inflamed they flie hastily and cheerfully to arms, but after the first emotions are over, to expect among such people as compose the bulk of an army that they are influenced by any other principles but those of interest is to look for what never did and I fear never will happen.

George Washington: Letter to the President of Congress, September 1776

I have not yet begun to fight!

John Paul Jones: On being asked if he had struck the colors of USS Bonhomme Richard *in her action against HMS* Serapis *off Flamborough Head, 23 September 1779*

They don't seem to know when they ought to die—indeed, these villains can hardly ever be killed. They are a people without the slightest idea of propriety.

Mikhail Yurevich Lermontov: of the Caucasian tribesmen, c. 1838

I have come from the West, where we have always seen the backs of our enemies—from an army whose business it has been to seek the adversary, and to beat him when found, whose policy has been attack and not defense.

Major General John Pope, USA: General Order to the Officers and Soldiers of the Army of the Potomac, 14 July 1862

Let your watchword be, fight, fight, fight . . .

Major General Joseph Hooker, USA: Order to Major General Stoneman, commanding the cavalry, Army of the Potomac, 1863

I propose to fight it out on this line if it takes all summer.

U.S. Grant: Despatch to Major General Halleck, from Spottsylvania Court House, 11 May 1864

I do not advise rashness, but I do desire resolute and actual fighting, with necessary casualties.

P.H. Sheridan: Message to Major General Averell, 23 September 1864

It is not simply the weapons one has in one's arsenal that give one flexibility, but the willingness and ability to use them.

Mao Tse-tung, 1893

I am short a cheekbone and an ear, but am able to whip all hell yet!

Colonel John M. Corse, USA: Report to General Sherman during the Confederate assault on Allatoona, Georgia, 6 October 1864

In battle, two moral forces, even more than two material forces, are in conflict. The stronger conquers. The victor has often lost . . . more men than the vanquished . . . With equal or even inferior power of destruction, he will win who is determined to advance.

Ardant du Picq, 1821–1870, Battle Studies

Come on, you sons of bitches! Do you want to live forever?

Gunnery Sergeant Daniel Daly, USMC: At Lucy-le-Bocage, as the 5th Marines assaulted Belleau Wood, 4 June 1918

The will to conquer is the first condition of victory.

Ferdinand Foch: The Principles of War, 1920

What men will fight for seems to be worth looking into.

H.L. Mencken: Prejudices, 1922

It is not the actual military structure of the moment that matters but rather the will and determination to use whatever military strength is available.

Adolf Hitler: Mein Kampf, 1925

The best arms are dead and useless materiel as long as the spirit is missing which is ready, willing, and determined to use them.

Adolf Hitler: Mein Kampf, 1925

In war the chief incalculable is the human will.

B.H. Liddell Hart: Strategy (Encyclopaedia Britannica, 1929 edition)

Errors toward the enemy (i.e., to fight) should be most leniently viewed, even if the consequences are not pleasant.
Winston Churchill: Memorandum for First Sea Lord, 24 September 1939

War is a contest of wills.
Winston Churchill: During Middle East planning conference, 27 April 1941

All the great struggles of history have been won by superior will-power wresting victory in the teeth of odds or upon the narrowest of margins.
Winston Churchill, 1874–1965

I will leave the beaches either a conqueror or a corpse.
George S. Patton, Jr.: To General Marshall before sailing for Casablanca, October 1942

Send them our latitude and longitude.
William F. Halsey: On receiving word that the day's Japanese propaganda broadcast had asked, "Where is the American Third Fleet?" October 1944

Fighting spirit is not primarily the result of a neat organization chart nor of a logical organizational setup. The former should never be sacrificed to the latter.
Ferdinand Eberstadt: Testimony on the Marine Corps Bill, Senate Armed Services Committee, 1951

It is fatal to enter any war without the will to win it.
Douglas MacArthur: Address to the Republican National Convention, 7 July 1952

The nation's combat potential is no stronger than the will of its people.
Hanson W. Baldwin: Critical Tomorrows, December 1962

It isn't the size of the dog in the fight that counts; it's the size of the fight in the dog.
Author unidentified

(*See also* Determination, Fight, Resolution, Tenacity.)

Wingate, Orde Charles (1903–1944)

. . . a man of genius who might well have become also a man of destiny.
Winston Churchill: To the House of Commons, 2 August 1944

To see Wingate urging action on some hesitant commander was to realize how a medieval baron felt when Peter the Hermit got after him to go crusading. Lots of barons found Peter the Hermit an uncomfortable fellow, but they went crusading all the same.
Sir William Slim 1891–

Winter Warfare

He smote the sledded Polack on the ice.
Shakespeare: Hamlet, i, 1, 1600

Every mile is two in winter.
George Herbert: Jacula Prudentum, 1651

Only absolute necessity and prospect of great advantages can excuse winter operations.
Frederick The Great: Instructions for His Generals, xxv, 1747

. . . a hard, dull bitterness of cold.
John Greenleaf Whittier, 1807–1892, Snow-Bound

Withdrawal

Instead of destroying that bridge, we should build another, that he may retire the more quickly from Europe.
Aristides: Disapproving a proposal to destroy Xerxes' bridge of boats across the Hellespont, c. 497 B.C.

There is no cause for despondency. *Battre en retraite* is a recognized operation of war, familiar to the Great Captains.
Sir Henry Wilson: Diary entry after the battle of Mons, August 1914

The withdrawal should be thought of as an offensive instrument, and exercises be framed to teach how the enemy can be lured into a trap, closed by a counter-stroke or devastating circle of fire.
B.H. Liddell Hart: Thoughts on War, xvi, 1944

(*See also* Retreat.)

Wolfe, James (1727–1759)

Oh! he is mad, is he? Then I wish he would bite some other of my generals.

George II: Reply to the remark that Wolfe was mad, 1758

Now, God be praised, I will die in peace.
Wolfe: Last words, on the Plains of Abraham, Quebec, 13 September 1759

(*See also* Quebec.)

Women

The sex is ever to a soldier kind.
Homer: The Odyssey, xiv, c. 1000 B.C.

A Spartan woman, as she handed her son his shield, exhorted him, saying, "Either with this or upon this."
Plutarch, 46–120 A.D., Moralia

No war without a woman.
John Clarke: Paroemiologia Anglo-Latina, 1639

Women adore a martial man.
William Wycherley: The Plain Dealer, ii, c. 1674

If upon service you have any ladies in your camp, be valiant in your conversation before them. There is nothing pleases the ladies more than to hear of storming breaches, attacking the covert-way sword in hand, and such like martial exploits.
Francis Grose: Advice to the Officers of the British Army, 1782

There being reasons to apprehend that a number of women have been clandestinely brought from England to several ships . . . the respective Captains are required by the Admiral to admonish those ladies upon the waste of water, and other disorders committed by them, and to make known to all, that on the first proof of water being obtained for washing from the scuttle-butt . . . every woman in the fleet . . . will be shipped home for England by the first convoy.
Lord St. Vincent: Mediterranean Fleet Order, 1796

Man shall be framed for War, and Woman for the entertainment of the Warrior. all else is folly.
F.W. Nietzche: Thus Spake Zarathrustra, i, 18, 1885

The young women, the young girls! Their unknowing eyes threw furtive glances at the medals and the uniforms. These sheeps' eyes almost made one believe he was a hero.
Captain Georges Thenault: Of service in the LaFayette Escadrille, 1916

The only war I ever approved of was the Trojan war; it was fought over a woman and the men knew what they were fighting for.
William Lyon Phelps: Sermon at Riverside Church, New York, 25 June 1933

(*See also* Dependents, Love and War.)

World War I (1914–1918)

If the iron dice roll, may God help us.
Theobald von Bethmann-Hollweg: To the German Reichstag, as Foreign Minister, 1 August 1914

The lamps are going out all over Europe; we shall not see them lit again in our lifetime.
Sir Edward Grey: On the evening of 4 August 1914, as war with Germany drew near.

. . . the struggle that will decide the course of history for the next hundred years.
Helmuth von Moltke ("The Younger"): Letter to Field Marshal Conrad von Hötzendorff, 5 August 1914

For all we have and are,
For all our children's fate,
Stand up and take the war.
The Hun is at the gate!
Rudyard Kipling: For All We Have and Are, 1914

The War That Will End War.
H.G. Wells: Title of book, 1914

The Great War and the Petty Peace.
H.G. Wells: Outline of History, xl, 1920

It is a fearful thing to lead this great peaceful people into war . . . We shall fight for the things we have always carried nearest our hearts—for democracy, for the right of those who submit to authority to have a voice in their own Governments, for the rights and liberties of small nations, for a universal dominion of right by such a concert of free peoples as shall bring peace and safety to all nations and make the world itself at last free.

Woodrow Wilson: War message to Congress, 2 April 1917

. . . and we won't come back till it's over, over there.

George M. Cohan: Song, Over There, 1917

. . . a straight duel between land-power and sea-power.

Halford J. Mackinder: Democratic Ideals and Reality, 1919 (referring to World War I)

The war to end wars has resulted in a peace to end peace.

Attributed to Kaiser Wilhelm II: On being apprised of the terms of the Treaty of Versailles, June 1919

Nobody wanted war . . . The nations backed their machines over the precipice.

Lloyd George, 1863–1945

Napoleon had said it was rare to find generals willing to fight battles. The curse of [World War I] was that so few could do anything else.

T.E. Lawrence: Science of Guerrilla Warfare, 1929

. . . a war of masses of men hurling masses of shells at each other.

Winston Churchill: Memorandum for the War Cabinet, 3 September 1940 (referring to World War I)

The First World War had causes but no objectives.

Correlli Barnett: The Swordbearers, i, 1963

(*See also* Arab Revolt, Chemical Warfare, Gallipoli, Trench Warfare, Western Front.)

Wound

A wound is nothing, be it ne'er so deep;
Blood is the god of war's rich livery.

Christopher Marlowe: Tamburlaine the Great, iii, 2, 1587

He jests at scars, that never felt a wound.

Shakespeare: Romeo and Juliet, ii, 2, 1594

ROMEO— Courage, man; the hurt cannot be much
MERCUTIO—No, 'tis not so deep as a well, nor so wide as a church door; but 'tis enough, 'twill serve.

Shakespeare: Romeo and Juliet, iii, 1 1594

. . . his cicatrice, an emblem of war, here on his sinister cheek.

Shakespeare: All's Well That Ends Well, ii, 1 1602

A scar nobly got, or a noble scar, is a good livery of honor.

Shakespeare: All's Well That Ends Well, iv, 5 1602

So well thy words become thee as thy
wounds;
They smack of honor both.

Shakespeare: Macbeth, i, 2, 1605

The history of a soldier's wound beguiles the pain of it.

Laurence Sterne: Tristram Shandy, i, 25, 1760

LORD UXBRIDGE— I've lost my leg, by God!
WELLINGTON—By God, sir, so you have!

Conversation at Waterloo, 18 June 1815

. . . for here I leave my second leg,
And the 42d Foot!

Thomas Hood: Faithless Nellie Gray, 1840

"You're wounded!" "Nay," the soldier's
pride
Touched to the quick, he said:
"I'm killed, Sire!" and his chief beside,
Smiling the boy fell dead.

Browning: Incident of the French Camp, 1846

It is nothing. For this, are we soldiers.

Captain Guy V. Henry, USA: After being shot through the face during the 3d Cavalry's action on the Rosebud River, 17 June 1876

(*See also* Casualties, Wounded.)

Wounded

"Fight on, my men," Sir Andrew days,

"A little I'm hurt, but not yet slain;
"I'll but lie down and bleed awhile,
"And then I'll rise and fight again."

Ballad of Sir Andrew Barton, author unknown, c. 1550

Wise men took refuge in the virtues of cold water, and kept the surgeons at a safe distance.

Sir John Fortescue: History of the British Army, I, 1899

Men, all I can say is, if I had been a better general, most of you would not be here.

George S. Patton, Jr.: To wounded soldiers at Walter Reed Hospital, Washington, 1945

(*See also* Casualties, Field Hospital, Wound.)

Yorktown (September–October 1781)

I propose a cessation of hostilities for twenty-four hours . . . to settle terms for the surrender.

Sir Charles Cornwallis: Note to Washington, 17 October 1781, asking for a truce to surrender the British army at Yorktown.

Oh God! It is all over!

Lord George Germain: On receiving word of Cornwallis's surrender, 25 November 1781

(*See also* American Revolution.)

Z

Zeal

A man with his heart in his profession imagines and finds resources where the worthless and lazy despair.
Frederick The Great: Instructions for His Generals, v, 1747

However services may be received, it is not right in an officer to slacken his zeal for his Country.
Nelson: Letter to his wife, June 1794

Above all, no zeal! (Surtout, point de zèle!)
Attributed to Talleyrand, 1754–1838

All zeal, Mr. Easy!
Frederick Marryat: Mr. Midshipman Easy, 1836

I can only plead my zeal to serve my country, and the chances of war.
David G. Farragut: Report to the Secretary of the Navy after the attack on Port Hudson, March 1863

No difficulty baffles great zeal.
Sir John Fisher: Memories, 1919

(*See also* Resolution.)

Text for this book was composed in 9-point Times Roman by a Photon 713 Textmaster.

Body stock is Warren's Olde Style wove offset 60-pound. Cover cloth is Columbia Mills Vynside "C" Vellum RVV-1750 on the spine, Lindemeyr-Schlosser Elephant Hide on Covers I and IV. Endpapers are Strathmore Rhododendron Pompeian Red Antique.

Composition and offset printing by the Science Press, Inc., Ephrata, Pennsylvania, a Subsidiary of Printing Corporation of America.
Binding by Albrecht Company, Baltimore, Maryland.